ARITHMETIC
A STRAIGHTFORWARD APPROACH

MARTIN M. ZUCKERMAN
CITY COLLEGE OF THE CITY UNIVERSITY OF NEW YORK

 ARDSLEY HOUSE PUBLISHERS, INC.   NEW YORK

Address orders and editorial
correspondence to:
Ardsley House, Publishers, Inc.
320 Central Park West
New York, NY 10025

Other Titles by Martin M. Zuckerman

Sets and Transfinite Numbers, Macmillan
Intermediate Algebra: A Straightforward Approach, 2nd Edition, John Wiley and Sons
Elementary Algebra: A Straightforward Approach, 2nd Edition, Allyn and Bacon
Geometry: A Straightforward Approach, 2nd Edition, Morton
Arithmetic without Trumpets or Drums, Allyn and Bacon
Algebra and Trigonometry: A Straightforward Approach, 2nd Edition, John Wiley and Sons
Basic Mathematics, Ardsley House
Intermediate Algebra: A Straightforward Approach, Alternate Edition, John Wiley and Sons
Passing the City University of New York Mathematics Skills Assessment Test, Ardsley House
Arithmetic with an Introduction to Algebra, Ardsley House
Aritmética con una Introducción al Algebra, Ardsley House
College Algebra: A Straightforward Approach, John Wiley and Sons

ISBN: 0-912675-07-1

Printed in the United States of America

10 9 8 7 6 5 4 3 2 1

CONTENTS

PREFACE

This book covers the basic topics in arithmetic with which every college student should be thoroughly familiar. It is written with the student in mind, in a style and at a level appropriate for student understanding. There are lots of illustrative examples and student exercises.

Major emphasis is placed on showing the student how to use the basic arithmetic operations in everyday life. There are numerous applications throughout the book—reading numbers in newspapers, cost and profit, approximating, taxes, reductions, percent increase or decrease, units of length, weight, and time, the Metric System, reading graphs, averaging. The most widely-used geometric figures are introduced, and the student is taught how to calculate area, perimeter, volume, and surface area. The book concludes with an introduction to algebra. The student is shown how to solve simple equations and how to apply these methods in solving proportions and problems involving costs, measures, and averaging.

How to use the book:

The format is extremely simple. Each section is divided into several topics. The student should first read Topic A and work through the illustrative examples presented. Immediately afterward, he or she should then attempt the Practice Exercises for Topic A which follow directly. (See pages 9 to 11.) Complete solutions to *all* Practice Exercises are provided in the back of the book, beginning on page 353. When the student has mastered these Practice Exercises, he or she should immediately try the Exercises for Topic A at the end of the section. (See page 13.) The student can then proceed to Topic B, followed by the Practice Exercises for Topic B and the Exercises for Topic B at the end of the section, and so on.

There is a diagnostic test at the beginning of the book to help the student pinpoint strengths and weaknesses. After each of the seven major units of the book, there are copious review exercises on the material of the unit, a practice exam on the unit, as well as review exercises on prior units. Two Practice Midterm Exams are provided after Unit IV and three Practice Final Exams are presented after Unit VII. In this way the student has several different opportunities to reinforce his or her knowledge of the techniques of the course.

Answers to the *odd-numbered exercises* at the end of each section are provided at the back of the book. Answers to *all* questions of the Diagnostic Test as well as to all Review Exercises are also given. Complete solutions are provided to the Practice Exam for each unit, and to all of the Practice Midterms and Practice Finals, as well as to the Practice Exercises within each section.

Acknowledgments:

I gratefully acknowledge the help of a number of people who have contributed to this endeavor. I would like to thank, in particular, the critics for their numerous useful suggestions:

Ignacio Bello, Hillsborough Community College
Ben P. Bockstege, Broward Community College
Kevin Reilley, Central Piedmont Community College
Harold Schoen, The University of Iowa
Heschel Shapiro, Los Angeles Trade Technical College
Lora Shapiro, City College of the City University of New York
Richard Spangler, Tacoma Community College

It was a pleasure to work with Karen Bernath, who designed the cover, Irving Rothman and John McAusland, who did the technical illustrations, Eileen Rosenfeld, who drew the cartoon section openings, Bill Boot, Nora Braverman, Ann Eisner, Laura Jones, and Robert E. Singleton, who handled composition and production, and Susan Hahn and Joan Alexander who checked the answers.

Martin M. Zuckerman

DIAGNOSTIC TEST

The purpose of this test is to help you understand which sections of the book you should emphasize in your studies. The question numbers correspond to the section numbers. For example, Question 1 corresponds to Section 1. If you are able to answer *all* parts of the question correctly, you may not need to study that section. But *if any part of your answer is wrong, you should study the section.* It is best to consult your instructor about your specific needs, but this test should help to pinpoint your strengths and weaknesses. When you have completed the test, turn to page 335 for the answers.

I. *BASIC CONCEPTS*

1. WORDS AND NUMERALS, page 9

   i. Fifty-three thousand eight hundred eight is written _____

   ii. Fifty-eight million two is written _____

   iii. Write 90,036 in verbal form. _____

2. BASIC ARITHMETIC, page 15

   i. 581 − 289 =   *ii.* 597 × 205 =   *iii.* 10,302 ÷ 17 =

   iv. 905 ÷ 42   *Answer:* quotient =   remainder =

3. **EXPONENTS**, page 27

   i. $5^3 =$ *ii.* $2^4 \times 3^2 =$

4. **PRIME FACTORS**, page 32

   i. Express 396 as the product of primes. *Answer:*

5. **DIVISIBILITY TESTS**, page 38

   i. Is 8687 divisible by 9? *Answer:*

   ii. Find the largest power of 10 that divides 5,050,000. *Answer:*

   iii. Express 48,600 as the product of primes. *Answer:*

6. **GREATEST COMMON DIVISORS AND LEAST COMMON MULTIPLES**, page 45

   i. gcd (54, 192) =

   ii. lcm (8, 10, 12) =

II. *FRACTIONS*

   7. **ADDING AND SUBTRACTING FRACTIONS**, page 55

      i. $\dfrac{1}{5} + \dfrac{3}{4} =$ *ii.* $\dfrac{5}{6} - \dfrac{1}{8} =$

   8. **MULTIPLYING AND DIVIDING FRACTIONS**, page 67

      i. $\dfrac{5}{8} \times \dfrac{2}{15} =$ *ii.* $\dfrac{3}{8} \div \dfrac{9}{4} =$

   9. **MIXED NUMBERS**, page 77

      i. $2\dfrac{2}{3} + 1\dfrac{1}{6} =$ *ii.* $2\dfrac{1}{2} \times 3\dfrac{1}{4} =$

      iii. $6\dfrac{1}{4} \div 3\dfrac{1}{8} =$

10. **SIZE OF FRACTIONS**, page 86

    i. Which of these fractions is the smallest?

    (A) $\dfrac{2}{5}$ (B) $\dfrac{1}{3}$ (C) $\dfrac{3}{8}$ (D) $\dfrac{2}{7}$ (E) $\dfrac{3}{10}$

III. *DECIMALS*

11. **DECIMAL NOTATION**, page 99

    i. Write forty-seven thousandths in decimal notation. *Answer:*

    ii. Change $\dfrac{7}{8}$ to a decimal. *Answer:*

    iii. Change $\dfrac{2}{11}$ to an infinite repeating decimal. *Answer:*

12. **SIZE OF DECIMALS**, page 108

    i. Which number is the smallest?

    (A) .404   (B) .411   (C) .4009   (D) .4010   (E) .4001

13. **ADDING AND SUBTRACTING DECIMALS**, page 115

    i. $5.905 + 7.08 + 12 =$ *ii.* $29.6 - 19.79 =$

14. **MULTIPLYING AND DIVIDING DECIMALS**, page 119

    i. Multiply. $4.4 \times .06$ *Answer:*

    ii. A gallon of paint costs $9.95. What is the cost of 39 gallons of that paint? *Answer:*

    iii. $720 \div .12 =$

    iv. Cherries sell for $1.40 per pound. How many pounds can be bought for $11.20? *Answer:*

15. **COST AND PROFIT**, page 127

    i. Find the total cost of 6 bottles of detergent at $2.25 per bottle and 8 rolls of paper towels at $1.04 per roll. *Answer:*

ii. At a concert, 632 tickets are sold at $7.50 each. It costs $1800 to rent the hall and $500 for other expenses. The profit is _____

iii. A limousine service charges $25 for the first hour and $15 for each additional hour. The charge for 6 hours of limousine service is _____

16. **ROUNDING,** page 132

i. Round 4,189,625 to the nearest 100,000. *Answer:*

ii. Round 17.059 to the nearest tenth. *Answer:*

iii. Estimate the product of 24.7 and 39.86 by first rounding each factor to the nearest integer. *Answer:*

IV. *PERCENTAGE*

17. **FRACTIONS, DECIMALS, AND PERCENT,** page 145

i. Change $\frac{4}{11}$ to a decimal rounded to the nearest hundredth.
Answer:

ii. Change .012 to a fraction in lowest terms. *Answer:*

iii. What is 56% expressed as a fraction in lowest terms? *Answer:*

18. **PERCENTAGE PROBLEMS,** page 152

i. What is 40% of 60? *Answer:*

ii. If 20% of a number is 15, what is the number? *Answer:*

iii. A pitcher throws 75 called strikes out of a total of 125 pitches in a game. What percent of his pitches are called strikes?
Answer:

iv. An alloy contains 40% copper. How much copper is there in 250 tons of the alloy?
Answer:

19. **TAXES, SALES, AND PERCENT INCREASE OR DECREASE**, page 160

    i. The price of gasoline, which was $1.20 per gallon, is increased by 5%. The new price per gallon is _____

    ii. A clock that was selling for $25 was reduced by 25%. The new price is _____

    iii. A suit sells for $180. There is also a 4% sales tax. What is the total price of the suit? *Answer:*

V. *MEASUREMENT AND GEOMETRY*

20. **MEASURE**, page 175

    i. A flight left at 7:52 A.M. and arrived at 11:25 A.M. that morning. How long did it last? *Answer:*

    ii.   9 pounds 9 ounces   *iii.* 4 feet 7 inches
          -4 pounds 10 ounces   +3 feet 9 inches

    iv. To the nearest kilometer, 10 miles = _____ kilometers.

21. **RECTANGLES AND SQUARES**, page 184

    i. How much does it cost to carpet a room that is 18 feet by 12 feet, if the carpeting costs $3 per square foot? *Answer:*

    ii. A woman wants to sew a silk border around a blanket that is 8 feet by 5 feet. How long a piece of silk does she need? *Answer:*

    iii. The area of a rectangle is 100 square inches and the length is 20 inches. Find the perimeter of the rectangle. *Answer:*

22. **TRIANGLES, CIRCLES, AND RECTANGULAR BOXES**, page 195

    i. A triangle has base 4 centimeters and height 8 centimeters. Its area is

_____

    ii. The circumference of a circle is 10π inches. Find its area to the nearest tenth of an inch. *Answer:*

    iii. Find the surface area of a rectangular box that is 5 feet by 2 feet by 3 feet. *Answer:*

23. **GRAPHS**, page 210

i. The graph at the right shows the number of students who graduated from Hoover High School each year from 1980 to 1984. The total number that graduated during the years 1980, 1981, and 1982 was closest to

(A) 10,000   (B) 11,000

(C) 11,500   (D) 12,000

(E) 13,000

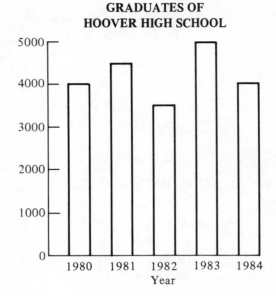

GRADUATES OF HOOVER HIGH SCHOOL

YEARLY INCOME: $25,000

ii. The graph at the left indicates a family's allotment of its yearly income of $25,000. For which item did the family spend $5000?

(A) food   (B) services

(C) clothing   (D) taxes

(E) rent

iii. The graph at the right indicates the annual profits of a textile manufacturer, in thousands of dollars. What was the decrease in profits from 1982 to 1983?

(A) $150   (B) $250

(C) $150,000   (D) $250,000

(E) $400,000

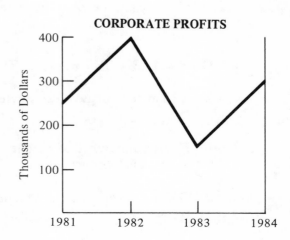

CORPORATE PROFITS

VI. *INTEGERS*

24. **ARITHMETIC WITH NEGATIVE INTEGERS**, page 223

   i. $10 - (-3) =$ *ii.* $(-5)(-6) =$ *iii.* $\dfrac{27}{-9} =$

25. **SEVERAL OPERATIONS**, page 239

   i. $4(2 + 3) - 5^2 =$ *ii.* $(-3)(-2)^2 - 5(-1)^2 =$

   iii. $\left(\dfrac{16 - 2^2}{3}\right) \times (8 - 3) =$

26. **AVERAGING**, page 246

   i. Find the average of 62, 84, 77, and 97. *Answer:*

   ii. On four chemistry exams, Judy's grades are 85, 65, 0, and 90. The average of her grades is _____

VII. *INTRODUCTION TO ALGEBRA*

27. **ALGEBRAIC EXPRESSIONS**, page 255

   i. If x represents a number, express 2 more than 3 times the number. *Answer:*

   ii. Find the value, in dollars, of x five-dollar bills and y ten-dollar bills. *Answer:*

28. **SUBSTITUTING NUMBERS FOR VARIABLES**, page 261

   i. Find the value of $2x^2 + 5x$ when $x = 4$. *Answer:*

   ii. Find the value of $4p^2 - 3pq$ when $p = -2$ and $q = 3$. *Answer:*

29. **ADDING AND SUBTRACTING POLYNOMIALS**, page 271

   i. Add $4x + 5y$ and $7x - 4y$. *Answer:*

   ii. Subtract $6x - 5$ from $8x^2 + 3$. *Answer:*

30. **SOLVING EQUATIONS**, page 280

    i. Solve for x:

$$5 - 3x = 2x - 5$$

    Answer:

31. **RATIOS AND PROPORTIONS**, page 289

    i. Find b if $\dfrac{6}{b} = \dfrac{-3}{5}$. *Answer:*

    ii. Find t if $\dfrac{t-3}{9} = \dfrac{t-5}{3}$. *Answer:*

    iii. If $\dfrac{y}{2} - 6 = \dfrac{y}{5}$, then $y =$

32. **APPLIED PROBLEMS**, page 297

    i. It costs a manufacturer \$2 to produce each wallet. In addition, there is a general overhead cost of \$5000. He decides to produce 8000 wallets. If he receives \$10 per wallet from a wholesaler, how many wallets must he sell to make a profit of \$10,000? *Answer:*

    ii. Three out of every 5 cars sold one month have radial tires. If an agency sells 15 cars with radial tires that month, how many cars do they sell that month? *Answer:*

    iii. Twenty percent of the 400 employees in a factory are women. In order to have 50% women employees in the factory, how many additional women must be hired? *Answer:*

33. **SQUARE ROOTS AND RIGHT TRIANGLES**, page 302

    i. $\sqrt{8^2 - 5^2} =$

    ii. In the right triangle shown here, find x.
    Answer:

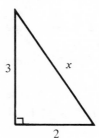

UNIT I BASIC CONCEPTS

1. Words and Numerals

The most commonly used numbers are the **counting numbers**

$$1, \ 2, \ 3, \ 4, \ 5, \ 6, \ 7, \ 8, \ 9, \ 10, \ 11, \ 12$$

and so on. The counting numbers, together with 0, are called the **whole numbers**. The first ten whole numbers, that is,

$$0, \ 1, \ 2, \ 3, \ 4, \ 5, \ 6, \ 7, \ 8, \ 9$$

are called **digits**. A number such as 35 is a 2-digit number. Here 3 is the 10's digit and 5 is the 1's digit.

The number 584 is a 3-digit number; here 5 is the 100's digit, 8 is the 10's digit, and 4 is the 1's digit.

A. WORDS TO NUMERALS

In books, newspapers, and magazine articles, numbers are often written in words. When working with these numbers, it is generally best to convert them to *numeral form*, that is, to write them in terms of digits. For example,

eight hundred fifty-six   is written   856

as a 3-digit number.

Note the correspondence:

1's ←——→ ones
10's ←——→ tens
100's ←——→ hundreds
1000's ←——→ thousands
10,000's ←——→ ten-thousands
100,000's ←——→ hundred-thousands
1,000,000's ←——→ millions

When writing **counting numbers** with *five or more digits*, begin at the right and use commas to separate the digits into groups of three. For example,

millions
  | *thousands*
  | |
12,607,259

This represents

12 million, 607 thousand, 259

Example 1 ▶ Write "seven thousand four hundred fifty-two" in numerals.
Solution.

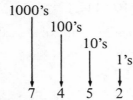

Thus write 7452.   ◀

Example 2 ▶ Write "forty thousand two hundred sixty" in numerals.
Solution.

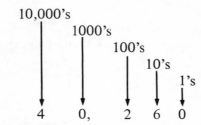

Thus write 40,260.   ◀

Example 3 ▶   Write "eight hundred thousand four hundred" in numerals.

Solution.

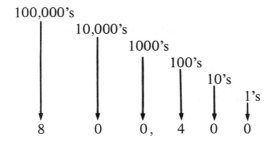

Thus write 800,400.   ◀

Example 4 ▶   Write "nine million twelve thousand seventy-eight" in numerals.

Solution.

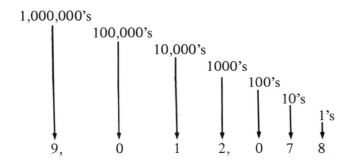

Thus write 9,012,078.   ◀

PRACTICE EXERCISES FOR TOPIC A

Complete Solutions to all Practice Exercises are given beginning on page 353.

Write in numerals.

1. fifty-three thousand five hundred four
2. six million ninety thousand fifteen
3. nine hundred forty-three thousand

Now try the exercises for Topic A on page 13.

B. NUMERALS TO WORDS

When writing an essay, numbers are generally expressed in words. It is important to know how to express numbers in *verbal form*.

Example 5 ▶   Write 5746 in verbal form.

Solution.

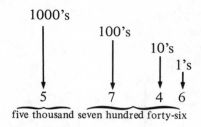

Example 6 ▶   Write 8,400,010 in verbal form.

Solution.

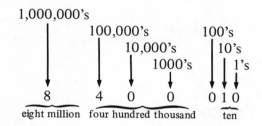

PRACTICE EXERCISES FOR TOPIC B

Write in verbal form.

4. 30,430
5. 12,000,529
6. 12,529,000

Now try the exercises for Topic B on page 13.

C. NUMBERS IN SENTENCES

Example 7 ▶   Suppose that six thousand four hundred fifty fans attended a football game. The number of fans at the game was

(A) 6405   **(B)** 645   **(C)** 6450   **(D)** 64,450   **(E)** 60,450

Solution. The number of fans was *six thousand four hundred fifty.*

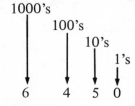

The correct choice is **(C).**

PRACTICE EXERCISES FOR TOPIC C

Draw a circle around the correct answer.

7. Suppose that a woman earns twenty thousand six hundred dollars a year. Which of these amounts expresses her yearly income?

    (A) $20,006   **(B)** $20,600   **(C)** $26,000   **(D)** $26,660   **(E)** $20,006

Now try the exercises for Topic C on page 14.

EXERCISES

Answers to Exercises are given beginning on page 335.

A. *Write in numerals*

1. two hundred forty-five   *Answer:*
2. six hundred nine   *Answer:*
3. seven hundred eighty   *Answer:*
4. one thousand two hundred thirty-six   *Answer:*
5. seven thousand six hundred two   *Answer:*
6. eight thousand forty-four   *Answer:*
7. six thousand five hundred   *Answer:*
8. twenty-four thousand five hundred   *Answer:*
9. sixty-three thousand fifty-six   *Answer:*
10. thirty-seven thousand nine   *Answer:*
11. four hundred thousand two hundred twelve   *Answer:*
12. three hundred fifty thousand six hundred   *Answer:*
13. seven hundred forty-seven thousand eight   *Answer:*
14. six hundred four thousand five hundred thirty   *Answer:*
15. one hundred ninety-eight thousand two hundred forty   *Answer:*
16. six million thirty-five thousand   *Answer:*
17. two million one thousand ninety-six   *Answer:*
18. five million nineteen thousand four hundred fifty   *Answer:*
19. ten million one hundred sixty-three thousand   *Answer:*
20. ninety million twenty thousand fourteen   *Answer:*

B. *Write in verbal form.*

21. 528   *Answer:*
22. 440   *Answer:*
23. 4800   *Answer:*
24. 7012   *Answer:*

25. 90,100   *Answer:*

26. 53,002   *Answer:*

27. 85,859   *Answer:*

28. 400,005   *Answer:*

29. 380,019   *Answer:*

30. 794,000   *Answer:*

31. 383,105   *Answer:*

32. 6,400,021   *Answer:*

33. 9,002,585   *Answer:*

34. 7,330,015   *Answer:*

35. 42,000,400   *Answer:*

36. 35,500,080   *Answer:*

C.   *Express the number in each sentence in numerals. Draw a circle around the correct letter choice, as illustrated in Exercise 37.*

37. There were twenty-nine students present.

   (A) 29   (B) 209   (C) 20   (D) 219   (E) 290

38. Four hundred fifty-three passengers boarded the ship.

   (A) 403   (B) 453   (C) 543   (D) 4053   (E) 400,053

39. Dinosaurs became extinct sixty million years ago.

   (A) 6,000,000   (B) 60,000,000   (C) 600,000

   (D) 600,000,000   (E) 1,000,060

40. The daily circulation of the *Globe* is over one hundred twenty-five thousand.

   (A) 120,005   (B) 120,125   (C) 125,000   (D) 1,250,000   (E) 1,000,125

41. The attendance at a world-series game was sixty-six thousand four hundred eighty.

   (A) 66,408   (B) 60,648   (C) 66,480   (D) 660,408   (E) 660,480

42. The novel contained one thousand two hundred ninety-two pages.

   (A) 192   (B) 1920   (C) 1292   (D) 1290   (E) 12,920

43. The population of Montgomery, Alabama is over one hundred seventy-seven thousand.

   (A) 107,000   (B) 177,000   (C) 1,770,000   (D) 170,007   (E) 170,077

44. The house sold for one hundred thirty-eight thousand five hundred dollars.

   (A) 138,000   (B) 135,000   (C) 138,555   (D) 130,580   (E) 138,500

45. There are twenty-two million six hundred thousand people in Yugoslavia.

   (A) 220,600   (B) 22,600   (C) 22,600,000

   (D) 22,660,000   (E) 22,000,600

46. Venus is about sixty-seven million miles from the sun.

    (A) 67,000,000   **(B)** 670,000,000   **(C)** 6,700,000

    (D) 67,000,000,000   **(E)** 67,000,067

2. Basic Arithmetic

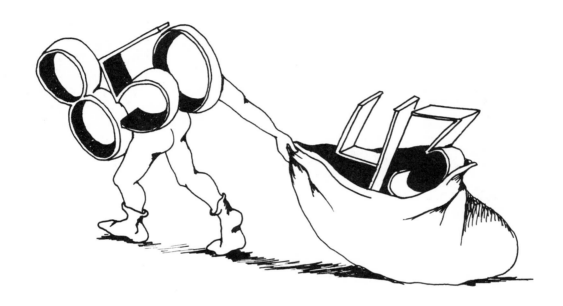

Let us briefly review the basic processes of adding, subtracting, multiplying, and dividing counting numbers.

A. ADDITION

Suppose Jerry has 3 books and decides to buy 2 more. He then has 5 books. This is illustrated as follows.

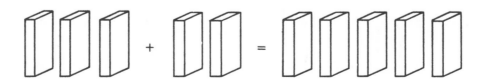

$3 + 2 = 5$ *(Read: "3 plus 2 equals 5.")*

Larger numbers are generally added in columns.

DEFINITION

> When two or more numbers are added, the result is called the **sum**.

Example 1 ▶ $276 + 798 =$

Solution. Arrange in columns.

Step 1.

100's	10's	1's
	1	
2	7	6
+ 7	9	8
		4

Note that in the 1's column, $6 + 8 = 14$. Carry the 1 to the 10's column.

Step 2.

100's	10's	1's
1	1	
2	7	6
+ 7	9	8
	7	4

Here in the 10's column, $1 + 7 + 9 = 17$. Carry the 1 to the 100's column.

Step 3.

100's	10's	1's
1	1	
2	7	6
+ 7	9	8
10	7	4

The sum is 1074.

PRACTICE EXERCISES FOR TOPIC A

Add.

1. $453 + 328 =$

2. 6029
    3894
    5190

Now try the exercises for Topic A on page 24.

B. SUBTRACTION

Subtraction *undoes* addition. Thus

$$5 - 2 = 3 \quad \text{because} \quad 2 + 3 = 5$$

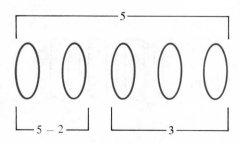
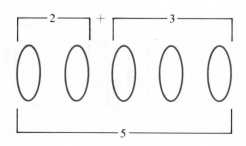

DEFINITION

> When one number is subtracted from another, the result is called the **difference**.

There are two methods of subtraction that are commonly used. Both are illustrated in Examples 2 and 3.

Example 2 ▶   315 − 178 =

Solution.

Step 1.

Borrow from the 10's column in order to increase 5 to 15 in the 1's column.

$$
\begin{array}{cc|c}
0 & & 15 \\
3\!\!\!/ & \!\!\!/5 \\
-\ 17 & 8 \\
\hline
 & & 7
\end{array}
\qquad \text{or} \qquad
\begin{array}{cc|c}
 & & 15 \\
31 & \!\!\!/5 \\
 & 8 \\
-\ 1\!\!\!/7 & 8 \\
\hline
 & & 7
\end{array}
$$

Step 2.

Borrow 10 tens from the 100's column.

$$
\begin{array}{c|c|c}
 & 10 \\
2 & \emptyset \\
\not{3} & \not{1} & 5 \\
-\ 1 & 7 & 8 \\
\hline
 & 3 & 7
\end{array}
\qquad \text{or} \qquad
\begin{array}{c|c|c}
 & 11 \\
3 & \not{1} & 5 \\
2 & 8 \\
-\ \not{1} & \not{1} & 8 \\
\hline
 & 3 & 7
\end{array}
$$

Step 3.

$$
\begin{array}{c|c}
2 \\
\not{3} & 15 \\
-\ 1 & 78 \\
\hline
1 & 37
\end{array}
\qquad \text{or} \qquad
\begin{array}{c|c}
3 & 15 \\
2 \\
-\ \not{1} & 78 \\
\hline
1 & 37
\end{array}
$$

The difference is 137.   ◀

Example 3 ▶   Find 504 − 169. Check your result.

Solution. Write

$$
\begin{array}{cc|c}
50 & 4 \\
-\ 16 & 9
\end{array}
$$

You want to borrow from the 10's column in order to subtract

$$
\begin{array}{r}
14 \\
-\ 9 \\
\hline
\end{array}
$$

in the 1's column. Because 0 is the 10's digit of 504, you must first borrow from the 100's column.

$$
\begin{array}{r}
4\ 10 \\
\cancel{5}\ \cancel{0}\ 4 \\
-1\ 6\ 9 \\
\hline
\end{array}
\qquad \text{or} \qquad
\begin{array}{r}
10 \\
5\ \cancel{0}\ 4 \\
2 \\
-\cancel{1}\ 6\ 9 \\
\hline
\end{array}
$$

Now proceed as before.

$$
\begin{array}{r}
9 \\
4\ \cancel{10}\ 14 \\
\cancel{5}\ \cancel{0}\ 4 \\
-1\ 6\ 9 \\
\hline
3\ 3\ 5
\end{array}
\qquad \text{or} \qquad
\begin{array}{r}
10\ 14 \\
5\ \cancel{0}\ \cancel{4} \\
2\ 7 \\
-\cancel{1}\ \cancel{6}\ 9 \\
\hline
3\ 3\ 5
\end{array}
$$

The difference is 335.

To *check* this result, add the difference, 335 to 169 (in either order). You should obtain 504.

Check.

$$
\begin{array}{r}
169 \\
335 \\
\hline
504
\end{array}
$$ ◀

Frequently, addition and subtraction are called for in the same example. Thus

$$
\underbrace{9 + 4}_{13} - 10 = 13 - 10 = 3
$$

and

$$
\underbrace{12 - 5}_{7} + 4 = 7 + 4 = 11
$$

PRACTICE EXERCISES FOR TOPIC B

Subtract. In Exercise 4 check your result.

3. 725 - 319 =

4. $\begin{array}{r} 607 \\ -218 \\ \hline \end{array}$ (*Check.*)

5. 7 + 9 - 6 =

Now try the exercises for Topic B on page 24.

C. MULTIPLICATION

Multiplication by a counting number amounts to repeated addition. Thus

$$5 \times 3 = \underbrace{5 + 5 + 5}_{\text{3 of these}} = 15$$

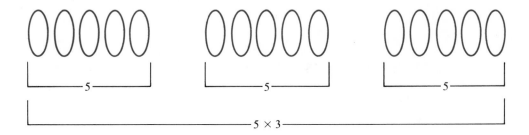

DEFINITION

> When two or more numbers are multiplied, each is called a **factor**, or a **divisor**, of the resulting **product**. For example, 2 and 3 are each factors (divisors) of 6 because the product of 2 and 3, or 2 × 3, is 6. Also, 6 is called a **multiple** of 2 and 3.

Example 4 ▶   485 × 329 =

Solution. Write in column form:

```
          4 8 5
        × 3 2 9
        4 3 6 5   - - - - - -   485 × 9
        9 7 0 0   - - - - -   485 × 2 0   =   970 0
    1 4 5 5 0 0   - - - - -   485 × 3 00 = 1455 00
    1 5 9 5 6 5
```

The product is 159,565.   ◀

Example 5 ▶   7831 × 5080 =

Solution. When one or more digits of the second factor, here 5080, is 0, the method can be shortened.

```
          7 8 3 1
        × 5 0 8 0
        6 2 6 4 8 0   - - - - - -   7831 ×   8 0   = 62648 0
    3 9 1 5 5 0 0 0   - - - - - -   7831 × 5 0 00 = 39155 0 00
    3 9 7 8 1 4 8 0
```

The product is 39,781,480.   ◀

Let n stand for a (counting) number. Then

$$n \times 1 = 1 \times n = n$$

For example,

$$6 \times 1 = 1 \times 6 \qquad \text{and} \qquad 3 \times 1 = 1 \times 3 = 3$$

Three or more numbers can be multiplied. For example,

$$2 \times 3 \times 4 = \underbrace{2 \times 3}_{6} \times 4 = 24$$

PRACTICE EXERCISES FOR TOPIC C

Multiply.

6. $485 \times 361 =$ 7. $6048 \times 3070 =$ 8. $5 \times 2 \times 7 =$

Now try the exercises for Topic C on page 24.

D. DIVISION

Division undoes multiplication, just as subtraction undoes addition. Thus

$$6 \div 2 = 3 \qquad \text{because} \qquad 3 \times 2 = 6$$

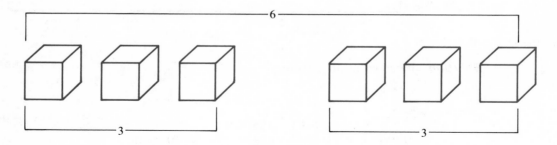

DEFINITION

> In the expression $6 \div 2 = 3$,   6 is called the **dividend**, 2 the **divisor**, and 3 the **quotient**.

To divide larger numbers, the following procedure can be used.

Example 6 ▶ Find $4902 \div 86$. Check your result.

Solution. Here 4902 is the dividend and 86 is the divisor. Write

$$86\overline{)4902}$$

The divisor, 86, is a 2-digit number. Now consider $\boxed{49}02$, whose first two digits are $\boxed{49}$. Because

$$86 > 49 \qquad \text{(86 is greater than 49)}$$

you must consider the first three digits $\boxed{490}$ and divide:

$$\frac{\boxed{49}\,0}{\boxed{8}\,6}$$

Because $\dfrac{49}{8} > 6$, *try* 6.

$$
\begin{array}{r}
6 \\
86\overline{)4902} \\
516 \\
\end{array}
\qquad \text{Too large!}
$$

Next try 5.

$$
\begin{array}{r}
5 \\
86\overline{)4902} \\
430 \\
\hline
60 \\
\end{array}
$$

Bring down the next digit of the divisor.

$$
\begin{array}{r}
5 \\
86\overline{)4902} \\
430 \\
\hline
602 \\
\end{array}
$$

Now divide 602 by the divisor, 86.

$$\frac{\boxed{60}\,2}{\boxed{8}\,6}$$

Because $\dfrac{60}{8} > 7$, *try* 7.

$$
\begin{array}{r}
57 \\
86\overline{)4902} \\
430 \\
\hline
602 \\
602 \\
\hline
\end{array}
$$

The quotient is 57.

You can *check* a division example by multiplying the divisor by the quotient (in either order). The resulting product should then be the dividend.

Check.

$$
\begin{array}{r}
86 \\
\times\ 57 \\
\hline
602 \\
430 \\
\hline
4902 \\
\end{array}
$$

◄

PRACTICE EXERCISES FOR TOPIC D

Divide. In Exercise 10, check your result.

9. $805 \div 23 =$ 10. $2622 \div 57 =$

(*Check.*)

Now try the exercises for Topic D on page 25.

E. DIVISION WITH A REMAINDER

Sometimes when you divide, there is a *remainder.* For example,

$$
\begin{array}{r}
3 \\
21\overline{)67} \\
\underline{63} \\
4
\end{array}
$$

We say that 3 is the **quotient** and that 4 is the **remainder.** The remainder is always smaller than the divisor.

Example 7 ▶   Find   *i.* the quotient and   *ii.* the remainder. Check your result.

$$3888 \div 19 =$$

Solution.

$$
\begin{array}{r}
204 \\
19\overline{)3888} \\
\underline{38} \\
88 \\
\underline{76} \\
12
\end{array}
$$

Observe that 12 is smaller than 19, and there are no more digits in the dividend to bring down.

i. The quotient is 204.   *ii.* The remainder is 12.

To check the result *when there is a remainder*, multiply the quotient and divisor (in either order); then add the remainder. The sum should be the dividend.

Check.
$$
\begin{array}{r}
204 \\
\times\ 19 \\
\hline
1836 \\
204 \\
\hline
3876 \\
+\ 12 \\
\hline
3888
\end{array}
$$

◀

PRACTICE EXERCISES FOR TOPIC E

Find *i.* the quotient and *ii.* the remainder. In Exercise 11 also check your result.

11. 418 ÷ 17 =   12. 5249 ÷ 58 =

  (*Check.*)

Now try the exercises for Topic E on page 25.

F. MULTIPLICATION BY 10, 100, 1000

The accompanying figure illustrates that

  3 × 10 = 30

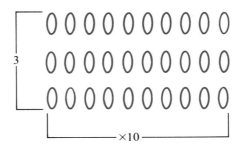

1. To multiply a counting number by 10, which ends in *one* 0, insert a 0 to the right of the 1's digit of the counting number.

2. To multiply by 100, which ends in *two* 0's, insert two 0's to the right of the 1's digit.

3. To multiply by 1000, which ends in *three* 0's, insert three 0's to the right of the 1's digit.

  How many 0's do you insert to the right of the 1's digit when multiplying a counting number by 1,000,000?

Example 8 ▶ *i.* 57 × 10 = 570   *ii.* 57 × 100 = 5700
  iii. 57 × 1000 = 57,000   *iv.* 57 × 1,000,000 = 57,000,000 ◀

PRACTICE EXERCISES FOR TOPIC F

Multiply.

13. 45 × 10 =   14. 63 × 100 =   15. 508 × 1000 =

Now try the exercises for Topic F on page 26.

G. APPLICATIONS

Example 9 ▶ A sum of four thousand eight hundred dollars is divided equally among six business partners. How much is each partner's share?

Solution. Divide 4800 by 6.

$$4800 \div 6 = 800$$

Each partner's share is $800.   ◀

In Example 9, a sum of money was *divided* into 8 equal shares. In other applied problems addition, subtraction, or multiplication are used.

PRACTICE EXERCISES FOR TOPIC G

16. A vendor sells programs for two dollars apiece. How much money does she obtain if she sells 86 programs?

Now try the exercises for Topic G on page 26.

EXERCISES

A. *Add.*

1. $818 + 362 =$ 2. $799 + 894 =$ 3. $8 + 5 + 9 + 3 =$

4. $12 + 15 + 30 + 9 + 18 =$

5. 634
517
209

6. 4830
1945
6685

7. 498
609
384
171

8. 2095
3129
7915
3810

B. *Subtract. In Exercises 19–22, check your result.*

9. $285 - 62 =$ 10. $329 - 92 =$ 11. $714 - 47 =$

12. $108 - 49 =$ 13. $594 - 263 =$ 14. $417 - 208 =$

15. $362 - 253 =$ 16. $743 - 248 =$ 17. $8 + 4 - 7 =$

18. $9 - 6 + 10 =$ 19. $1989 - 998 =$ 20. $5084 - 3989 =$
 (*Check.*)   (*Check.*)

21. $594 - 388 =$ 22. $883 - 466 =$
 (*Check.*)   (*Check.*)

C. *Multiply.*

23. $522 \times 43 =$ 24. $207 \times 81 =$ 25. $345 \times 213 =$

26. $486 \times 306 =$ 27. $786 \times 590 =$ 28. $872 \times 565 =$

29. 8043 × 621 =   **30.** 7205 × 106 =   **31.** 6043 × 3908 =

32. 4925 × 3479 =   **33.** 5 × 2 × 6 =   **34.** 3 × 2 × 4 × 10 =

D. *Divide. In Exercises 45–48, check your result.*

35. 713 ÷ 23 =   **36.** 1428 ÷ 42 =   **37.** 936 ÷ 36 =

38. 893 ÷ 47 =   **39.** 1024 ÷ 64 =   **40.** 2592 ÷ 48 =

41. 8060 ÷ 26 =   **42.** 1530 ÷ 34 =   **43.** 3876 ÷ 68 =

44. 5913 ÷ 27 =   **45.** 3498 ÷ 33 =   **46.** 6431 ÷ 59 =

             (Check.)   *(Check.)*

47. 2835 ÷ 63 =   **48.** 7480 ÷ 68 =

(Check.)   *(Check.)*

E. *Find i. the quotient and ii. the remainder. In Exercises 53–56, iii. check your result.*

49. 89 ÷ 13

Answer: *i.* quotient =   *ii.* remainder =

50. 148 ÷ 17

Answer: *i.* quotient =   *ii.* remainder =

51. 536 ÷ 24

Answer: *i.* quotient =   *ii.* remainder =

52. 805 ÷ 42

Answer: *i.* quotient =   *ii.* remainder =

53. 597 ÷ 39

Answer: *i.* quotient =   *ii.* remainder =

       iii. Check.

54. 3083 ÷ 66

Answer: *i.* quotient =   *ii.* remainder =

       iii. Check.

55. $8902 \div 78$

Answer: i.   quotient =   ii.   remainder =

iii.   Check.

56. $9043 \div 89$

Answer: i.   quotient =   ii.   remainder =

iii.   Check.

F. *Multiply.*

57. $28 \times 10 =$ **58.** $15 \times 100 =$ **59.** $620 \times 100 =$

60. $191 \times 1000 =$ **61.** $12 \times 10,000 =$

62. $287 \times 100,000 =$

G. **63.** A worker earns $53 on Monday, $46 on Tuesday, $49 on Wednesday, $38 on Thursday, and $58 on Friday. How much does he earn for this 5-day week? *Answer:*

64. A woman buys six sundresses for $27 each. How much does she spend for these dresses? *Answer:*

65. Nancy pays for a $23 item with a $50 bill. How much change does she receive? *Answer:*

66. A 12-ounce bottle of diet cola contains 84 calories. How many calories are there per ounce? *Answer:*

67. A man drives 150 miles before lunch and an additional 125 miles afterward. Altogether, how many miles does he drive? *Answer:*

68. How many days are there in 9 weeks? *Answer:*

69. An 18-floor building contains 270 apartments. If each floor contains the same number of apartments, how many apartments are there on a floor? *Answer:*

70. Each of the 42 rows in an auditorium contains 36 seats. How many seats are there in this auditorium? *Answer:*

3. Exponents

Sometimes it is necessary to multiply a number by itself. In fact, the number in question can be used as a factor several times.

A. SQUARE OF A NUMBER

DEFINITION

> The **square** of a number a means
>
> $$a \times a$$

For example, the square of 5 means 5×5, or 25. Write

5^2 for   the square of 5

Thus

$$5^2 = 5 \times 5 = 25$$

Similarly,

$$10^2 = 10 \times 10 = 100$$

In general, write

a^2 for   the square of a

Example 1 ▶   Find:

i. 8^2 *ii.* 12^2

Solution.

i. $8^2 = 8 \times 8 = 64$ *ii.* $12^2 = 12 \times 12 = 144$ ◄

PRACTICE EXERCISES FOR TOPIC A

Find each square.

1. 3^2 2. 20^2

Now try the exercises for Topic A on page 31.

B. CUBE OF A NUMBER

DEFINITION

> The **cube** of a number *a* means
>
> $a \times a \times a$

Thus the cube of 2 means $\underbrace{2 \times 2}_{4} \times 2$, or 8. Write

2^3 for   the cube of 2

so that

$2^3 = 2 \times 2 \times 2 = 8$

Similarly,

$5^3 = \underbrace{5 \times 5}_{25} \times 5 = 125$

In general, write

a^3 for   the cube of *a*

Example 2 ▶ Find:

i. 3^3 *ii.* 10^3

Solution.

i. $3^3 = \underbrace{3 \times 3}_{9} \times 3 = 27$ *ii.* $10^3 = \underbrace{10 \times 10}_{100} \times 10 = 1000$ ◄

PRACTICE EXERCISES FOR TOPIC B

Find each cube.

3. 4^3 4. 100^3

Now try the exercises for Topic B on page 31.

C. HIGHER POWERS

For any number a,

a^4 means $a \times a \times a \times a$ (with 4 factors)

a^5 means $a \times a \times a \times a \times a$ (with 5 factors)

and so on. Thus

$$2^4 = \underbrace{2 \times 2 \times 2}_{8} \underset{\times\ 2}{} \times 2 = 16$$

$$2^5 = \underbrace{2 \times 2 \times 2}_{8} \underset{\times}{} \underbrace{\times 2 \times 2}_{4} = 32$$

Example 3 ▶ Find:

   i. 10^4 *ii.* 10^5

Solution.

   i. $10^4 = 10 \times 10 \times 10 \times 10 = 10,000$

   ii. $10^5 = \underbrace{10 \times 10 \times 10}_{1000} \underset{\times}{} \underbrace{\times 10 \times 10}_{100} = 100,000$ ◀

DEFINITION

> In the expression 10^4, 10 is called the **base** and 4, the **exponent**. We speak of 10^4 as raising 10 **to the fourth power.**

Exponents are introduced to express repeated multiplication. For example, in the expression 10^4, the *exponent* 4 indicates that the *base* 10 is used as a factor 4 times. Thus

$$10^4 = 10 \times 10 \times 10 \times 10$$

Example 4 ▶ In the expression 5^8: *i.* what is the base? *ii.* what is the exponent? *iii.* to what power is 5 raised?

Solution.

   i. 5   *ii.* 8   *iii.* the eighth power   ◀

Observe that

$$1^2 = 1 \times 1 = 1$$

and

$$1^3 = 1 \times 1 \times 1 = 1$$

In fact, every power of 1 is 1. Thus

$$1^{20} = 1 \quad \text{and} \quad 1^{50} = 1$$

Next observe that

$$10^2 = 10 \times 10 = 100 \quad \text{(with 2 zeros)}$$

$$10^3 = 10 \times 10 \times 10 = 1000 \quad \text{(with 3 zeros)}$$

In general, if n is any counting number, then 10^n has n zeros following the digit 1. For example,

$$10^6 = 1,000,000 \quad \text{(6 zeros)}$$

$$10^9 = 1,000,000,000 \quad \text{(9 zeros)}$$

PRACTICE EXERCISES FOR TOPIC C

In Exercises 5–7, find the resulting power.

5. 3^4 6. 10^7 7. 5^4

8. In the expression 3^9 :
   i. What is the base?
   ii. What is the exponent?
   iii. To what power is 3 raised?

Now try the exercises for Topic C on page 31.

D. PRODUCTS OF POWERS

Often you will consider products such as

$$2^3 \times 5^2$$

Here you multiply

the cube of 2   by   the square of 5

Thus

$$2^3 \times 5^2 = 8 \times 25 = 200$$

Example 5 ▶ Find $2^4 \times 3^2$.

Solution.

$$2^4 = 2 \times 2 \times 2 \times 2 = 16$$

$$3^2 = 3 \times 3 = 9$$

Thus

$$2^4 \times 3^2 = 16 \times 9 = 144$$ ◀

Example 6 ▶ Find $2 \times 5 \times 7^2$.

Solution.

$$\underbrace{2 \times 5}_{10} \times \underbrace{7^2}_{49} = 10 \times 49 = 490$$ ◀

PRACTICE EXERCISES FOR TOPIC D

Find each product.

9. $2^3 \times 7^2$ 10. $3 \times 5^2 \times 10^3$

Now try the exercises for Topic D on this page.

EXERCISES

A. *Find each square.*

1. $2^2 =$	2. $4^2 =$	3. $6^2 =$	4. $7^2 =$
5. $9^2 =$	6. $1^2 =$	7. $11^2 =$	8. $15^2 =$
9. $30^2 =$	10. $100^2 =$	11. $16^2 =$	12. $25^2 =$

B. *Find each cube.*

13. $6^3 =$	14. $7^3 =$	15. $1^3 =$	16. $8^3 =$
17. $9^3 =$	18. $20^3 =$	19. $11^3 =$	20. $1000^3 =$

C. *In Exercises 21–30, find each power.*

21. $2^5 =$	22. $2^6 =$	23. $2^7 =$	24. $2^8 =$
25. $4^4 =$	26. $4^5 =$	27. $20^4 =$	28. $20^5 =$
29. $1^{17} =$	30. $1^{99} =$		

31. In the expression 5^6:
    i. What is the base?
    ii. What is the exponent?
    iii. To what power is 5 raised?

32. In the expression 2^{10}:
    i. What is the base?
    ii. What is the exponent?
    iii. To what power is 2 raised?

D. *Find each product.*

33. $2^2 \times 7 =$	34. $3^2 \times 10 =$	35. $2^2 \times 3^2 =$
36. $2^3 \times 3^2 =$	37. $5^2 \times 11 =$	38. $3^4 \times 10^2 =$
39. $7^2 \times 3^4 =$	40. $5^3 \times 10^6 =$	41. $2^2 \times 3 \times 7 =$

42. $5^2 \times 7 \times 10 =$ **43.** $2^3 \times 3^2 \times 5 =$ **44.** $2^4 \times 3^3 \times 7 =$

45. $2^3 \times 7 \times 11 =$ **46.** $2^4 \times 5^2 \times 7^2 =$

4. Prime Factors

The process of writing a counting number as a product of simpler numbers is called *factoring*. This will be helpful when combining fractions.

A. FACTORS

Observe that

$$14 = 2 \times 7$$

Thus 14 can be expressed as the product of two counting numbers, 2 and 7, which are called *factors*, or *divisors*, of 14. We will also say that 14 is a **multiple of** 2 and 7, and that 14 is **divisible by** 2 and 7.

Example 1▶   Show that 7 is a factor of 21.

Solution.

7 is a factor of 21 because $21 = 7 \times 3$.   ◀

Clearly, 3 is also a factor of 21. And 21 is *divisible by* both 3 and 7.

Example 2▶   Find all factors of 10.

Solution.

$$10 = 5 \times 2$$

and also

$$10 = 10 \times 1$$

There is no other way of writing 10 as the product of two counting numbers. The factors of 10 are thus

1, 2, 5, and 10 ◄

The *multiples* of 10 are

10 × 1 (or 10)   10 × 2 (or 20)   10 × 3 (or 30)
10 × 4 (or 40)

and so on. In general, a multiple of 10 can be written as $10 \times c$ for some counting number c. Thus 80 is a multiple of 10 because $80 = 10 \times 8$. On the other hand, the only numbers that are *divisors* (factors) *of* 10 are

1, 2, 5, and 10

because for *counting numbers b* and c,

$$10 = b \times c$$

only when b and c are among the numbers 1, 2, 5, and 10.
    Finally, observe that

50 is *divisible by* 10

because $50 = 10 \times 5$, but that

5 is a *divisor of* 10

because $10 = 5 \times 2$.

Example 3 ▶ *i.* Which of the following are *multiples* of 6? *ii.* Which are *divisors* of 6?

(A) 2   (B) 3   (C) 6   (D) 12   (E) 30

Solution.

i. A *multiple* of 6 must have the form $6 \times c$ for some counting number c. Note that

$$6 = 6 \times 1$$
$$12 = 6 \times 2$$
$$30 = 6 \times 5$$

Therefore (C), (D), and (E) are multiples of 6.
    ii. On the other hand, a counting number c is a *divisor of* 6 if $6 = b \times c$ for some counting number b. Because

$$6 = 3 \times 2 = 2 \times 3$$

6 is divisible by both 2 and 3. And because

$$6 = 1 \times 6$$

6 is divisible by itself. Thus 6 is divisible by choices (A), (B), and (C). ◄

Just as 6 is both a multiple of itself and is divisible *by* itself, so too

> every counting number a is a multiple of a
> and is divisible by a.

PRACTICE EXERCISES FOR TOPIC A

In Exercises 1–3 find all factors of the given counting number.

1. 4   2. 11   3. 12

4. *i.* Which of the following numbers are multiples of 8? *ii.* Which are divisors of 8?

   (A) 4   (B) 8   (C) 12   (D) 16   (E) 48

5. *i.* Which of the following numbers are multiples of 10? *ii.* Which are divisors of 10?

   (A) 1   (B) 5   (C) 25   (D) 50   (E) 230

Now try the exercises for Topic A on page 37.

B. PRIMES AND COMPOSITES

The only way you can express 5 as the product of counting numbers is

$$5 = 5 \times 1 = 1 \times 5$$

Thus 1 and 5 are the only factors of 5.

"Prime" numbers serve as "atoms" among the counting numbers. By multiplying primes, you can obtain all other counting numbers except 1.

DEFINITION

> A counting number p, other than 1, is said to be a **prime** if the only factors of p are p itself and 1.

Clearly, 5 is a prime. Here are some other examples of primes.

Example 4 ▶   *i.* Show that 2 is a prime.
              ii. Show that 11 is a prime.

Solution.

i. 2 is a prime because the only factors of 2 are 1 and 2.
ii. 11 is a prime because the only factors of 11 are 1 and 11. ◄

The first few primes are 2, 3, 5, 7, 11, and 13.

DEFINITION

> Counting numbers other than 1 and primes are known as **composites**.

A composite is thus a counting number that has a factor other than itself or 1.

Example 5 ▶ Show that the following are composites:

i. 4   *ii.* 6   *iii.* 10

Solution.

i. 4 = 2 × 2
Thus 4 is a composite because it has a factor, 2, other than itself or 1.

ii. 6 = 2 × 3
Thus 6 is a composite because it has the factors 2 and 3.

iii. 10 = 2 × 5
Thus 10 is a composite.   ◄

Example 6 ▶ Which of the following are primes? Which are composites?

i. 15   *ii.* 17   *iii.* 33

Solution.

i. 15 is a composite because

15 = 3 × 5

ii. 17 is a prime. Its only factors are 1 and 17.

iii. 33 is a composite because

33 = 3 × 11   ◄

Observe that 1 is not classified as either a prime or a composite. Thus *every counting number is either*

i. 1,
ii. a prime, *or*
iii. a composite

PRACTICE EXERCISES FOR TOPIC B

Which numbers are primes and which are composites?

6. 21   7. 23   8. 31   9. 39

Now try the exercises for Topic B on page 37.

C. PRIME FACTORS

Clearly,

$$12 = 4 \times 3$$

But the factor 4 can be further simplified because $4 = 2 \times 2$. Thus,

$$12 = 2 \times 2 \times 3$$

Every composite can be expressed as the product of primes. Thus,

$$40 = 8 \times 5$$
$$= (2 \times 2 \times 2) \times 5$$
$$= 2^3 \times 5$$

Use exponents to express repeated prime factors. Except for the order of the factors, there is only one way of expressing a composite as the product of primes. No matter how you begin factoring, the *prime* factors are the same. For example,

$$40 = 4 \times 10$$
$$= (2 \times 2) \times (2 \times 5)$$
$$= 2^3 \times 5$$

Example 7 ▶   Express each composite as the product of primes.

   i. 24   *ii.* 30   *iii.* 50

Solution.

   i. $24 = 8 \times 3 = 2^3 \times 3$
  ii. $30 = 6 \times 5 = 2 \times 3 \times 5$
 iii. $50 = 5 \times 10 = 5 \times 2 \times 5 = 2 \times 5^2$ ◀

Example 8 ▶   Express each composite as the product of primes.

   i. 64   *ii.* 72   *iii.* 96

Solution.

   i. $64 = 8 \times 8 = (2 \times 2 \times 2) \times (2 \times 2 \times 2) = 2^6$
  ii. $72 = 8 \times 9 = 2^3 \times 3^2$
 iii. $96 = 3 \times 32 = 3 \times 2^5$ ◀

PRACTICE EXERCISES FOR TOPIC C

Express each composite as the product of primes.

10. 32   11. 44   12. 84

Now try the exercises for Topic C on page 38.

EXERCISES

A. *In Exercises 1–10, find all factors of each number.*

1. 6 *Answer:*   2.   7 *Answer:*

3. 8 *Answer:*   4. 18 *Answer:*

5. 19 *Answer:*   6. 20 *Answer:*

7. 22 *Answer:*   8. 28 *Answer:*

9. 30 *Answer:*   10. 32 *Answer:*

11. *i.* Which of the following are multiples of 5? *ii.* Which are divisors of 5?
  (A) 5   (B) 10   (C) 25   (D) 50   (E) 52
  i. Answer: *ii. Answer:*

12. *i.* Which of the following are multiples of 7? *ii.* Which are divisors of 7?
  (A) 1   (B) 14   (C) 23   (D) 35   (E) 49
  i. Answer: *ii. Answer:*

13. *i.* Which of the following are multiples of 20? *ii.* Which are divisors of 20?
  (A) 4   (B) 5   (C) 10   (D) 40   (E) 50
  i. Answer: *ii. Answer:*

14. *i.* Which of the following are multiples of 3? *ii.* Which are divisors of 3?
  (A) 1   (B) 3   (C) 6   (D) 15   (E) 39
  i. Answer: *ii. Answer:*

15. *i.* Which of the following are multiples of 9? *ii.* Which are divisors of 9?
  (A) 3   (B) 6   (C) 9   (D) 15   (E) 36
  i. Answer: *ii. Answer:*

16. *i.* Which of the following are multiples of 12? *ii.* Which are divisors of 12?
  (A) 3   (B) 6   (C) 96   (D) 144   (E) 288
  i. Answer: *ii. Answer:*

17. Find a number that is a multiple of both 2 and 3. *Answer:*

18. Find a number that is a multiple of both 5 and 6. *Answer:*

19. Find a number that is a multiple of both 6 and 9. *Answer:*

20. Find a number that is a multiple of 2, 3, and 5.   *Answer:*

21. Find a number that is a divisor of both 4 and 6.   *Answer:*

22. Find a number that is a divisor of both 4 and 8.   *Answer:*

B. *Which numbers are primes and which are composites?*

23. 5   *Answer:*   24. 7   *Answer:*   25. 9   *Answer:*

26. 13   *Answer:*   27. 18   *Answer:*   28. 19   *Answer:*

29. 23   *Answer:*   30. 30   *Answer:*   31. 31   *Answer:*

32. 39   *Answer:*   33. 41   *Answer:*   34. 43   *Answer:*

35. 45   *Answer:*   36. 47   *Answer:*

C. *Express each composite as the product of primes.*

37. 6 =	**38.** 8 =	**39.** 14 =
40. 15 =	**41.** 20 =	**42.** 25 =
43. 36 =	**44.** 42 =	**45.** 48 =
46. 54 =	**47.** 60 =	**48.** 80 =
49. 81 =	**50.** 90 =	**51.** 100 =
52. 144 =	**53.** 169 =	**54.** 196 =
55. 288 =	**56.** 720 =	

5. Divisibility Tests

There are very simple tests to determine whether a counting number is divisible by 2, 3, 5, 9, or by a power of 10. These tests are particularly useful when you are factoring a large number, such as 11,700.

A. DIVISIBILITY BY 2

DEFINITION

> A number is **even** if it is divisible by 2; otherwise, it is **odd**.

The following are even:

0:    0 = 0 × 2
2:    2 = 1 × 2
4:    4 = 2 × 2
6:    6 = 3 × 2
8:    8 = 4 × 2

On the other hand,

1, 3, 5, 7, and 9

are odd.

A number is even if its 1's digit is even, and is odd if its 1's digit is odd.

Example 1 ▶   *i.* 456 is even because 6, its 1's digit, is even.
ii. 285 is odd because 5, its 1's digit, is odd.   ◀

PRACTICE EXERCISES FOR TOPIC A

In Exercises 1 and 2, state whether the given number is even or odd.

1. 558   2. 6943

In Exercises 3 and 4, which numbers are divisible by 2?

3. 7530   4. 68,421

Now try the exercises for Topic A on page 44.

B. DIVISIBILITY BY 5

Observe the pattern:

 5 = 1 × 5
10 = 2 × 5
15 = 3 × 5
20 = 4 × 5
          •
          •
          •
75 = 15 × 5
80 = 16 × 5

Each of the numbers, 5, 10, 15, 20, . . . , 75, 80 is divisible by 5.

A number is divisible by 5 if its 1's digit is 5 or 0; otherwise, it is not divisible by 5.

Example 2 ▶   Which of the following numbers are divisible by 5?

i. 435   *ii.* 7490   *iii.* 362

Solution.

    i. 435 is divisible by 5 because its 1's digit is 5.
   ii. 7490 is divisible by 5 because its 1's digit is 0.
  iii. 362 is *not* divisible by 5 because its 1's digit is 2 (rather than 5 or 0).   ◀

PRACTICE EXERCISES FOR TOPIC B

Which numbers are divisible by 5?

5. 105   6. 559   7. 8040

Now try the exercises for Topic B on page 44.

C. DIVISIBILITY BY POWERS OF 10

A number is divisible by 10 if its 1's digit is 0.
A number is divisible by 100 if its 1's digit and 10's digit are *both* 0.
A number is divisible by 1000 if its 1's digit, 10's digit, and 100's digit are *all* 0.

Example 3 ▶   *i.* 760 is divisible by 10.   *ii.* 8100 is divisible by 100.
           iii. 702,000 is divisible by 1000.   ◀

If a number is divisible by 10, then because

$$10 = 5 \times 2$$

the number is also divisible by both 5 and 2. Thus, 760, whose 1's digit is 0, is divisible by both 5 and 2.
    Finally, observe that

$10 = 10^1$	10 ends in 1 zero.
$100 = 10^2$	100 ends in 2 zeros.
$1000 = 10^3$	1000 ends in 3 zeros.

In general, a number is divisible by 10^n if the number ends in n zeros. (Here, n is a counting number.)

Example 4 ▶ Find the highest power of 10 that divides 67,000,000.

Solution.

67,000,000, which ends in 6 zeros, is divisible by 10^6, or 1,000,000. This is the highest power of 10 that divides 67,000,000.   ◀

PRACTICE EXERCISES FOR TOPIC C

8. Which of these numbers are divisible by 10?

     (A) 60   (B) 3002   (C) 4400   (D) 104   (E) 1010

9. Which of these numbers are divisible by 100?

    (A) 540   (B) 5400   (C) 54,000   (D) 101   (E) 1001

10. Which of these numbers are divisible by 1000?

    (A) 200   (B) 2001   (C) 201,000   (D) 1000   (E) 10,001

11. Find the highest power of 10 that divides 5,800,000.

Now try the exercises for Topic C on page 44.

D. DIVISIBILITY BY 3

Observe that

$$36 = 12 \times 3$$

Thus, 36 is divisible by 3. Notice that if you add the digits of 36, you obtain

$$3 + 6 = 9$$

Thus the sum of the digits is also divisible by 3. In general, a number is divisible by 3 if the sum of its digits is divisible by 3. Otherwise, the number is not divisible by 3.

Example 5 ▶   Which of the following numbers are divisible by 3?

    (A) 51   (B) 75   (C) 83   (D) 477   (E) 6020

Solution.

(A) $5 + 1 = 6$

    Thus, 51 is divisible by 3 because 6, the sum of the digits, is divisible by 3. Note that

    $$51 = 17 \times 3$$

(B) $7 + 5 = 12$

    Thus, 75 is divisible by 3 because 12 is divisible by 3. In fact,

    $$75 = 25 \times 3$$

(C) $8 + 3 = 11$

    Here, 83 is *not* divisible by 3 because 11 is not divisible by 3.

(D) $4 + 7 + 7 = 18$

    and furthermore,

    $$1 + 8 = 9$$

    Thus, 477 is divisible by 3 because 18 is divisible by 3. Divide 477 by 3 and note that

$$477 = 159 \times 3$$

(E) $6 + 0 + 2 + 0 = 8$

Thus 6020 is *not* divisible by 3 because 8 is not divisible by 3.

The correct choices are (A), (B), and (D).   ◄

PRACTICE EXERCISES FOR TOPIC D

12. Which of the following numbers are divisible by 3?

   (A) 72   (B) 96   (C) 112   (D) 8076   (E) 2030

Now try the exercises for Topic D on page 45.

E. DIVISIBILITY BY 9

Is there a similar test for divisibility by 9?

$$189 = 21 \times 9$$

and

$$1 + 8 + 9 = 18$$

Thus, 189 is divisible by 9, and 18, the sum of its digits, is divisible by 9.
   A number is divisible by 9 if the sum of its digits is divisible by 9. Otherwise, the number is not divisible by 9.

Example 6 ▶   Which of the following numbers are divisible by 9?

(A) 54   (B) 6084   (C) 8319

Solution.

(A) $5 + 4 = 9$

   Thus, 54 is divisible by 9 because 9, the sum of its digits, is divisible by 9. Here,

   $$54 = 6 \times 9$$

(B) $6 + 0 + 8 + 4 = 18$

   Thus, 6084 is divisible by 9 because 18 is divisible by 9. Divide 6084 by 9 to obtain

   $$6084 = 676 \times 9$$

(C) $8 + 3 + 1 + 9 = 21$

   Thus, 8319 is *not* divisible by 9 because 21 is *not* divisible by 9. Observe that 8319 *is* divisible by 3 because 21 is divisible by 3.   ◄

PRACTICE EXERCISES FOR TOPIC E

13. Which of the following numbers are divisible by 9?

    (A) 171   (B) 567   (C) 829   (D) 6777   (E) 12,345

Now try the exercises for Topic E on page 45.

F. FACTORING

Recall that

    a is *divisible by b*

means that

    b is a factor of *a*

Because 402 is divisible by 3, it follows that 3 is a factor of 402. A major reason for studying divisibility is to aid in factoring large numbers. (And you will sometimes have to factor large numbers when you add fractions.)

Example 7 ▶ Express 275 as the product of primes.

Solution.

The 1's digit of 275 is 5. Divide 275 by 5.

$$275 = 55 \times 5$$

The 1's digit of 55 is 5. Divide 55 by 5 to obtain

$$55 = \underset{\text{prime}}{11} \times 5$$

Thus,

$$275 = \overbrace{11 \times 5}^{55} \times 5$$
$$= 5^2 \times 11$$

◀

Example 8 ▶ Express 11,700 as the product of primes.

Solution.

$$11{,}700$$

ends in 2 zeros. Thus, 11,700 is divisible by 100.

$$11{,}700 = 117 \times 10^2$$

Next,

$$10 = 2 \times 5$$

Therefore,

$$10^2 = 10 \times 10 = (2 \times 5) \times (2 \times 5)$$
$$= 2^2 \times 5^2$$

Also,
$$1 + 1 + 7 = 9$$

Divide 117 by 9 to obtain
$$117 = 13 \times 9$$
$$= 13 \times 3^2$$

Thus,
$$11,700 = 117 \times 10^2$$
$$= (13 \times 3^2) \times (2^2 \times 5^2)$$
$$= 2^2 \times 3^2 \times 5^2 \times 13$$

◀

PRACTICE EXERCISES FOR TOPIC F

Express each composite as the product of primes.

14. 625   15. 4050   16. 147

Now try the exercises for Topic F on page 45.

EXERCISES

A. *In Exercises 1–4, state whether the given number is even or odd.*

  1. 98   *Answer:*   **2.** 89   *Answer:*

  3. 146   *Answer:*   **4.** 7510   *Answer:*

In Exercises 5–8, which numbers are divisible by 2?

  5. 118   *Answer:*   **6.** 793   *Answer:*

  7. 2229 *Answer:*   **8.** 30,410 *Answer:*

B. *Which numbers are divisible by 5?*

  9. 554   *Answer:*   **10.** 375   *Answer:*

11. 1190 *Answer:*   **12.** 55,551 *Answer:*

13. 31,770 *Answer:*   **14.** 50,009 *Answer:*

C. *In Exercises 15–18, which numbers are divisible by 10?*

15. 360   *Answer:*   **16.** 3006   *Answer:*

17. 5080   *Answer:*   **18.** 77,000 *Answer:*

In Exercises 19–22, which numbers are divisible by 100?

19. 2200   *Answer:*   **20.** 2460   *Answer:*

21. 36,000 *Answer:*   **22.** 5,050,000 *Answer:*

In Exercises 23–26, which numbers are divisible by 1000?

23. 9000   *Answer:*   **24.** 60,001   *Answer:*

25. 994,000 *Answer:*   **26.** 8,040,700 *Answer:*

27. Is 67,000 divisible by 10,000?   *Answer:*

28. Is 100,600,000 divisible by 100,000?   *Answer:*

In Exercises 29–32, find the largest power of 10 that divides each number.

29. 9050   *Answer:*   30. 35,900,000 *Answer:*

31. 47,000,000 *Answer:*   32. 6,000,010   *Answer:*

D. *Which numbers are divisible by 3?*

33. 87   *Answer:*   34. 133   *Answer:*

35. 279   *Answer:*   36. 5006   *Answer:*

37. 81,243   *Answer:*   38. 752,907   *Answer:*

E. *Which numbers are divisible by 9?*

39. 126   *Answer:*   40. 849   *Answer:*

41. 7434   *Answer:*   42. 6,000,309   *Answer:*

43. 52,854   *Answer:*   44. 555,555,555 *Answer:*

F. *Express each composite as the product of primes.*

45. 256 =   46. 243 =   47. 3125 =

48. 7000 =   49. 4500 =   50. 441 =

51. 135 =   52. 153 =   53. 216 =

54. 585 =   55. 567 =   56. 6561 =

---

6. Greatest Common Divisors and Least Common Multiples

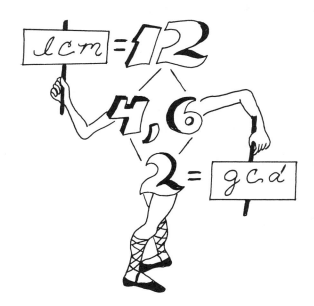

Greatest common divisors are used when simplifying fractions (Section 7).
Least common multiples are used when adding fractions (Section 8).

A. GREATEST COMMON DIVISORS

Recall that when two or more numbers are multiplied, each is called a *factor*, or a *divisor*, of the product. For example,

$$10 = 2 \times 5$$

so that 2 and 5 are each factors (or divisors) of 10. Note that 5 is also a factor of 15 because

$$15 = 3 \times 5$$

We say that 5 is a *common factor* of 10 and 15, according to the following definition:

DEFINITION

> Let m and n be any counting numbers. Then a is called a **common factor** (or a **common divisor**) **of m and n** if a is a factor of *both m and n*. The *largest* common factor of m and n is called the **greatest common divisor of m and n**.

Write

$$gcd\ (m, n) \quad \text{for} \quad \text{the greatest common divisor of } m \text{ and } n$$

Example 1 ▶ Find $gcd\ (6, 9)$.

Solution.

1. The factors of 6 are 1, 2, 3, and 6.
2. The factors of 9 are 1, 3, and 9.
3. The *common* factors of 6 and 9 are 1 and 3.
4. The larger of these common factors is 3. Thus

$$gcd\ (6, 9) = 3$$ ◀

Example 2 ▶ Find $gcd\ (8, 12)$.

Solution.

1. The factors of 8 are 1, 2, 4, and 8.
2. The factors of 12 are 1, 2, 3, 4, 6, and 12.
3. The *common* factors of 8 and 12 are 1, 2, and 4.
4. The largest of the common factors is 4. Thus

$$gcd\ (8, 12) = 4$$ ◀

PRACTICE EXERCISES FOR TOPIC A

Find each *gcd*.

1. $gcd\ (4, 8) =$ 2. $gcd\ (10, 15) =$ 3. $gcd\ (21, 28) =$

Now try the exercises for Topic A on page 51.

B. *gcd* BY FACTORING

When the numbers are comparatively simple, their greatest common divisor can be determined at sight, as in Examples 1 and 2. Otherwise, use prime factoring, as in Example 3.

Example 3 ▶ Find *gcd* (56, 140).

Solution.

$$56 = 8 \times 7 = 2^3 \times 7$$

$$140 = 14 \times 10$$
$$= 2 \times 7 \times 2 \times 5$$
$$= 2^2 \times 5 \times 7$$

To obtain the *gcd*, find the *smaller* power of each *common prime factor*. Then multiply these smaller powers. Thus the common prime factors are 2 and 7. The *smaller* power of 2 is 2^2 (in the prime factorization of 140). And the smaller power of 7 is 7 itself (in both factorizations). Thus

$$gcd\ (56, 140) = 2^2 \times 7$$
$$= 4 \times 7$$
$$= 28$$

◀

To summarize:

> To find the *gcd* of two or more numbers by factoring:
>
> 1. Express them in terms of prime factors.
> 2. Find which primes occur in *all* factorizations.
> 3. For each common prime factor, select the *smallest* power that occurs.
> 4. Multiply these smallest powers to obtain the *gcd*.

PRACTICE EXERCISES FOR TOPIC B

Find each *gcd* by the factoring method.

4. *gcd* (24, 42) =   5. *gcd* (72, 132) =

6. *gcd* (108, 144) =

Now try the exercises for Topic B on page 52.

C. *gcd* OF THREE NUMBERS

Often, the *gcd* of three numbers can be found at sight, as outlined in Example 4.

Example 4 ▶   Find *gcd* (4, 6, 10).

> ***Solution.*** 1. The factors of 4 are 1, 2, and 4.
> 2. The factors of 6 are 1, 2, 3, and 6.
> 3. The factors of 10 are 1, 2, 5, and 10.
> 4. The *common* factors of 4, 6, and 10 are 1 and 2.
> 5. The larger of these common factors is 2.
>
> Thus
>
> $$gcd\ (4, 6, 10) = 2$$ ◀

With more complicated numbers, it is better to use the factoring method.

Example 5 ▶   Find *gcd* (24, 36, 60).

> ***Solution.***
>
> $$24 = 8 \times 3 = 2^3 \times 3$$
> $$36 = 4 \times 9 = 2^2 \times 3^2$$
> $$60 = 4 \times 3 \times 5 = 2^2 \times 3 \times 5$$

The common prime factors are 2 and 3. The *smallest* power of 2 that occurs is 2^2 and the smallest power of 3 that occurs is 3 itself. Thus

$$gcd\ (24, 36, 60) = 2^2 \times 3 = 4 \times 3 = 12$$ ◀

PRACTICE EXERCISES FOR TOPIC C

Find each *gcd*.

7. *gcd* (10, 25, 50) =   8. *gcd* (16, 40, 96) =   9. *gcd* (28, 25, 50) =

Now try the exercises for Topic C on page 52.

D. *lcm* BY LISTING

Recall that 6 is a *multiple* of both 2 and 3 because $2 \times 3 = 6$.

DEFINITION

> The **least common multiple** of two (or more) counting numbers is the *smallest* number that is a multiple of each of them.

Clearly, none of the numbers 1, 2, 3, 4, or 5 is a multiple of *both 2 and 3*. Thus 6 is the smallest multiple of both 2 and 3. As such, it is the least common multiple of 2 and 3.

> *To find the least common multiple of two (relatively simple) counting numbers*, first list the *smallest* multiples of each of them. Then find the smallest number that is on both lists. This is the *least common multiple.*

For example, to find the least common multiple of 4 and 6, observe that the multiples of 4 are

4, 8, $\boxed{12}$, 16, 20, 24, and so on

and the multiples of 6 are

6, $\boxed{12}$, 18, 24, and so on

The smallest number that is on both lists is 12; thus 12 is the least common multiple of 4 and 6.

Write

lcm (4, 6)

for the least common multiple of 4 and 6. Thus

lcm (4, 6) = 12

Example 6 ▶   Find *lcm* (12, 16).

Solution. List the multiples of 12 and 16.

12, 24, 36, $\boxed{48}$, 60, and so on

16, 32, $\boxed{48}$, 64, and so on

The smallest number that is on both lists is 48. Thus

lcm (12, 16) = 48   ◀

PRACTICE EXERCISES FOR TOPIC D

Find each *lcm.*

10. *lcm* (3, 4) =   11. *lcm* (10, 12) =

12. *lcm* (6, 15) =

Now try the exercises for Topic D on page 52.

E. *lcm* BY FACTORING

Sometimes, the least common multiple can be found almost immediately, as in the preceding examples. At other times, it is best to consider the prime factors of the numbers.

> To find the *lcm* of two or more numbers:
>
> 1. Express them in terms of prime factors.
> 2. The *lcm* is the product of the *highest* powers of all primes that occur in any of these factorizations.

Example 7 ▶   Find *lcm* (25, 30).

Solution.

$$25 = 5^2$$
$$30 = 2 \times 3 \times 5$$

The highest power of 2 that occurs is 2 itself.
The highest power of 3 that occurs is 3 itself.
The highest power of 5 that occurs is 5^2.

$$lcm\ (25, 30) = 2 \times 3 \times 5^2$$
$$= 6 \times 25$$
$$= 150$$

◀

Example 8 ▶   Find *lcm* (72, 96).

Solution.

$$72 = 8 \times 9 = 2^3 \times 3^2$$
$$96 = 32 \times 3 = 2^5 \times 3$$

The highest power of 2 that occurs is 2^5.
The highest power of 3 that occurs is 3^2.

$$lcm\ (72, 96) = 2^5 \times 3^2$$
$$= 32 \times 9$$
$$= 288$$

◀

PRACTICE EXERCISES FOR TOPIC E

Find each *lcm* by the factoring method.

13. *lcm* (20, 32) =   14. *lcm* (18, 45) =

15. *lcm* (48, 54) =

Now try the exercises for Topic E on page 52.

F. *lcm* OF THREE NUMBERS

To find the *lcm* of three numbers, the factoring method is generally preferred.

Example 9 ▶ Find *lcm* (40, 50, 64).

Solution.

$$40 = 2^3 \times 5$$
$$50 = 2 \times 5^2$$
$$64 = 2^6$$

The highest power of 2 that occurs is 2^6.
The highest power of 5 that occurs is 5^2. Thus,

$$lcm\ (40, 50, 64) = 2^6 \times 5^2$$
$$= 64 \times 25$$
$$= 1600$$
◀

Example 10 ▶ Find *lcm* (12, 28, 49).

Solution.

$$12 = 2^2 \times 3$$
$$28 = 2^2 \times 7$$
$$49 = 7^2$$

The highest power of 2 that occurs is 2^2.
The highest power of 3 that occurs is 3.
The highest power of 7 that occurs is 7^2. Thus,

$$lcm\ (12, 28, 49) = 2^2 \times 3 \times 7^2$$
$$= \underbrace{4 \times 3}_{12} \times 49$$
$$= 588$$
◀

PRACTICE EXERCISES FOR TOPIC F

Find each *lcm*.

16. *lcm* (9, 12, 16) =   17. *lcm* (20, 25, 30) =

18. *lcm* (24, 42, 48) =

Now try the exercises for Topic F on page 52.

EXERCISES

A. *Find each gcd.*

1. *gcd* (6, 12) =   2. *gcd* (7, 14) =   3. *gcd* (4, 6) =

4. *gcd* (8, 20) =   5. *gcd* (9, 15) =   6. *gcd* (12, 16) =

7. *gcd* (15, 20) =   8. *gcd* (20, 30) =   9. *gcd* (12, 21) =

10. *gcd* (24, 32) =   11. *gcd* (3, 5) =   12. *gcd* (30, 45) =

B. *Find each gcd by the factoring method.*

13. *gcd* (24, 60) =   14. *gcd* (45, 125) =   15. *gcd* (77, 84) =

16. *gcd* (54, 96) =   17. *gcd* (64, 160) =   18. *gcd* (112, 392) =

19. *gcd* (189, 297) =   20. *gcd* (144, 240) =

C. *Find each gcd.*

21. *gcd* (5, 10, 15) =   22. *gcd* (12, 16, 20) =   23. *gcd* (18, 24, 42) =

24. *gcd* (40, 50, 80) =   25. *gcd* (72, 84, 96) =   26. *gcd* (48, 108, 144) =

D. *Find each lcm.*

27. *lcm* (2, 3) =   28. *lcm* (5, 7) =   29. *lcm* (4, 8) =

30. *lcm* (5, 15) =   31. *lcm* (6, 9) =   32. *lcm* (10, 15) =

33. *lcm* (22, 33) =   34. *lcm* (9, 12) =   35. *lcm* (15, 20) =

36. *lcm* (14, 21) =   37. *lcm* (15, 25) =   38. *lcm* (20, 25) =

E. *Find each lcm by the factoring method.*

39. *lcm* (6, 10) =   40. *lcm* (9, 21) =   41. *lcm* (16, 24) =

42. *lcm* (20, 75) =   43. *lcm* (12, 30) =   44. *lcm* (50, 125) =

45. *lcm* (27, 45) =   46. *lcm* (32, 50) =   47. *lcm* (28, 48) =

48. *lcm* (40, 64) =   49. *lcm* (90, 100) =   50. *lcm* (72, 96) =

51. *lcm* (44, 99) =   52. *lcm* (132, 144) =   53. *lcm* (98, 105) =

54. *lcm* (108, 162) =   55. *lcm* (160, 176) =   56. *lcm* (242, 693) =

F. *Find each lcm.*

57. *lcm* (2, 3, 5) =   58. *lcm* (4, 8, 16) =   59. *lcm* (3, 6, 12) =

60. *lcm* (2, 3, 4) =   61. *lcm* (6, 9, 16) =   62. *lcm* (8, 12, 20) =

63. *lcm* (50, 75, 100) =   64. *lcm* (25, 40, 100) =   65. *lcm* (48, 64, 72) =

66. *lcm* (44, 48, 132) =   67. *lcm* (45, 81, 99) =   68. *lcm* (56, 64, 196) =

Review Exercises for Unit I

In Exercises 1–3, write in numerals:

1. five thousand eight hundred forty   *Answer:*

2. nine million six hundred thousand   *Answer:*

3. twenty-seven thousand fifty-two   *Answer:*

In Exercises 4 and 5, write in verbal form:

4. 6208   *Answer:*

5. 150,090   *Answer:*

6. Express the number in the following sentence in numerals by drawing a circle around the correct letter.

There were over three thousand five hundred people in the audience.

(A) 305   (B) 3050   (C) 3005   (D) 3500   (E) 3,000,500

7. Add. 849 + 592 =

8. Add. 3851
          7209
          5010
          ‾‾‾‾

9. Subtract. 6039 – 1041 =   **10.** Multiply. 584 × 409 =

11. Multiply. 409 × 1000 =   **12.** Divide. 2184 ÷ 26 =

13. Find *i.* the quotient and *ii.* the remainder: 927 ÷ 48

Answer: *i.* quotient =   *ii.* remainder =

14. A woman pays for a $36 dress with a $50 bill. How much change does she receive?

Answer:

15. How many envelopes are there in 12 boxes, each of which contains 125 envelopes?

Answer:

16. 8^2 =   **17.** 5^3 =   **18.** 3^5 =   **19.** 10^7 =

20. In the expression 4^6: *i.* what is the base? *ii.* what is the exponent? *iii.* to what power is 4 raised?

Answer: *i.*   *ii.*   *iii.*

21. $2^4 × 3^2$ =   **22.** $5^2 × 7 × 10^3$ =

23. Find all factors of 24.   *Answer:*

24. *i.* Which of the following are multiples of 8? *ii.* Which are divisors of 8?

(A) 1   (B) 4   (C) 8   (D) 20   (E) 40

Answer: *i.*   *ii.*

25. Find a number that is a multiple of both 2 and 7. *Answer:*

26. Find a number greater than 1 that is a divisor of both 20 and 25. *Answer:*

In Exercises 27–29, which numbers are primes and which are composites?

27. 17 *Answer:*   **28.** 29 *Answer:*   **29.** 39 *Answer:*

In Exercises 30–33, express each composite as the product of primes.

30. 28 =   **31.** 54 =   **32.** 96 =   **33.** 240 =

34. Is 47 even or odd? *Answer:*   **35.** Is 2210 divisible by 2? *Answer:*

36. Is 5405 divisible by 5?   *Answer:*

37. Find the largest power of 10 that divides 450,200.   *Answer:*

38. Is 3913 divisible by 3?   *Answer:*   39. Is 4743 divisible by 9?   *Answer:*

In Exercises 40 and 41, express each composite as the product of primes.

40. 4480 =   41. 6993 =

In Exercises 42–47, find each gcd.

42. gcd (15, 20) =   43. gcd (36, 60) =

44. gcd (49, 77) =   45. gcd (108, 135) =

46. gcd (20, 40, 70) =   47. gcd (84, 105, 147) =

In Exercises 48–53, find each lcm.

48. lcm (9, 15) =   49. lcm (28, 40) =

50. lcm (52, 56) =   51. lcm (120, 144) =

52. lcm (6, 15, 20) =   53. lcm (32, 36, 90) =

Practice Exam on Unit I

1. Write in numerals: forty-two thousand one hundred six   *Answer:*

2. Write 220,015 in verbal form: *Answer:*

3. Multiply. 4085 $\times$ 669   *Answer:*

4. Find *i.* the quotient and *ii.* the remainder: 589 $\div$ 36

      Answer: i. quotient =   *ii.* remainder =

5. How many $4 items can be purchased for $72?   *Answer:*

6. 11^2 =   7. 4^4 =   8. $5^2 \times 7$ =

9. Express 84 as the product of primes.   *Answer:*

10. Is 8405 divisible by 5?   *Answer:*

11. Find the largest power of 10 that divides 204,000.   *Answer:*

12. Is 5589 divisible by 9?   *Answer:*

13. Express 17,820 as the product of primes.   *Answer:*

14. Find gcd (48, 64). *Answer:*   15. Find lcm (12, 20, 25). *Answer:*

UNIT II FRACTIONS

7. Adding and Subtracting Fractions

In this section the notion of a fraction is introduced. You will learn how to add and subtract fractions with the same denominator and with different denominators.

A. LOWEST TERMS

DEFINITION

Fractions are numbers, such as $\frac{1}{5}$ and $\frac{3}{10}$, that are written as quotients. For the fraction $\frac{1}{5}$, the *upper* number, 1, is its **numerator** and the *lower* number, 5, is its **denominator**. A fraction, such as $\frac{2}{3}$, in which the numerator is smaller than the denominator, is called a **proper fraction**. And a fraction, such as $\frac{3}{2}$ or $\frac{4}{4}$, in which the numerator is larger than or the same as the denominator, is called an **improper fraction**.

Every counting number can be written as a fraction with denominator 1. For example, $4 = \frac{4}{1}$. Also, the number 1 can be written as a fraction in various ways.

$$1 = \frac{1}{1} = \frac{2}{2} = \frac{3}{3} = \frac{4}{4}, \text{ and so on}$$

A fraction represents a *part* of a whole quantity. For example, the fraction $\frac{3}{5}$ can be represented by the following "pie diagram" in which there are 3 shaded regions out of a total of 5 *equal-sized* regions in all.

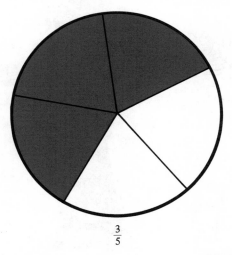

$$\frac{3}{5}$$

A fraction such as $\frac{2}{4}$ is not in *simplest form*. In the accompanying pie diagram observe that $\frac{2}{4}$ of the pie amounts to the same as $\frac{1}{2}$ of the pie.

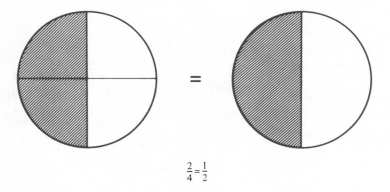

$$\frac{2}{4} = \frac{1}{2}$$

Now observe that the numerator 2 and the denominator 4 of $\frac{2}{4}$ have the factor 2 in common. The fraction $\frac{2}{4}$ can be **reduced to lowest terms** by dividing the numerator and denominator by 2. Thus

$$\frac{2}{4} = \frac{\overset{1}{\cancel{2}} \times 1}{\cancel{2} \times 2} = \frac{1}{2}$$

Note that $\frac{2}{4}$ can also be obtained from $\frac{1}{2}$ by *multiplying* the numerator *and* denominator of $\frac{1}{2}$ by 2. The fractions $\frac{2}{4}$ and $\frac{1}{2}$ are called **equivalent fractions**.

In this example, $\frac{2}{4}$ was reduced to lowest terms by dividing the numerator, 2, and denominator, 4, by 2, which is the *greatest common divisor* of 2 and 4. For the fraction $\frac{6}{9}$,

$$gcd\,(6, 9) = 3$$

Thus to reduce $\frac{6}{9}$ to lowest terms, divide numerator and denominator by 3.

$$\frac{6}{9} = \frac{\overset{1}{\cancel{3}} \times 2}{\underset{1}{\cancel{3}} \times 3} = \frac{1}{3}$$

> To reduce a fraction to lowest terms, divide the numerator and denominator by their greatest common divisor.

Example 1 ▶ Reduce each fraction to lowest terms.

  i. $\dfrac{12}{16}$ ii. $\dfrac{25}{40}$

Solution. Look for the greatest common divisor of the numerator and denominator of each fraction.

  i. $gcd\,(12, 16) = 4$ ii. $gcd\,(25, 40) = 5$

$$\frac{12}{16} = \frac{\overset{1}{\cancel{4}} \times 3}{\underset{1}{\cancel{4}} \times 4} = \frac{3}{4} \qquad\qquad \frac{25}{40} = \frac{\overset{1}{\cancel{5}} \times 5}{\underset{1}{\cancel{5}} \times 8} = \frac{5}{8}$$ ◄

PRACTICE EXERCISES FOR TOPIC A

Reduce each fraction to lowest terms.

1. $\dfrac{5}{20}$ 2. $\dfrac{12}{15}$ 3. $\dfrac{24}{44}$

Now try the exercises for Topic A on page 64.

B. BUILDING FRACTIONS

Usually, it is easiest to work with fractions that have been reduced to lowest terms. However, when adding fractions, a fraction may have to be expressed as an equivalent fraction that is not in lowest terms. The new fraction is *built* from the given one by multiplying the numerator and denominator by a certain counting number. For example, $\frac{1}{2}$ may have to be expressed with denominator 12. The new denominator 12 is then 6 times the given denominator, 2.

$$12 = 6 \times 2$$

Thus multiply the numerator and denominator of $\frac{1}{2}$ by 6.

$$\frac{1}{2} = \frac{6 \times 1}{6 \times 2} = \frac{6}{12}$$

Example 2 ▶   Express $\frac{3}{4}$ *i.* with denominator 8   *ii.* with denominator 24.

Solution.

i. $8 = 2 \times 4$ *ii.* $24 = 6 \times 4$

   Thus   Thus

$$\frac{3}{4} = \frac{2 \times 3}{2 \times 4} = \frac{6}{8}$$ $$\frac{3}{4} = \frac{6 \times 3}{6 \times 4} = \frac{18}{24}$$ ◀

PRACTICE EXERCISES FOR TOPIC B

4. Express $\frac{2}{3}$ with denominator 12.

5. Express $\frac{7}{10}$ *i.* with denominator 20   *ii.* with denominator 50.

Now try the exercises for Topic B on page 64.

C. ADDING AND SUBTRACTING FRACTIONS WITH THE SAME DENOMINATOR

> *To add or subtract fractions with the same denominator D:*
>
> 1. Add or subtract the numerators.
> 2. The denominator is D.
> 3. Reduce to lowest terms, if necessary.

This procedure is illustrated in the accompanying figure.

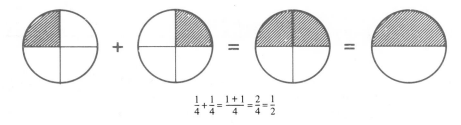

$$\frac{1}{4} + \frac{1}{4} = \frac{1+1}{4} = \frac{2}{4} = \frac{1}{2}$$

Example 3 ▶ Add. $\frac{1}{5} + \frac{2}{5}$

Solution. Both fractions have the same denominator, 5.

Add the numerators.

$$\frac{1}{5} + \frac{2}{5} = \frac{\overbrace{1+2}}{5} = \frac{3}{5}$$

Write the denominator 5.

Here $\frac{3}{5}$ is already in lowest terms because there are no (positive) factors (other than 1) common to the numerator and denominator. ◀

Example 4 ▶ Subtract. $\frac{3}{8} - \frac{1}{8}$

Solution.

$$\frac{3}{8} - \frac{1}{8} = \frac{3-1}{8} = \frac{2}{8} = \frac{1}{4}$$

Here $\frac{2}{8} = \frac{\overset{1}{\cancel{2}} \times 1}{\underset{1}{\cancel{2}} \times 4} = \frac{1}{4}$ ◀

PRACTICE EXERCISES FOR TOPIC C

Add or subtract.

6. $\frac{3}{7} + \frac{2}{7}$ 7. $\frac{5}{9} - \frac{2}{9}$ 8. $\frac{5}{12} + \frac{7}{12}$

Now try the exercises for Topic C on page 64.

D. LEAST COMMON DENOMINATORS

In order to add fractions with *different* denominators, you must first express each of them with the same denominator.

DEFINITION

> The **least common denominator** (*lcd*) of two (or more) fractions is the least common *multiple* of the individual denominators.

For example, the least common denominator of $\frac{2}{3}$ and $\frac{1}{4}$ is the least common multiple of 3 and 4, or 12. To add or subtract $\frac{2}{3}$ and $\frac{1}{4}$, first write them as *equivalent* fractions with 12, the *lcd*, as denominator. To express $\frac{2}{3}$ with denominator 12, note that $12 = 4 \times 3$. Thus multiply the numerator and denominator of $\frac{2}{3}$ by 4.

$$\frac{2}{3} = \frac{4 \times 2}{4 \times 3} = \frac{8}{12}$$

To express $\frac{1}{4}$ with denominator 12, note that $12 = 3 \times 4$, and multiply the numerator and denominator of $\frac{1}{4}$ by 3.

$$\frac{1}{4} = \frac{3 \times 1}{3 \times 4} = \frac{3}{12}$$

Example 5 ▶   *i.* Find the *lcd* of $\frac{5}{12}$ and $\frac{2}{15}$.

ii. Write equivalent fractions with this *lcd* as denominator.

Solution.

i. First find the least common multiple of the individual denominators, 12 and 15.

$$12 = 4 \times 3 = 2^2 \times 3$$
$$15 = 3 \times 5$$

$$\begin{aligned} lcm\ (12, 15) &= 2^2 \times 3 \times 5 \\ &= 4 \times 15 \\ &= 60 \end{aligned}$$

Therefore 60 is the *lcd* of $\frac{5}{12}$ and $\frac{2}{15}$.

ii. To write $\frac{5}{12}$ with denominator 60, multiply the numerator and denominator by 5.

$$\frac{5}{12} = \frac{5 \times 5}{5 \times 12} = \frac{25}{60}$$

To write $\frac{2}{15}$ with denominator 60, multiply the numerator and denominator by 4.

$$\frac{2}{15} = \frac{4 \times 2}{4 \times 15} = \frac{8}{60}$$

◀

PRACTICE EXERCISES FOR TOPIC D

9. i. Find the *lcd* of $\frac{3}{4}$ and $\frac{1}{6}$.

    ii. Write equivalent fractions with this *lcd* as denominator.

10. i. Find the *lcd* of $\frac{5}{8}$ and $\frac{3}{20}$.

    ii. Write equivalent fractions with this *lcd* as denominator.

Now try the exercises for Topic D on page 64.

E. ADDING AND SUBTRACTING FRACTIONS WITH DIFFERENT DENOMINATORS

> *To add or subtract fractions with <u>different</u> denominators:*
> 1. Find their *lcd*.
> 2. Then find equivalent fractions with this *lcd* as denominator.
> 3. Add or subtract the numerators.
> 4. Reduce the resulting fraction to lowest terms, if necessary.

This procedure is illustrated in the accompanying figure.

$$\frac{1}{2} + \frac{1}{4} = \frac{2}{4} + \frac{1}{4} = \frac{3}{4}$$

Example 6 ▶ Add. $\frac{5}{6} + \frac{3}{5}$

Solution. The *lcd* of $\frac{5}{6}$ and $\frac{3}{5}$ is 30, the least common multiple of 6 and 5.

$$\frac{5}{6} = \frac{5 \times 5}{5 \times 6} = \frac{25}{30}$$

$$\frac{3}{5} = \frac{6 \times 3}{6 \times 5} = \frac{18}{30}$$

$$\frac{5}{6} + \frac{3}{5} = \frac{25}{30} + \frac{18}{30} = \frac{25 + 18}{30} = \frac{43}{30}$$

The numerator and denominator of $\frac{43}{30}$ have no factor in common. Thus

$$\frac{5}{6} + \frac{3}{5} = \frac{43}{30} \qquad \blacktriangleleft$$

Example 7 ▶ Subtract. $\frac{11}{20} - \frac{5}{12}$

Solution. First find the least common multiple of 20 and 12.

$$20 = 2^2 \times 5$$
$$12 = 2^2 \times 3$$

$$\begin{aligned} lcm \, (20, 12) &= 2^2 \times 3 \times 5 \\ &= 4 \times 15 \\ &= 60 \end{aligned}$$

Thus 60 is also the *lcd* of the given fractions.

$$\frac{11}{20} = \frac{3 \times 11}{3 \times 20} = \frac{33}{60}$$

$$\frac{5}{12} = \frac{5 \times 5}{5 \times 12} = \frac{25}{60}$$

$$\frac{11}{20} - \frac{5}{12} = \frac{33}{60} - \frac{25}{60}$$

$$= \frac{33 - 25}{60}$$

$$= \frac{8}{60} \qquad \text{(Divide numerator and denominator by 4.)}$$

$$= \frac{2}{15} \qquad \blacktriangleleft$$

PRACTICE EXERCISES FOR TOPIC E

Add or subtract.

11. $\frac{2}{3} + \frac{1}{4}$ 12. $\frac{9}{10} - \frac{3}{4}$ 13. $\frac{3}{10} + \frac{2}{25}$

Now try the exercises for Topic E on page 65.

F. SEVERAL FRACTIONS

Example 8 ▶ $\dfrac{1}{2} + \dfrac{3}{8} - \dfrac{1}{6} =$

(A) $\dfrac{3}{4}$ (B) $\dfrac{3}{8}$ (C) $\dfrac{7}{12}$ (D) $\dfrac{17}{24}$ (E) $\dfrac{25}{24}$

Solution.

$$2 = 2$$
$$8 = 2^3$$
$$6 = 2 \times 3$$
$$lcm\,(2, 8, 6) = 2^3 \times 3$$
$$= 8 \times 3$$
$$= 24$$

Thus the *lcd* of the fractions is 24.

$$\frac{1}{2} = \frac{12 \times 1}{12 \times 2} = \frac{12}{24}$$

$$\frac{3}{8} = \frac{3 \times 3}{3 \times 8} = \frac{9}{24}$$

$$\frac{1}{6} = \frac{4 \times 1}{4 \times 6} = \frac{4}{24}$$

$$\frac{1}{2} + \frac{3}{8} - \frac{1}{6} = \frac{12}{24} + \frac{9}{24} - \frac{4}{24}$$

$$= \frac{12 + 9 - 4}{24}$$

$$= \frac{17}{24}$$

This is in lowest terms. The correct choice is (D).   ◀

PRACTICE EXERCISES FOR TOPIC F

14. $\dfrac{2}{5} + \dfrac{1}{9} + \dfrac{1}{3} =$

15. $\dfrac{7}{12} - \dfrac{1}{2} + \dfrac{5}{6} =$

(A) $\dfrac{1}{12}$ (B) $\dfrac{11}{16}$ (C) $\dfrac{5}{12}$ (D) $\dfrac{11}{12}$ (E) $\dfrac{3}{4}$

Now try the exercises for Topic F on page 65.

EXERCISES

A. *Reduce each fraction to lowest terms.*

1. $\dfrac{7}{14} =$

2. $\dfrac{4}{20} =$

3. $\dfrac{15}{27} =$

4. $\dfrac{35}{50} =$

5. $\dfrac{36}{54} =$

6. $\dfrac{84}{98} =$

7. $\dfrac{108}{144} =$

8. $\dfrac{128}{192} =$

B. 9. Express $\dfrac{1}{5}$ with denominator 10. *Answer:*

10. Express $\dfrac{3}{7}$ with denominator 21. *Answer:*

11. Express $\dfrac{4}{9}$ with denominator 36. *Answer:*

12. Express $\dfrac{9}{11}$ with denominator 55. *Answer:*

13. Express $\dfrac{5}{8}$ *i.* with denominator 16   *ii.* with denominator 24.

    Answer: *i.*   *ii.*

14. Express $\dfrac{7}{10}$ *i.* with denominator 30   *ii.* with denominator 50.

    Answer: *i.*   *ii.*

15. Express $\dfrac{5}{12}$ *i.* with denominator 60   *ii.* with denominator 72.

    Answer: *i.*   *ii.*

16. Express $\dfrac{11}{25}$ *i.* with denominator 100   *ii.* with denominator 125.

    Answer: *i.*   *ii.*

C. *Add or subtract.*

17. $\dfrac{2}{7} + \dfrac{3}{7} =$

18. $\dfrac{5}{9} + \dfrac{1}{9} =$

19. $\dfrac{4}{11} + \dfrac{7}{11} =$

20. $\dfrac{5}{8} - \dfrac{3}{8} =$

21. $\dfrac{7}{6} - \dfrac{5}{6} =$

22. $\dfrac{9}{10} + \dfrac{3}{10} =$

23. $\dfrac{17}{20} - \dfrac{3}{20} =$

24. $\dfrac{31}{42} - \dfrac{11}{42} =$

25. $\dfrac{25}{64} + \dfrac{7}{64} =$

26. $\dfrac{53}{98} - \dfrac{5}{98} =$

D.   *i. Find the lcd of the given fractions.*
    ii. Write equivalent fractions with this lcd as denominator.

27. $\dfrac{1}{2}$ and $\dfrac{1}{4}$ *Answer:*   *i.*   *ii.*

28. $\frac{2}{3}$ and $\frac{1}{5}$ *Answer:*   *i.*   *ii.*

29. $\frac{3}{4}$ and $\frac{5}{6}$ *Answer:*   *i.*   *ii.*

30. $\frac{3}{10}$ and $\frac{4}{15}$ *Answer:*   *i.*   *ii.*

31. $\frac{2}{9}$ and $\frac{1}{12}$ *Answer:*   *i.*   *ii.*

32. $\frac{4}{5}$ and $\frac{9}{20}$ *Answer:*   *i.*   *ii.*

33. $\frac{7}{12}$ and $\frac{3}{16}$ *Answer:*   *i.*   *ii.*

34. $\frac{1}{18}$ and $\frac{5}{24}$ *Answer:*   *i.*   *ii.*

35. $\frac{1}{30}$ and $\frac{7}{40}$ *Answer:*   *i.*   *ii.*

36. $\frac{1}{48}$ and $\frac{5}{36}$ *Answer:*   *i.*   *ii.*

E. *Add or subtract.*

37. $\frac{1}{2} + \frac{1}{3} =$ 38. $\frac{1}{4} + \frac{1}{8} =$ 39. $\frac{3}{5} - \frac{1}{3} =$

40. $\frac{7}{10} - \frac{1}{2} =$ 41. $\frac{2}{7} + \frac{1}{9} =$ 42. $\frac{3}{8} - \frac{1}{6} =$

43. $\frac{5}{8} + \frac{3}{10} =$ 44. $\frac{7}{9} - \frac{1}{6} =$ 45. $\frac{7}{12} - \frac{1}{10} =$

46. $\frac{3}{8} + \frac{2}{3} =$ 47. $\frac{11}{12} - \frac{5}{8} =$ 48. $\frac{7}{10} + \frac{3}{20} =$

49. $\frac{9}{25} - \frac{3}{10} =$ 50. $\frac{1}{36} + \frac{7}{24} =$ 51. $\frac{5}{18} + \frac{5}{12} =$

52. $\frac{9}{40} - \frac{2}{25} =$ 53. $\frac{1}{4} + \frac{2}{3} =$ 54. $\frac{3}{4} + \frac{1}{8} =$

55. $\frac{1}{10} - \frac{1}{20} =$

F. *Combine.*

56. $\frac{1}{12} + \frac{1}{12} + \frac{5}{12} =$ 57. $\frac{4}{9} - \frac{1}{9} + \frac{2}{9} =$ 58. $\frac{1}{2} + \frac{1}{4} + \frac{1}{6} =$

59. $\frac{3}{8} + \frac{1}{4} + \frac{5}{12} =$ 60. $\frac{7}{10} + \frac{2}{5} - \frac{13}{20} =$ 61. $\frac{5}{8} + \frac{3}{4} + \frac{7}{16} =$

62. $\frac{5}{6} + \frac{1}{12} + \frac{2}{9} =$ 63. $\frac{7}{10} + \frac{1}{25} - \frac{1}{50} =$

In the remaining exercises, draw a circle around the correct letter.

64. $\frac{1}{12} + \frac{2}{9} + \frac{3}{4} =$

(A) $\frac{7}{12}$ (B) $\frac{9}{108}$ (C) $\frac{19}{18}$ (D) $\frac{35}{36}$ (E) $\frac{37}{36}$

65. $\frac{3}{10} + \frac{2}{25} + \frac{7}{50} =$

(A) $\frac{13}{25}$ (B) $\frac{12}{25}$ (C) $\frac{13}{50}$ (D) $\frac{23}{50}$ (E) $\frac{27}{100}$

66. $\frac{7}{10} + \frac{4}{15} + \frac{1}{20} =$

(A) $\frac{12}{45}$ (B) $\frac{12}{25}$ (C) $\frac{17}{20}$ (D) $\frac{61}{60}$ (E) 1

67. Ed eats $\frac{1}{4}$ of an apple pie and Joe eats $\frac{3}{8}$ of the pie. How much of the pie remains?

(A) $\frac{1}{2}$ (B) $\frac{5}{8}$ (C) $\frac{3}{8}$ (D) $\frac{1}{8}$ (E) $\frac{1}{4}$

68. A stock goes up $\frac{3}{4}$ of a point one day and it falls $\frac{5}{8}$ of a point the next day. What is its gain or loss for the two days?

(A) It gains $\frac{1}{8}$ of a point.   (B) It loses $\frac{1}{8}$ of a point.   (C) It loses $\frac{1}{4}$ of a point.

(D) It gains one point.   (E) It gains $\frac{3}{8}$ of a point.

69. A 1-yard pole is immersed in a pond so that $\frac{1}{5}$ of the pole is in bottom sand and $\frac{1}{10}$ is in water. What fraction of the pole is above water?

(A) $\frac{3}{20}$ (B) $\frac{3}{10}$ (C) $\frac{2}{5}$ (D) $\frac{3}{5}$ (E) $\frac{7}{10}$

70. On a $\frac{3}{4}$-mile trip, Judy bikes for $\frac{2}{3}$ of a mile. How much of the trip remains?

(A) $\frac{1}{4}$ of a mile   (B) $\frac{1}{3}$ of a mile   (C) $\frac{1}{12}$ of a mile

(D) $\frac{7}{12}$ of a mile   (E) $\frac{1}{2}$ of a mile

8. Multiplying and Dividing Fractions

It is probably easier to multiply or divide fractions than it is to add or subtract them.

A. MULTIPLYING TWO FRACTIONS

The accompanying figure illustrates how fractions are multiplied.

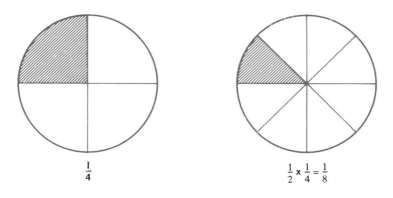

$\frac{1}{4}$ $\frac{1}{2} \times \frac{1}{4} = \frac{1}{8}$

> *To multiply two fractions:*
>
> 1. Divide by factors common to the numerators and denominators.
> 2. Multiply the numerators.
> 3. Multiply the denominators.

Example 1 ▶   Multiply. $\frac{3}{5} \times \frac{1}{4}$

Solution. The numerators and denominators have no factors in common. Thus multiply the numerators and multiply the denominators.

$$\frac{3}{5} \times \frac{1}{4} = \frac{3 \times 1}{5 \times 4} = \frac{3}{20}$$ ◀

When the numerators and denominators have no factors in common, as in Example 1, the rule for multiplying fractions can be expressed as follows:

$$\frac{a}{b} \times \frac{c}{d} = \frac{a \times c}{b \times d}$$

When the numerators and denominators have factors in common, it is best to divide by these common factors before multiplying. You will then be working with smaller numbers.

Example 2 ▶   Multiply. $\frac{3}{10} \times \frac{4}{9}$

Solution. The numerator 3 and the denominator 9 have the factor 3 in common. Divide 3 and 9 by 3. Similarly, the numerator 4 and the denominator 10 have the factor 2 in common. Divide 4 and 10 by 2.

$$\overset{1}{\underset{5}{\cancel{\frac{3}{10}}}} \times \overset{2}{\underset{3}{\cancel{\frac{4}{9}}}} = \frac{1 \times 2}{5 \times 3} = \frac{2}{15}$$ ◀

PRACTICE EXERCISES FOR TOPIC A

Multiply.

1. $\frac{3}{7} \times \frac{2}{5}$ 2. $\frac{5}{8} \times \frac{4}{15}$ 3. $\frac{9}{10} \times \frac{5}{6}$

Now try the exercises for Topic A on page 73.

B. MULTIPLYING THREE FRACTIONS

The same rule applies when multiplying three or more fractions.

Example 3 ▶   Multiply. $\frac{5}{8} \times \frac{12}{7} \times \frac{7}{10}$

Solution. The numerator 5 and the denominator 10 have the factor 5 in common. Divide 5 and 10 by 5. The numerator 12 and the denominator 8 have the factor 4 in common. Divide 12 and 8 by 4. Divide the remaining numerator 7 by the remaining denominator 7.

$$\frac{\overset{1}{\cancel{5}}}{\underset{2}{\cancel{8}}} \times \frac{\overset{3}{\cancel{12}}}{\underset{1}{\cancel{7}}} \times \frac{\overset{1}{\cancel{7}}}{\underset{2}{\cancel{10}}} = \frac{1 \times 3 \times 1}{2 \times 1 \times 2} = \frac{3}{4}$$ ◄

PRACTICE EXERCISES FOR TOPIC B

Multiply.

4. $\frac{2}{9} \times \frac{1}{4} \times \frac{3}{5}$ 5. $\frac{5}{6} \times \frac{1}{10} \times 9$ 6. $\frac{1}{2} \times \frac{1}{3} \times \frac{4}{5} \times \frac{3}{8}$

Now try the exercises for Topic B on page 74.

C. POWERS OF FRACTIONS

To raise a fraction to a power:
1. Raise the numerator to that power.
2. Raise the denominator to that power.

In symbols:

$$\left(\frac{a}{b}\right)^n = \frac{a^n}{b^n}$$

Example 4 ► Find $\left(\frac{3}{5}\right)^2$.

Solution.

$$\left(\frac{3}{5}\right)^2 = \frac{3^2}{5^2} = \frac{9}{25}$$ ◄

Example 5 ► Find $\left(\frac{1}{2}\right)^3$.

Solution.

$$\left(\frac{1}{2}\right)^3 = \frac{1^3}{2^3} = \frac{1}{8}$$ ◄

PRACTICE EXERCISES FOR TOPIC C

Find each power.

7. $\left(\frac{3}{4}\right)^2$ 8. $\left(\frac{1}{10}\right)^3$ 9. $\left(\frac{1}{2}\right)^4$

Now try the exercises for Topic C on page 74.

D. DIVIDING FRACTIONS

In order to discover the rule for dividing by a fraction, consider the pattern in the right-hand column of the following table.

$8 \div 8 =$	1
$8 \div 4 =$	2
$8 \div 2 =$	4
$8 \div 1 =$	8
$8 \div \dfrac{1}{2} =$	
$8 \div \dfrac{1}{4} =$	
$8 \div \dfrac{1}{8} =$	

The divisors are 8, 4, 2, 1, and so on. Each subsequent divisor beginning with 4 is one-half of the preceding divisor. The first four quotients, on the other hand, are 1, 2, 4, and 8. Each subsequent quotient beginning with 2 is twice the preceding quotient. This pattern suggests that

$$8 \div \frac{1}{2} = 16 = 8 \times \frac{2}{1} \qquad \text{(Invert } \tfrac{1}{2} \text{ and multiply.)}$$

$$8 \div \frac{1}{4} = 32 = 8 \times \frac{4}{1} \qquad \text{(Invert } \tfrac{1}{4} \text{ and multiply.)}$$

$$8 \div \frac{1}{8} = 64 = 8 \times \frac{8}{1} \qquad \text{(Invert } \tfrac{1}{8} \text{ and multiply.)}$$

Note that the division at the beginning of the table can also be expressed in terms of multiplication. For example, $8 \div 4 = 2$ can be expressed by

$$\frac{8}{1} \div \frac{4}{1} = \frac{8}{1} \times \frac{1}{4} = 2 \qquad \text{(Invert } \tfrac{4}{1} \text{ and multiply.)}$$

To divide fractions,

$$\frac{a}{b} \div \frac{c}{d}$$

invert the divisor $\frac{c}{d}$) to obtain $\frac{d}{c}$, and then multiply. Thus

$$\frac{a}{b} \div \frac{c}{d} = \frac{a}{b} \times \frac{d}{c} = \frac{a \times d}{b \times c}$$

Example 6 ▶   Divide. $\dfrac{1}{5} \div \dfrac{2}{3}$

Solution.

$$\dfrac{1}{5} \div \dfrac{2}{3} = \dfrac{1}{5} \times \dfrac{3}{2}$$

$$= \dfrac{1 \times 3}{5 \times 2}$$

$$= \dfrac{3}{10}$$ ◀

Example 7 ▶   Divide. $\dfrac{9}{16} \div \dfrac{3}{8}$

Solution.

$$\dfrac{9}{16} \div \dfrac{3}{8} = \dfrac{\overset{3}{\cancel{9}}}{\underset{2}{\cancel{16}}} \times \dfrac{\overset{1}{\cancel{8}}}{\underset{1}{\cancel{3}}} = \dfrac{3}{2}$$ ◀

PRACTICE EXERCISES FOR TOPIC D

Divide.

10. $\dfrac{2}{3} \div \dfrac{4}{5}$ 11. $\dfrac{3}{4} \div \dfrac{7}{12}$ 12. $\dfrac{12}{25} \div \dfrac{3}{20}$

Now try the exercises for Topic D on page 74.

E. FURTHER EXAMPLES

Example 8 ▶ $\dfrac{4}{15} \div \dfrac{2}{5} =$

(A) $\dfrac{8}{75}$ (B) $\dfrac{2}{3}$ (C) $\dfrac{2}{15}$ (D) $\dfrac{2}{75}$ (E) $\dfrac{20}{15}$

Solution.

$$\dfrac{4}{15} \div \dfrac{2}{5} = \dfrac{\overset{2}{\cancel{4}}}{\underset{3}{\cancel{15}}} \times \dfrac{\overset{1}{\cancel{5}}}{\underset{1}{\cancel{2}}} = \dfrac{2}{3}$$

The correct choice is **(B)**. ◀

PRACTICE EXERCISES FOR TOPIC E

13. $\dfrac{5}{9} \times \dfrac{3}{5} =$

(A) 3   (B) $\dfrac{1}{3}$ (C) $\dfrac{1}{9}$ (D) $\dfrac{1}{5}$ (E) $\dfrac{1}{45}$

14. $\frac{7}{12} \div \frac{3}{4} =$

(A) $\frac{1}{2}$ (B) $\frac{7}{16}$ (C) $\frac{16}{7}$ (D) $\frac{7}{9}$ (E) $\frac{9}{7}$

Now try the exercises for Topic E on page 74.

F. APPLICATIONS

Example 9▶   Two-thirds of the radios sold by an appliance store are FM. Of the FM radios sold, $\frac{3}{5}$ are table models. What fraction of the radios sold at that store are FM table models?

(A) $\frac{2}{5}$ (B) $\frac{5}{8}$ (C) $\frac{9}{10}$ (D) $\frac{10}{9}$ (E) $\frac{1}{15}$

Solution. The word "of" frequently indicates multiplication. Let us suppose, for the moment, that 30 radios are sold, in all. If $\frac{2}{3}$ of these are FM then,

$$\frac{2}{\cancel{3}} \times \frac{\cancel{30}^{10}}{1} = \frac{20}{1} = 20$$

20 FM radios are sold. Furthermore, $\frac{3}{5}$ *of* these 20 are table models, so that

$$\frac{3}{\cancel{5}} \times \frac{\cancel{20}^{4}}{1} = \frac{12}{1} = 12$$

12 of the radios sold are FM table models. Here we multiplied the total number of radios sold, 30, by $\frac{2}{3}$, the fraction that are FM and then by $\frac{3}{5}$, the fraction of the FM's that are table models. Note that

$$\frac{\cancel{30}^{6}}{1} \times \frac{2}{\cancel{3}} \times \frac{\cancel{3}^{1}}{\cancel{5}} = \frac{12}{1} = 12$$

No matter what number of radios are sold, to find the fraction that are both FM and table models, multiply by $\frac{2}{3}$ *and* by $\frac{3}{5}$.

$$\frac{2}{\cancel{3}} \times \frac{\cancel{3}^{1}}{5} = \frac{2}{5}$$

Clearly, $\frac{2}{5}$ of the radios sold are FM table models. The correct choice is (A). ◀

Example 10 ▶ The recipe for one batch of cookies requires $\frac{1}{4}$ of a pound of chopped walnuts. How many batches of these cookies can be made from $\frac{1}{2}$ of a pound of chopped walnuts?

(A) $\frac{1}{4}$ (B) $\frac{1}{2}$ (C) $\frac{3}{4}$ (D) 2 (E) 8

Solution. Let us first consider the same kind of problem without fractions. Suppose that one batch of cookies requires 2 pounds of walnuts. Then, from 6 pounds of walnuts you could make

$$6 \div 2, \text{ or } 3$$

batches of cookies. The present recipe calls for $\frac{1}{4}$ of a pound of walnuts. From $\frac{1}{2}$ pound of walnuts you can make $\frac{1}{2} \div \frac{1}{4}$ batches of cookies.

$$\frac{1}{2} \div \frac{1}{4} = \frac{1}{2} \times \frac{4}{1} = 2$$

The correct choice is (D). ◀

PRACTICE EXERCISES FOR TOPIC F

15. One-sixth of a family's income is spent on food. One-ninth of this amount goes for produce. What fraction of this family's income is spent on produce?

(A) $\frac{1}{3}$ (B) $\frac{2}{3}$ (C) $\frac{1}{54}$ (D) $\frac{5}{18}$ (E) $\frac{1}{18}$

16. One-eighth of the students at a school can fit into one bus. How many buses are required to transport $\frac{3}{4}$ of the students?

(A) 2 (B) 3 (C) 4 (D) 5 (E) 6

Now try the exercises for Topic F on page 75.

EXERCISES

A. *Multiply.*

1. $\frac{1}{2} \times \frac{1}{3} =$ 2. $\frac{3}{4} \times \frac{1}{4} =$ 3. $\frac{2}{5} \times \frac{1}{2} =$ 4. $\frac{3}{10} \times \frac{1}{3} =$

5. $\frac{5}{7} \times \frac{2}{5} =$ 6. $\frac{7}{9} \times \frac{3}{7} =$ 7. $\frac{5}{8} \times \frac{4}{15} =$ 8. $\frac{7}{10} \times \frac{5}{14} =$

9. $\frac{8}{9} \times \frac{3}{4} =$ 10. $\frac{9}{10} \times \frac{5}{3} =$ 11. $\frac{3}{7} \times \frac{5}{6} =$ 12. $\frac{7}{12} \times \frac{3}{14} =$

13. $\frac{5}{12} \times \frac{1}{15} =$ 14. $\frac{7}{20} \times \frac{3}{14} =$ 15. $\frac{9}{100} \times \frac{10}{27} =$ 16. $\frac{3}{32} \times \frac{12}{13} =$

B. *Multiply.*

17. $\frac{1}{5} \times \frac{1}{3} \times \frac{1}{2} =$ 18. $\frac{2}{9} \times \frac{1}{2} \times \frac{3}{5} =$ 19. $\frac{1}{12} \times \frac{2}{3} \times \frac{3}{4} =$

20. $\frac{5}{3} \times \frac{1}{10} \times \frac{9}{10} =$ 21. $8 \times \frac{1}{4} \times \frac{1}{2} =$ 22. $\frac{2}{3} \times \frac{1}{5} \times 15 =$

23. $\frac{1}{2} \times \frac{1}{3} \times \frac{1}{4} \times \frac{1}{5} =$ 24. $\frac{2}{3} \times \frac{3}{4} \times \frac{8}{9} \times 27 =$

C. *Find each power.*

25. $\left(\frac{1}{2}\right)^2 =$ 26. $\left(\frac{2}{5}\right)^2 =$ 27. $\left(\frac{3}{10}\right)^2 =$ 28. $\left(\frac{4}{9}\right)^2 =$

29. $\left(\frac{5}{12}\right)^2 =$ 30. $\left(\frac{1}{3}\right)^3 =$ 31. $\left(\frac{2}{3}\right)^3 =$ 32. $\left(\frac{3}{10}\right)^4 =$

D. *Divide.*

33. $\frac{1}{3} \div \frac{1}{2} =$ 34. $\frac{1}{2} \div \frac{1}{3} =$ 35. $\frac{1}{4} \div \frac{1}{2} =$ 36. $\frac{2}{5} \div \frac{4}{7} =$

37. $\frac{3}{8} \div \frac{9}{10} =$ 38. $\frac{5}{6} \div \frac{15}{16} =$ 39. $\frac{7}{12} \div \frac{7}{9} =$ 40. $\frac{3}{8} \div \frac{9}{2} =$

41. $\frac{4}{9} \div \frac{2}{15} =$ 42. $\frac{2}{3} \div \frac{6}{7} =$ 43. $\frac{4}{5} \div \frac{8}{25} =$ 44. $9 \div \frac{1}{6} =$

45. $\frac{1}{9} \div 6 =$ 46. $12 \div \frac{2}{3} =$ 47. $\frac{5}{144} \div \frac{25}{12} =$ 48. $\frac{9}{64} \div \frac{27}{32} =$

E. *Draw a circle around the correct letter.*

49. $\frac{2}{7} \times \frac{5}{8} =$

(A) $\frac{7}{15}$ (B) $\frac{5}{28}$ (C) $\frac{1}{4}$ (D) $\frac{25}{78}$ (E) $\frac{16}{35}$

50. $\frac{7}{10} \times \frac{2}{21} =$

(A) $\frac{7}{100}$ (B) $\frac{9}{210}$ (C) $\frac{1}{15}$ (D) $\frac{147}{20}$ (E) $\frac{9}{31}$

51. $\frac{9}{14} \times \frac{2}{3} =$

(A) $\frac{3}{7}$ (B) $\frac{7}{3}$ (C) $\frac{27}{28}$ (D) $\frac{9}{42}$ (E) $\frac{11}{42}$

52. $\frac{5}{9} \times \frac{3}{10} =$

(A) $\frac{1}{6}$ (B) 6 (C) $\frac{50}{27}$ (D) $\frac{27}{50}$ (E) $\frac{8}{90}$

53. $\frac{2}{5} \div \frac{4}{9} =$

    (A) $\frac{8}{45}$ (B) $\frac{9}{10}$ (C) $\frac{10}{9}$ (D) $\frac{45}{8}$ (E) $\frac{38}{45}$

54. $\frac{3}{7} \div \frac{9}{10} =$

    (A) $\frac{10}{21}$ (B) $\frac{21}{10}$ (C) $\frac{27}{70}$ (D) $\frac{10}{7}$ (E) $\frac{7}{30}$

55. $\frac{5}{6} \div \frac{1}{12} =$

    (A) $\frac{5}{72}$ (B) 10   (C) $\frac{1}{10}$ (D) 2   (E) $\frac{11}{12}$

56. $\frac{7}{10} \div \frac{3}{5} =$

    (A) $\frac{1}{10}$ (B) $\frac{7}{6}$ (C) $\frac{6}{7}$ (D) $\frac{21}{50}$ (E) $\frac{50}{21}$

57. $\frac{5}{12} \times \frac{3}{8} =$

    (A) $\frac{5}{32}$ (B) $\frac{11}{8}$ (C) $\frac{10}{9}$ (D) $\frac{9}{10}$ (E) $\frac{19}{24}$

58. $\frac{6}{7} \div 14 =$

    (A) $\frac{3}{49}$ (B) 12   (C) $\frac{1}{12}$ (D) 3   (E) $\frac{2}{49}$

59. $\frac{9}{10} \times 15 =$

    (A) $\frac{3}{2}$ (B) $\frac{27}{2}$ (C) $\frac{2}{3}$ (D) $\frac{3}{50}$ (E) $\frac{9}{2}$

60. $\frac{3}{32} \div \frac{4}{9} =$

    (A) $\frac{3}{8}$ (B) $\frac{8}{3}$ (C) $\frac{1}{32}$ (D) $\frac{27}{128}$ (E) $\frac{128}{27}$

F. *Draw a circle around the correct letter.*

61. Jerry finds three-fourths of an apple pie in the refrigerator and eats one-sixth of this portion. What fraction of the entire pie does he eat?

    (A) $\frac{1}{24}$ (B) $\frac{3}{8}$ (C) $\frac{1}{12}$ (D) $\frac{1}{8}$ (E) $\frac{5}{24}$

62. Three-fifths of the students at a college are women. One-tenth of the women are psychology majors. What fraction of the students at this college are women who major in psychology?

    (A) $\frac{3}{10}$ (B) $\frac{3}{50}$ (C) $\frac{1}{6}$ (D) $\frac{1}{2}$ (E) $\frac{7}{10}$

63. A recipe calls for $\frac{2}{3}$ of a cup of sugar to make a pound of cookies. How much sugar is there in $\frac{1}{4}$ of a pound of these cookies?

    (A) $\frac{1}{3}$ cup   (B) $\frac{3}{7}$ cup   (C) $\frac{5}{12}$ cup   (D) $\frac{2}{7}$ cup   (E) $\frac{1}{6}$ cup

64. One-half of the graduating class at Central High School goes on to college. Among these college-bound students, $\frac{3}{5}$ graduate from college. What fraction of the graduates of Central High School graduate from college?

    (A) $\frac{3}{10}$ (B) $\frac{4}{7}$ (C) $\frac{3}{7}$ (D) $\frac{5}{6}$ (E) $\frac{1}{10}$

65. Three-fifths of the students at Southwestern Tech are foreign-born. Among these, half are women. Among these foreign-born women, $\frac{5}{6}$ are Spanish-speaking. What fraction of the students at Southwestern Tech are foreign-born, Spanish-speaking women?

    (A) $\frac{5}{12}$ (B) $\frac{3}{10}$ (C) $\frac{1}{5}$ (D) $\frac{1}{4}$ (E) $\frac{1}{8}$

66. One portion of a meat sauce requires $\frac{1}{8}$ of a pound of meat. How many portions of that meat sauce can be made from $\frac{3}{4}$ of a pound of meat?

    (A) $\frac{3}{32}$ (B) $\frac{7}{8}$ (C) 6   (D) $\frac{1}{6}$ (E) $\frac{5}{8}$

67. A box contains $\frac{7}{8}$ cup of bird seed. If a bird eats $\frac{1}{16}$ of a cup each day, how many days will the box last?

    (A) 7   (B) 14   (C) 3   (D) 6   (E) $\frac{7}{128}$

68. Four-fifths of the 250 automobiles an agency rents one month have automatic transmission. Nine-tenths of these cars with automatic transmission are new. How many new cars with automatic transmission are rented by the agency that month?

    (A) 150   (B) 180   (C) 200   (D) 220   (E) 225

9. Mixed Numbers

Mixed numbers involve a counting number plus a proper fraction. Here we consider the basic arithmetic operations on mixed numbers.

A. MIXED NUMBERS TO IMPROPER FRACTIONS

The number

$$4\frac{1}{2} \quad \text{stands for} \quad 4 + \frac{1}{2}$$

This is a **mixed number** (a counting number plus a proper fraction). Here 4 is called the **whole part** and $\frac{1}{2}$ is called the **fractional part**. Observe that

$$4\frac{1}{2} = 4 + \frac{1}{2}$$

$$= \frac{4}{1} + \frac{1}{2}$$

$$= \frac{8}{2} + \frac{1}{2}$$

$$= \frac{9}{2}$$

Thus the mixed number $4\frac{1}{2}$ can also be expressed as the improper fraction $\frac{9}{2}$.

> *To express a mixed number as an improper fraction:*
>
> 1. Write the whole part as a fraction with denominator 1.
> 2. Express this fraction as an equivalent fraction whose denominator is that of the fractional part.
> 3. Add the resulting fractions.

Example 1 ▶   Express $6\frac{2}{3}$ as an improper fraction.

Solution.

$$6\frac{2}{3} = 6 + \frac{2}{3}$$

$$= \frac{6}{1} + \frac{2}{3}$$

$$= \frac{18}{3} + \frac{2}{3}$$

$$= \frac{20}{3}$$

◀

PRACTICE EXERCISES FOR TOPIC A

Express each mixed number as an improper fraction.

1. $3\frac{1}{2}$ 2. $2\frac{3}{4}$ 3. $8\frac{2}{3}$

Now try the exercises for Topic A on page 83.

B. IMPROPER FRACTIONS TO MIXED NUMBERS

How do we convert improper fractions to mixed numbers? Consider the improper fraction $\frac{16}{5}$. Divide the numerator 16 by the denominator 5.

$$
\begin{array}{r}
3 \\
5\overline{)16} \\
15 \\
\hline
1
\end{array}
$$

The quotient, 3, is the *whole part* of the resulting mixed number. The remainder is 1 and the divisor is 5, so that $\frac{1}{5}$ is the fractional part. Thus

$$\frac{16}{5} = 3\frac{1}{5}$$

> *To express an improper fraction $\dfrac{N}{D}$ as a mixed number:*
>
> 1. Divide the numerator, N, by the denominator, D.
> 2. The quotient, Q, is the whole part of the resulting mixed number.
> 3. If R is the remainder, then the fractional part is $\dfrac{R}{D}$.
> 4. Reduce the fractional part $\dfrac{R}{D}$ to lowest terms, if necessary. (If the given improper fraction is in lowest terms, this step is unnecessary.)

Example 2 ▶   Express $\dfrac{23}{6}$ as a mixed number.

Solution. Divide the numerator, 23, by the denominator, 6.

$$\begin{array}{r} 3 \\ 6\,\overline{)23} \\ \underline{18} \\ 5 \end{array}$$

The whole part of the mixed number is the quotient, 3. The remainder is 5. The fractional part is thus $\dfrac{5}{6}$.

Therefore,

$$\frac{23}{6} = 3\,\frac{5}{6}$$ ◀

PRACTICE EXERCISES FOR TOPIC B

Express each improper fraction as a mixed number.

4. $\dfrac{9}{2}$ 5. $\dfrac{11}{3}$ 6. $\dfrac{20}{7}$

Now try the exercises for Topic B on page 84.

C. ADDING AND SUBTRACTING MIXED NUMBERS

To add or subtract mixed numbers:

1. Add or subtract the fractional parts.

2. Add or subtract the whole parts.

3. Simplify, if necessary.

Example 3 ▶ Add. $4\frac{1}{3} + 2\frac{2}{3}$

Solution.

$$
\begin{array}{c}
4 + \dfrac{1}{3} \\
+\, 2 + \dfrac{2}{3} \\
\hline
6 + \dfrac{3}{3} = 6 + 1 \qquad \left(\text{because } \dfrac{3}{3} = 1\right) \\
= 7
\end{array}
$$

◀

Sometimes one or both of the mixed numbers must be changed to an equivalent form before adding or subtracting.

Example 4 ▶ Subtract. $5\frac{1}{2} - 2\frac{3}{4}$

Solution. The fractional parts have different denominators. The *lcd* of $\frac{1}{2}$ and $\frac{3}{4}$ is 4. Thus write $\frac{1}{2} = \frac{2}{4}$ and consider

$$
\begin{array}{r}
5\dfrac{2}{4} \\
-\, 2\dfrac{3}{4} \\
\hline
\end{array}
$$

First subtract the fractional parts. Because $\frac{2}{4}$ is smaller than $\frac{3}{4}$, borrow 1 from the whole part, 5. Thus

$$5\frac{2}{4} = 4 + 1 + \frac{2}{4}$$

$$= 4 + \frac{4}{4} + \frac{2}{4}$$

$$= 4 + \frac{6}{4}$$

Now subtract.

$$4\frac{6}{4}$$

$$-\ 2\frac{3}{4}$$

$$2\frac{3}{4}$$

Thus

$$5\frac{1}{2} - 2\frac{3}{4} = 2\frac{3}{4}$$

Alternate Solution. Convert to fractions and subtract. Then write the resulting fraction as a mixed number.

$$5\frac{1}{2} - 2\frac{3}{4} = \frac{11}{2} - \frac{11}{4}$$

$$= \frac{22}{4} - \frac{11}{4}$$

$$= \frac{11}{4}$$

$$= \frac{8}{4} + \frac{3}{4}$$

whole part / fractional part

$$= 2\frac{3}{4}$$

◀

PRACTICE EXERCISES FOR TOPIC C

Add or subtract.

7. $5\frac{1}{7} + 2\frac{4}{7}$ 　　　　　8. $3\frac{1}{4} - 1\frac{3}{8}$ 　　　　　9. $6\frac{2}{3} + 4\frac{3}{4}$

Now try the exercises for Topic C on page 84.

D. MULTIPLYING AND DIVIDING MIXED NUMBERS

> *To multiply or divide mixed numbers:*
>
> 1. Convert to equivalent improper fractions.
> 2. Multiply or divide these fractions.
> 3. Write the resulting fraction as a mixed number.

Example 5 ▶ Multiply. $4\frac{1}{5} \times 2\frac{1}{2}$

Solution.

$$4\frac{1}{5} = 4 + \frac{1}{5} = \frac{20}{5} + \frac{1}{5} = \frac{21}{5}$$

$$2\frac{1}{2} = 2 + \frac{1}{2} = \frac{4}{2} + \frac{1}{2} = \frac{5}{2}$$

$$4\frac{1}{5} \times 2\frac{1}{2} = \frac{21}{\cancel{5}} \times \frac{\cancel{5}^{\,1}}{2}$$

$$= \frac{21}{2}$$

$$= \frac{20}{2} \overset{\text{whole part}}{\big/} + \frac{1}{2} \longleftarrow \text{fractional part}$$

$$= 10\frac{1}{2}$$

Example 6 ▶ Divide. $5 \div 3\frac{1}{4}$

Solution. First observe that

$$3\frac{1}{4} = 3 + \frac{1}{4} = \frac{12}{4} + \frac{1}{4} = \frac{13}{4}$$

so that

$$5 \div 3\frac{1}{4} = \frac{5}{1} \div \frac{13}{4}$$

$$= \frac{5}{1} \times \frac{4}{13}$$

$$= \frac{20}{13}$$

$$= \frac{13}{13} + \frac{7}{13}$$

$$= 1\frac{7}{13}$$

PRACTICE EXERCISES FOR TOPIC D

Multiply or divide.

10. $2\frac{1}{4} \times 3\frac{1}{3}$ 11. $4\frac{1}{2} \div 2\frac{3}{8}$ 12. $8 \times 4\frac{1}{2} \times 2\frac{3}{4}$

Now try the exercises for Topic D on page 84.

E. FURTHER EXAMPLES

Example 7 ▶ From $3\frac{1}{4}$ cups of flour, a chef uses $1\frac{2}{3}$ cups in preparing a recipe. How many cups of flour remain?

(A) $\frac{29}{12}$ (B) $\frac{31}{12}$ (C) $\frac{19}{12}$ (D) 2   (E) $\frac{7}{12}$

Solution. Subtract: $3\frac{1}{4} - 1\frac{2}{3}$. The *lcd* of $\frac{1}{4}$ and $\frac{2}{3}$ is 12.

$$\frac{1}{4} = \frac{3 \times 1}{3 \times 4} = \frac{3}{12}$$

$$\frac{2}{3} = \frac{4 \times 2}{4 \times 3} = \frac{8}{12}$$

Because $\frac{3}{12}$ is smaller than $\frac{8}{12}$, borrow 1 from the whole part, 3.

$$3\frac{1}{4} = 3\frac{3}{12} = 2 + \frac{12}{12} + \frac{3}{12} = 2\frac{15}{12}$$

$$\begin{array}{r} 2\frac{15}{12} \\ - 1\frac{8}{12} \\ \hline 1\frac{7}{12} \end{array}$$

Now observe that $1\frac{7}{12} = \frac{12}{12} + \frac{7}{12} = \frac{19}{12}$. The correct choice is (C).   ◀

PRACTICE EXERCISES FOR TOPIC E

13. A woman has $2\frac{1}{2}$ meat loaves in her refrigerator. If she is able to obtain 4 portions from each (whole) meat loaf, how many portions can she serve?

Now try the exercises for Topic E on page 85.

EXERCISES

A. *Express as an improper fraction.*

1. $1\frac{1}{2} =$ 2. $2\frac{1}{4} =$ 3. $4\frac{1}{5} =$ 4. $1\frac{3}{4} =$

5. $5\frac{2}{5} =$ 6. $10\frac{3}{4} =$ 7. $7\frac{2}{3} =$ 8. $9\frac{5}{8} =$

B. *Express as a mixed number.*

9. $\frac{8}{5} =$ 10. $\frac{16}{7} =$ 11. $\frac{31}{4} =$ 12. $\frac{55}{8} =$

13. $\frac{89}{6} =$ 14. $\frac{77}{12} =$

C. *Add or subtract. Express the result either as a counting number, as a mixed number, or as a proper fraction.*

15. $2\frac{1}{4} + 3\frac{1}{4} =$ 16. $6\frac{2}{3} - 4\frac{1}{3} =$ 17. $4\frac{1}{2} + 1\frac{1}{2} =$

18. $5\frac{2}{5} - 2\frac{2}{5} =$ 19. $5\frac{1}{2} + 2\frac{1}{4} =$ 20. $6\frac{3}{4} - 1\frac{1}{2} =$

21. $2\frac{2}{5} + 1\frac{1}{2} =$ 22. $5\frac{2}{3} + 1\frac{1}{6} =$ 23. $3\frac{3}{4} + 5\frac{3}{8} =$

24. $10\frac{1}{2} + 7\frac{3}{4} =$ 25. $3\frac{1}{2} - 1\frac{3}{4} =$ 26. $4\frac{1}{4} - 2\frac{1}{2} =$

27. $7\frac{1}{10} - 4\frac{3}{5} =$ 28. $4\frac{2}{9} - 2\frac{2}{3} =$ 29. $3\frac{1}{8} - 2\frac{3}{4} =$

30. $5\frac{1}{12} - 4\frac{3}{8} =$ 31. $2\frac{1}{4} + 5\frac{1}{2} + 4\frac{1}{8} =$ 32. $6\frac{1}{2} + 3\frac{1}{3} - 2\frac{2}{3} =$

D. *Multiply or divide. Express the result either as a counting number, as a mixed number, or as a proper fraction.*

33. $4 \times 2\frac{1}{2} =$ 34. $8 \times 2\frac{2}{3} =$ 35. $2\frac{1}{4} \times 3\frac{1}{2} =$

36. $1\frac{2}{5} \times 1\frac{1}{2} =$ 37. $3\frac{3}{4} \times 1\frac{1}{8} =$ 38. $2\frac{1}{5} \times 4\frac{1}{4} =$

39. $5 \times 2\frac{1}{2} \times 1\frac{1}{4} =$ 40. $2\frac{1}{4} \times 3\frac{1}{3} \times 1\frac{1}{7} =$ 41. $6 \div 1\frac{1}{2} =$

42. $8 \div 2\frac{3}{4} =$ 43. $9\frac{1}{2} \div 3 =$ 44. $2\frac{1}{4} \div 1\frac{3}{4} =$

45. $5\frac{1}{2} \div 1\frac{5}{8} =$ 46. $6\frac{1}{4} \div \frac{3}{8} =$ 47. $3\frac{2}{3} \div 1\frac{5}{6} =$

48. $3\frac{3}{5} \div 1\frac{7}{10} =$

Multiply or divide. Express the result as a fraction.

49. $2\frac{1}{6} \times 5\frac{1}{4} =$ 50. $7\frac{1}{2} \div 3\frac{1}{4} =$ 51. $12\frac{1}{2} \div 10\frac{5}{8} =$

52. $7\frac{3}{4} \div 2\frac{1}{12} =$

E. *Draw a circle around the correct letter.*

53. $4\frac{1}{2} + 2\frac{3}{4} =$

   (A) $6\frac{1}{4}$ (B) $7\frac{1}{4}$ (C) $7\frac{1}{2}$ (D) $6\frac{3}{4}$ (E) $6\frac{7}{8}$

54. $10 - 2\frac{1}{7} =$

   (A) $8\frac{1}{7}$ (B) 8 (C) $7\frac{9}{10}$ (D) $7\frac{6}{7}$ (E) 7

55. $2\frac{1}{4} \times 4\frac{1}{2} =$

   (A) $8\frac{1}{8}$ (B) $6\frac{3}{4}$ (C) 10 (D) $\frac{81}{8}$ (E) $\frac{1}{2}$

56. $3\frac{1}{6} \div 3\frac{1}{3} =$

   (A) 1 (B) $\frac{1}{2}$ (C) $\frac{1}{6}$ (D) $\frac{1}{18}$ (E) $\frac{19}{20}$

57. $6 \div 2\frac{2}{3} =$

   (A) 16 (B) $\frac{9}{4}$ (C) $\frac{10}{3}$ (D) $\frac{9}{8}$ (E) $\frac{3}{10}$

58. $2\frac{1}{8} + 5\frac{3}{4} =$

   (A) 8 (B) $8\frac{1}{8}$ (C) $7\frac{7}{8}$ (D) $8\frac{1}{4}$ (E) $7\frac{3}{32}$

59. $2\frac{2}{3} \div 6 =$

   (A) $4\frac{1}{3}$ (B) $3\frac{1}{3}$ (C) $12\frac{2}{3}$ (D) 16 (E) $\frac{4}{9}$

60. $2\frac{1}{4} \div 1\frac{2}{5} =$

   (A) $1\frac{1}{20}$ (B) $\frac{1}{20}$ (C) $\frac{63}{20}$ (D) $\frac{45}{28}$ (E) $1\frac{5}{8}$

61. An apartment house orders a shipment of $2\frac{1}{4}$ tons of coal in December and another shipment of $1\frac{1}{2}$ tons of coal in January. How many tons of coal does it order?

   (A) $3\frac{1}{2}$ (B) $3\frac{3}{4}$ (C) $3\frac{5}{8}$ (D) $3\frac{7}{8}$ (E) 4

62. On a highway map $\frac{1}{4}$ of an inch represents 1 mile. How many miles are represented by $1\frac{1}{2}$ inches?

   (A) 6 miles   (B) $1\frac{1}{4}$ miles   (C) $1\frac{3}{4}$ miles   (D) $\frac{3}{8}$ mile   (E) 8 miles

63. A restaurant divides each pie into 6 equal portions. If $4\frac{1}{2}$ pies are left, how many portions of pie remain?

(A) 24   (B) $24\frac{1}{2}$ (C) 26   (D) 27   (E) 28

64. On three plays a football team gains $1\frac{1}{2}$ yards, $\frac{3}{4}$ yard, and $4\frac{1}{2}$ yards. How many yards have been gained?

(A) $6\frac{3}{4}$ yards   (B) $5\frac{3}{4}$ yards   (C) 6 yards   (D) 7 yards   (E) $7\frac{1}{4}$ yards

65. A $10\frac{1}{2}$-yard rope is divided into 4 equal parts. How long is each piece?

(A) $2\frac{1}{2}$ yards   (B) $2\frac{1}{8}$ yards   (C) $2\frac{3}{8}$ yards   (D) $2\frac{5}{8}$ yards   (E) $2\frac{7}{8}$ yards

66. A stock which sold at $64\frac{1}{8}$ (points) on Monday fell to $61\frac{3}{4}$ on Tuesday. How much did it fall?

(A) $2\frac{1}{8}$ points   (B) $2\frac{3}{8}$ points   (C) $2\frac{7}{8}$ points   (D) $3\frac{3}{8}$ points   (E) $3\frac{5}{8}$ points

10. Size of Fractions

Which weighs less?

$\frac{3}{4}$ of a ton of coal   or $\frac{4}{5}$ of a ton of coal

Which is the shortest distance?

$$\frac{3}{4} \text{ mile} \quad \text{or} \quad \frac{5}{8} \text{ mile} \quad \text{or} \quad \frac{2}{3} \text{ mile}$$

It is important to recognize which of several fractions is the smallest and which is the largest.

A. FRACTIONS WITH THE SAME DENOMINATOR

Write

$$a < b \qquad \text{if } a \text{ is } \textit{smaller than } b$$

When this is the case, then b is *larger than a.* This is written

$$b > a$$

For example,

$$2 < 5 \qquad (2 \text{ is smaller than } 5)$$

and also,

$$5 > 2 \qquad (5 \text{ is larger than } 2)$$

Observe that the symbols

$$<, \quad >$$

each point to the *smaller* number.

Before comparing the sizes of fractions with the *same* denominator, first note that

$$\frac{20}{5} = 4 \qquad \text{whereas} \qquad \frac{30}{5} = 6$$

Think of $\frac{20}{5}$ and $\frac{30}{5}$ as (improper) fractions. Thus when 20 is divided into 5 parts, the resulting quotient is less than when 30 is divided into 5 parts because

$$20 < 30$$

Similarly, for proper fractions, when 2 is divided into 5 parts, the resulting fraction is less than when 3 is divided into 5 parts. In symbols,

$$\frac{2}{5} < \frac{3}{5}$$

In general, when two fractions[†] have the *same* denominator, the one with

---

[†] Here we assume the fractions are "positive."

the smaller numerator is the smaller fraction. Thus, if a, b, and c are counting numbers,

$$\frac{a}{b} < \frac{c}{b} \quad \text{if} \quad a < c$$

For example,

$$\frac{1}{4} < \frac{3}{4} \quad \text{because} \quad 1 < 3$$

Example 1 ▶ Which of these fractions is the smallest?

(A) $\frac{5}{12}$ (B) $\frac{1}{12}$ (C) $\frac{7}{12}$ (D) $\frac{9}{12}$ (E) $\frac{13}{12}$

Solution. Each of these fractions has denominator 12. Among these, the smallest numerator is 1. Thus the smallest fraction is $\frac{1}{12}$. The correct choice is (B). ◀

PRACTICE EXERCISE FOR TOPIC A

1. Which of these fractions is the smallest?

(A) $\frac{2}{7}$ (B) $\frac{5}{7}$ (C) $\frac{4}{7}$ (D) $\frac{6}{7}$ (E) $\frac{10}{7}$

Now try the exercises for Topic A on page 91.

B. TWO FRACTIONS WITH DIFFERENT DENOMINATORS

To compare the size of two (positive) fractions with *different* denominators, first find equivalent fractions with the same denominator. Then compare the numerators, as before.

When the numbers are small, it is often easiest to multiply the denominators to obtain a common denominator, even though this may not be the *least common denominator.*

Example 2 ▶ Which is smaller? $\frac{3}{4}$ or $\frac{5}{6}$

Solution. Write equivalent fractions with denominator 24 (or 4 × 6).

$$\frac{3}{4} = \frac{6 \times 3}{6 \times 4} = \frac{18}{24}$$

$$\frac{5}{6} = \frac{4 \times 5}{4 \times 6} = \frac{20}{24}$$

Thus $\frac{3}{4} < \frac{5}{6}$ because 18 < 20. ◀

Observe that you could also obtain this by "cross-multiplying":

$$\frac{3}{4} \diagdown \frac{5}{6}$$

$$3 \times 6 < 5 \times 4$$

and therefore

$$\frac{3}{4} < \frac{5}{6}$$

In general, for counting numbers a, b, c, and d,

$$\frac{a}{b} < \frac{c}{d} \quad \text{if} \quad a \times d < b \times c$$

PRACTICE EXERCISES FOR TOPIC B

Which is smaller?

2. $\frac{2}{3}$ or $\frac{3}{4}$ 3. $\frac{5}{8}$ or $\frac{7}{10}$ 4. $\frac{5}{12}$ or $\frac{9}{20}$

Now try the exercises for Topic B on page 92.

C. FINDING THE SMALLEST FRACTION

Example 3 ▶ Which of these fractions is the smallest?

(A) $\frac{3}{4}$ (B) $\frac{5}{8}$ (C) $\frac{7}{8}$ (D) $\frac{3}{5}$ (E) $\frac{5}{7}$

Solution. Compare (A) with (B); then compare the smaller of these with (C), and so on.

(A), (B): $\frac{5}{8} < \frac{3}{4}$ because $5 \times 4 < 3 \times 8$ or $20 < 24$

(A), (B), (C): $\frac{5}{8} < \frac{7}{8}$ because $5 < 7$

(A), (B), (C), (D): $\frac{3}{5} < \frac{5}{8}$ because $3 \times 8 < 5 \times 5$ or $24 < 25$

(A), (B), (C), (D), (E): $\frac{3}{5} < \frac{5}{7}$ because $3 \times 7 < 5 \times 5$ or $21 < 25$

The smallest of these fractions is $\frac{3}{5}$. The correct choice is (D).   ◀

PRACTICE EXERCISES FOR TOPIC C

Which fraction is the smallest? Draw a circle around the correct letter.

5. (A) $\dfrac{1}{6}$ (B) $\dfrac{5}{6}$ (C) $\dfrac{3}{10}$ (D) $\dfrac{7}{10}$ (E) $\dfrac{9}{10}$

6. (A) $\dfrac{1}{2}$ (B) $\dfrac{2}{5}$ (C) $\dfrac{3}{5}$ (D) $\dfrac{4}{7}$ (E) $\dfrac{5}{9}$

7. (A) $\dfrac{7}{10}$ (B) $\dfrac{2}{3}$ (C) $\dfrac{5}{8}$ (D) $\dfrac{3}{5}$ (E) $\dfrac{3}{4}$

Now try the exercises for Topic C on page 92.

D. FINDING THE LARGEST FRACTION

Example 4 ▶ Which of these fractions is the largest?

(A) $\dfrac{2}{7}$ (B) $\dfrac{3}{7}$ (C) $\dfrac{5}{7}$ (D) $\dfrac{3}{8}$ (E) $\dfrac{5}{8}$

Solution. Fractions (A), (B), and (C) all have denominator 7. Among these, clearly $\dfrac{5}{7}$ is the largest. Fractions (D) and (E) have denominator 8. The larger of fractions (D) and (E) is $\dfrac{5}{8}$. Finally,

$$\frac{5}{8} < \frac{5}{7} \quad \text{because} \quad 5 \times 7 < 8 \times 5 \quad \text{or} \quad 35 < 40$$

Observe that when two (positive) fractions have the same numerator, the *smaller fraction* corresponds to the *larger denominator*. Thus the *largest* of the five fractions is $\dfrac{5}{7}$ [choice (C)]. ◀

PRACTICE EXERCISES FOR TOPIC D

Which fraction is the largest? Draw a circle around the correct letter.

8. (A) $\dfrac{1}{10}$ (B) $\dfrac{7}{10}$ (C) $\dfrac{9}{10}$ (D) $\dfrac{7}{12}$ (E) $\dfrac{11}{12}$

9. (A) $\dfrac{4}{5}$ (B) $\dfrac{7}{8}$ (C) $\dfrac{3}{4}$ (D) $\dfrac{5}{6}$ (E) $\dfrac{7}{10}$

Now try the exercises for Topic D on page 93.

E. APPLICATIONS

Example 5 ▶   Which is a better buy? a 2-ounce box of raisins for 25 cents or a 3-ounce box for 37 cents

Solution. We want to increase the amount of raisins we get for each cent spent. Thus form the fractions

$$\frac{2 \text{ (ounces)}}{25 \text{ (cents)}} \quad \text{and} \quad \frac{3 \text{ (ounces)}}{37 \text{ (cents)}}$$

Compare $\frac{2}{25}$ with $\frac{3}{37}$. Clearly

$$\underbrace{2 \times 37}_{74} < \underbrace{3 \times 25}_{75}$$

Thus

$$\frac{2}{25} < \frac{3}{37}$$

Therefore, the 3-ounce box for 37 cents is the better buy.   ◀

PRACTICE EXERCISES FOR TOPIC E

10. Which weighs less? $\frac{3}{4}$ of a ton of coal or $\frac{4}{5}$ of a ton of coal

11. Which is the shortest distance? $\frac{3}{4}$ mile, $\frac{5}{8}$ mile, or $\frac{2}{3}$ mile

Now try the exercises for Topic E on page 94.

EXERCISES

A. *In each exercise, which fraction is the smallest? Draw a circle around the correct letter.*

1. (A) $\frac{2}{5}$ (B) $\frac{3}{5}$ (C) $\frac{4}{5}$

2. (A) $\frac{4}{9}$ (B) $\frac{2}{9}$ (C) $\frac{5}{9}$ (D) $\frac{8}{9}$

3. (A) $\frac{7}{10}$ (B) $\frac{9}{10}$ (C) $\frac{1}{10}$ (D) $\frac{3}{10}$

4. (A) $\frac{4}{13}$ (B) $\frac{7}{13}$ (C) $\frac{3}{13}$ (D) $\frac{11}{13}$ (E) $\frac{12}{13}$

5. (A) $\frac{4}{15}$ (B) $\frac{11}{15}$ (C) $\frac{7}{15}$ (D) $\frac{2}{15}$ (E) $\frac{13}{15}$

6. (A) $\frac{7}{20}$ (B) $\frac{9}{20}$ (C) $\frac{11}{20}$ (D) $\frac{3}{20}$ (E) $\frac{23}{20}$

B. *In each exercise, which is smaller?*

7. $\frac{1}{2}$ or $\frac{3}{5}$ *Answer:*   8. $\frac{3}{7}$ or $\frac{4}{9}$ *Answer:*

9. $\frac{2}{5}$ or $\frac{1}{3}$ *Answer:*   10. $\frac{3}{10}$ or $\frac{2}{7}$ *Answer:*

11. $\frac{7}{8}$ or $\frac{5}{6}$ *Answer:*   12. $\frac{7}{9}$ or $\frac{4}{5}$ *Answer:*

13. $\frac{9}{11}$ or $\frac{11}{13}$ *Answer:*   14. $\frac{3}{14}$ or $\frac{5}{13}$ *Answer:*

15. $\frac{9}{16}$ or $\frac{7}{13}$ *Answer:*   16. $\frac{10}{19}$ or $\frac{11}{24}$ *Answer:*

C. *In each exercise, which fraction is the smallest? Draw a circle around the correct letter.*

17. (A) $\frac{1}{4}$ (B) $\frac{3}{4}$ (C) $\frac{1}{5}$ (D) $\frac{3}{5}$ (E) $\frac{4}{5}$

18. (A) $\frac{7}{8}$ (B) $\frac{3}{8}$ (C) $\frac{1}{8}$ (D) $\frac{1}{9}$ (E) $\frac{7}{9}$

19. (A) $\frac{1}{10}$ (B) $\frac{1}{9}$ (C) $\frac{1}{8}$ (D) $\frac{1}{7}$ (E) $\frac{1}{6}$

20. (A) $\frac{3}{8}$ (B) $\frac{3}{10}$ (C) $\frac{3}{13}$ (D) $\frac{1}{9}$ (E) $\frac{1}{11}$

21. (A) $\frac{2}{5}$ (B) $\frac{1}{3}$ (C) $\frac{3}{7}$ (D) $\frac{2}{9}$ (E) $\frac{2}{11}$

22. (A) $\frac{7}{10}$ (B) $\frac{2}{3}$ (C) $\frac{3}{5}$ (D) $\frac{5}{8}$ (E) $\frac{8}{11}$

23. (A) $\frac{4}{9}$ (B) $\frac{1}{2}$ (C) $\frac{2}{5}$ (D) $\frac{5}{11}$ (E) $\frac{3}{7}$

24. (A) $\frac{9}{10}$ (B) $\frac{8}{9}$ (C) $\frac{9}{11}$ (D) $\frac{7}{8}$ (E) $\frac{19}{20}$

25. (A) $\frac{1}{6}$ (B) $\frac{1}{3}$ (C) $\frac{2}{11}$ (D) $\frac{1}{10}$ (E) $\frac{2}{9}$

26. (A) $\frac{3}{4}$ (B) $\frac{7}{10}$ (C) $\frac{4}{5}$ (D) $\frac{7}{9}$ (E) $\frac{5}{8}$

27. (A) $\frac{1}{10}$ (B) $\frac{9}{100}$ (C) $\frac{91}{1000}$ (D) $\frac{3}{20}$ (E) $\frac{1}{5}$

28. (A) $\frac{3}{5}$ (B) $\frac{5}{8}$ (C) $\frac{5}{9}$ (D) $\frac{7}{10}$ (E) $\frac{2}{3}$

29. (A) $\frac{7}{6}$ (B) $\frac{5}{4}$ (C) $\frac{4}{3}$ (D) $\frac{6}{5}$ (E) $\frac{8}{7}$

30. (A) $\frac{3}{10}$ (B) $\frac{2}{7}$ (C) $\frac{4}{9}$ (D) $\frac{3}{8}$ (E) $\frac{5}{12}$

31. (A) $\frac{7}{10}$ (B) $\frac{4}{5}$ (C) $\frac{7}{9}$ (D) $\frac{7}{8}$ (E) $\frac{9}{11}$

32. (A) $\frac{5}{12}$ (B) $\frac{7}{16}$ (C) $\frac{1}{2}$ (D) $\frac{2}{5}$ (E) $\frac{3}{8}$

33. (A) $\frac{8}{11}$ (B) $\frac{3}{4}$ (C) $\frac{7}{9}$ (D) $\frac{5}{6}$ (E) $\frac{4}{5}$

34. (A) $\frac{2}{9}$ (B) $\frac{1}{4}$ (C) $\frac{3}{8}$ (D) $\frac{2}{7}$ (E) $\frac{3}{11}$

35. (A) $\frac{7}{20}$ (B) $\frac{1}{3}$ (C) $\frac{5}{11}$ (D) $\frac{5}{12}$ (E) $\frac{4}{15}$

36. (A) $\frac{8}{13}$ (B) $\frac{7}{11}$ (C) $\frac{5}{7}$ (D) $\frac{13}{20}$ (E) $\frac{7}{12}$

D. *In each exercise, which fraction is the largest? Draw a circle around the correct letter.*

37. (A) $\frac{1}{2}$ (B) $\frac{1}{5}$ (C) $\frac{1}{6}$ (D) $\frac{1}{7}$ (E) $\frac{1}{8}$

38. (A) $\dfrac{7}{9}$ (B) $\dfrac{2}{9}$ (C) $\dfrac{5}{9}$ (D) $\dfrac{2}{3}$ (E) $\dfrac{1}{3}$

39. (A) $\dfrac{5}{8}$ (B) $\dfrac{4}{7}$ (C) $\dfrac{3}{5}$ (D) $\dfrac{2}{3}$ (E) $\dfrac{7}{10}$

40. (A) $\dfrac{2}{5}$ (B) $\dfrac{1}{3}$ (C) $\dfrac{4}{11}$ (D) $\dfrac{3}{8}$ (E) $\dfrac{5}{12}$

41. (A) $\dfrac{7}{9}$ (B) $\dfrac{3}{4}$ (C) $\dfrac{8}{11}$ (D) $\dfrac{13}{16}$ (E) $\dfrac{7}{10}$

42. (A) $\dfrac{1}{8}$ (B) $\dfrac{2}{11}$ (C) $\dfrac{3}{20}$ (D) $\dfrac{2}{15}$ (E) $\dfrac{1}{10}$

43. (A) $\dfrac{5}{3}$ (B) $\dfrac{7}{5}$ (C) $\dfrac{5}{4}$ (D) $\dfrac{8}{7}$ (E) $\dfrac{11}{9}$

44. (A) $\dfrac{7}{15}$ (B) $\dfrac{11}{20}$ (C) $\dfrac{1}{2}$ (D) $\dfrac{4}{7}$ (E) $\dfrac{6}{11}$

E. 45. Which weighs more? $\dfrac{1}{2}$ pound of feathers or $\dfrac{7}{16}$ pound of lead? *Answer:*

46. Henry's portion is $\dfrac{3}{8}$ of the pie, Walter's is $\dfrac{2}{5}$. Who receives the larger portion?
Answer:

47. One side of a kerchief measures $\dfrac{13}{20}$ of a meter. An adjacent side measures $\dfrac{7}{10}$ of a
meter. Which side is longer? *Answer:*

48. A salad dressing calls for $\dfrac{3}{4}$ cup of wine and $\dfrac{2}{3}$ cup of vinegar. Is there more wine or
more vinegar in the dressing? *Answer:*

49. Which is a better buy? a 6-ounce bar of soap for 75 cents or an 8-ounce bar for
92 cents *Answer:*

50. Which is a better buy? a 12-ounce can of string beans for 64 cents or a pound can
for a dollar *Answer:*

Review Exercises for Unit II

1. Reduce $\frac{12}{20}$ to lowest terms. *Answer:*   2. Reduce $\frac{45}{75}$ to lowest terms. *Answer:*

3. Express $\frac{3}{8}$ *i.* with denominator 24   *ii.* with denominator 56.

   Answer: i.   *ii.*

4. Add. $\frac{5}{12} + \frac{1}{12} =$ 5. Subtract. $\frac{7}{15} - \frac{2}{15} =$

In Exercises 6 and 7, i. find the lcd of the given fractions. ii. Write equivalent fractions with the lcd as denominator.

6. $\frac{5}{6}$ and $\frac{4}{9}$ *Answer: i.*   *ii.*

7. $\frac{7}{12}$ and $\frac{3}{20}$ *Answer: i.*   *ii.*

In Exercises 8–12, combine.

8. $\frac{3}{4} + \frac{1}{10} =$ 9. $\frac{7}{8} - \frac{1}{5} =$ 10. $\frac{7}{15} + \frac{3}{25} =$

11. $\frac{3}{4} + \frac{5}{6} + \frac{1}{8} =$ 12. $\frac{1}{2} + \frac{9}{10} - \frac{3}{5} =$

13. Bob completes $\frac{1}{2}$ of a job and Joan completes $\frac{1}{3}$ of this job. How much of the job remains?

   (A) $\frac{1}{5}$ (B) $\frac{4}{5}$ (C) $\frac{1}{6}$ (D) $\frac{5}{6}$ (E) $\frac{1}{4}$

In Exercises 14–18, multiply.

14. $\frac{2}{7} \times \frac{3}{5} =$ 15. $\frac{4}{9} \times \frac{3}{10} =$ 16. $\frac{3}{16} \times \frac{5}{12} =$

17. $\frac{2}{7} \times \frac{3}{10} \times \frac{21}{4} =$ 18. $12 \times \frac{5}{18} \times \frac{9}{10} =$

In Exercises 19–22, divide.

19. $\frac{1}{6} \div \frac{2}{3} =$ 20. $\frac{3}{8} \div \frac{9}{4} =$ 21. $10 \div \frac{2}{5} =$ 22. $\frac{9}{56} \div \frac{7}{12} =$

In Exercises 23–25, draw a circle around the correct letter.

23. $\frac{5}{12} \times \frac{9}{10} =$

   (A) $\frac{3}{8}$ (B) $\frac{7}{11}$ (C) $\frac{25}{49}$ (D) $\frac{1}{2}$ (E) $\frac{3}{4}$

24. $\frac{12}{25} \div \frac{3}{50} =$

    (A) $\frac{1}{8}$ (B) 8   (C) $\frac{9}{25}$ (D) $\frac{18}{625}$ (E) $\frac{1}{5}$

25. At a matinee performance, four-fifths of the people are retired. Two-thirds of these retired people like the show. What fraction of those attending the matinee are retired people who like the show?

    (A) $\frac{8}{15}$ (B) $\frac{5}{6}$ (C) $\frac{3}{5}$ (D) $\frac{2}{5}$ (E) $\frac{2}{15}$

26. Express $2\frac{3}{5}$ as an improper fraction.   *Answer:*

27. Express $10\frac{1}{4}$ as an improper fraction.   *Answer:*

28. Express $\frac{29}{6}$ as a mixed number.   *Answer:*

In Exercises 29–36, combine. Express the result either as a counting number, as a mixed number, or as a proper fraction.

29. $4\frac{1}{2} + 3\frac{1}{2} =$ 30. $10\frac{5}{6} - 3\frac{1}{6} =$ 31. $1\frac{3}{4} + 2\frac{5}{8} =$

32. $5\frac{2}{3} - 4\frac{1}{2} =$ 33. $5\frac{2}{5} \times 3 =$ 34. $2\frac{3}{4} \times 4\frac{1}{6} =$

35. $1\frac{3}{10} \div 4 =$ 36. $5\frac{1}{2} \div 2\frac{7}{8} =$

In Exercises 37 and 38, multiply and express the result as a fraction.

37. $2\frac{3}{8} \times 4\frac{1}{4} =$ 38. $5\frac{1}{4} \div 1\frac{3}{10} =$

In Exercises 39–41, draw a circle around the correct letter.

39. $5\frac{1}{6} + 3\frac{1}{4} =$

    (A) $8\frac{1}{10}$ (B) $8\frac{1}{12}$ (C) $15\frac{1}{24}$ (D) $8\frac{5}{12}$ (E) $8\frac{1}{3}$

40. If a pound and three-quarters of sugar has been poured from a 5-pound box, how much sugar remains in the box?

    (A) $6\frac{3}{4}$ pounds   (B) $4\frac{1}{4}$ pounds   (C) $3\frac{1}{4}$ pounds   (D) $8\frac{3}{4}$ pounds

    (E) $2\frac{6}{7}$ pounds

41. Which fraction is the smallest? Draw a circle around the correct letter.

    (A) $\frac{3}{16}$ (B) $\frac{9}{16}$ (C) $\frac{5}{16}$ (D) $\frac{7}{16}$ (E) $\frac{13}{16}$

42. Which is smaller? $\frac{3}{5}$ or $\frac{7}{10}$ *Answer:*   **43.** Which is smaller? $\frac{13}{18}$ or $\frac{2}{3}$ *Answer:*

In Exercises 44–46, which fraction is the smallest? Draw a circle around the correct letter.

44. (A) $\frac{1}{9}$ (B) $\frac{7}{9}$ (C) $\frac{3}{10}$ (D) $\frac{1}{10}$ (E) $\frac{7}{10}$

45. (A) $\frac{1}{4}$ (B) $\frac{2}{9}$ (C) $\frac{1}{3}$ (D) $\frac{3}{10}$ (E) $\frac{5}{12}$

46. (A) $\frac{2}{25}$ (B) $\frac{1}{10}$ (C) $\frac{3}{50}$ (D) $\frac{9}{100}$ (E) $\frac{11}{100}$

In Exercises 47 and 48, which fraction is the largest? Draw a circle around the correct letter.

47. (A) $\frac{3}{4}$ (B) $\frac{5}{8}$ (C) $\frac{1}{2}$ (D) $\frac{3}{8}$ (E) $\frac{7}{16}$

48. (A) $\frac{4}{15}$ (B) $\frac{5}{12}$ (C) $\frac{2}{9}$ (D) $\frac{5}{9}$ (E) $\frac{8}{15}$

49. Which weighs more? $\frac{3}{10}$ of a pound of chicken or $\frac{1}{4}$ of a pound of liver *Answer:*

50. Which is a better buy? a 12-ounce box of cornflakes for 80 cents or a 14-ounce box for 90 cents *Answer:*

Review Exercises on Unit I

---

  1. Write in numerals: seventy-two thousand forty-five

  2. Write in verbal form: 1,004,802

  3. Subtract 384 from 491.

  4. Multiply: 585 × 478

  5. Find *i.* the quotient and *ii.* the remainder: 797 ÷ 28

  6. If a pad of paper contains 45 sheets, how many sheets are there in 12 pads?

  7. $2^5 \times 3^2 =$

  8. Express 132 as the product of primes.

  9. Find the largest power of 10 that divides 505,000.

 10. Find *lcm* (25, 30).

 11. Find *gcd* (18, 105).

 12. Express 1440 as the product of primes.

Practice Exam on Unit II

1. Reduce $\frac{20}{32}$ to lowest terms. *Answer:*   2. $\frac{5}{12} + \frac{2}{9} =$

3. $\frac{3}{8} - \frac{5}{16} =$ 4. $\frac{7}{10} - \frac{1}{5} + \frac{1}{2} =$ 5. $\frac{3}{10} \times \frac{20}{27} =$ 6. $\frac{5}{12} \div \frac{10}{21} =$

7. Express $5\frac{3}{4}$ as an improper fraction. *Answer:*

8. Express $\frac{10}{3}$ as a mixed number. *Answer:*

9. $6\frac{1}{2} - 4\frac{3}{4} =$ 10. $4\frac{3}{5} \div 1\frac{1}{2} =$

11. Which fraction is the smallest?

   (A) $\frac{3}{4}$ (B) $\frac{3}{8}$ (C) $\frac{5}{16}$ (D) $\frac{7}{20}$ (E) $\frac{13}{22}$

12. Which fraction is the largest?

   (A) $\frac{1}{2}$ (B) $\frac{3}{5}$ (C) $\frac{2}{3}$ (D) $\frac{5}{9}$ (E) $\frac{6}{13}$

13. On a highway map $\frac{1}{3}$ of an inch represents 1 mile. How many miles are represented by $2\frac{2}{3}$ inches?

   (A) 5   (B) 6   (C) 7   (D) 8   (E) 9

14. A stock which sold at $56\frac{1}{8}$ (points) gains a point and a half. What is its new price?

   (A) $57\frac{1}{8}$ (B) $57\frac{3}{8}$ (C) $57\frac{5}{8}$ (D) $57\frac{7}{8}$ (E) $57\frac{3}{4}$

UNIT III DECIMALS

11. Decimal Notation

Decimals provide a convenient way of writing fractions in digital form. It is often easier to add, subtract, multiply, or divide decimals than it is to work with fractions. Here you will learn how to read numbers in decimal form, and how to express fractions as decimals and vice versa.

A. DECIMALS AND POWERS OF TEN

A fraction, such as

$$\frac{3}{10}, \quad \text{or} \quad \frac{79}{100}, \quad \text{or} \quad \frac{541}{1000}$$

whose denominator is a *power* of 10, can be expressed in **decimal notation** as follows:

$$\frac{3}{10} = .3$$

decimal point

tenths

Read this decimal as "three tenths" or as "point three."

99

$$\frac{79}{100} = .79$$

Read this decimal as "seventy-nine hundredths" or as "point seven, nine."

tenths
hundredths

$$\frac{79}{100} = \frac{70 + 9}{100} = \frac{70}{100} + \frac{9}{100} = \frac{7}{10} + \frac{9}{100}$$

Example 1 ▶ Fill in the blanks.

$$.27 = \frac{\boxed{}}{10} + \frac{\boxed{}}{100}$$

Solution.

$$.27 = \frac{27}{100}$$

$$= \frac{20 + 7}{100}$$

$$= \frac{20}{100} + \frac{7}{100}$$

$$= \frac{\boxed{2}}{10} + \frac{\boxed{7}}{100}$$ ◀

The fraction $\frac{541}{1000}$ is expressed as .541.

$$\frac{541}{1000} = .541$$

Read this decimal as "five hundred forty-one thousandths" or as "point five, four, one."

tenths
hundredths
thousandths

$$\frac{5}{10} + \frac{4}{100} + \frac{1}{1000} = \frac{500}{1000} + \frac{40}{1000} + \frac{1}{1000} = \frac{500 + 40 + 1}{1000} = \frac{541}{1000}$$

Example 2 ▶ Fill in the blanks.

$$.813 = \frac{\boxed{}}{10} + \frac{\boxed{}}{100} + \frac{\boxed{}}{1000}$$

Solution.

$$.813 = \frac{813}{1000}$$

$$= \frac{800 + 10 + 3}{1000}$$

$$= \frac{800}{1000} + \frac{10}{1000} + \frac{3}{1000}$$

$$= \frac{\boxed{8}}{10} + \frac{\boxed{1}}{100} + \frac{\boxed{3}}{1000}$$ ◀

PRACTICE EXERCISES FOR TOPIC A

Fill in the blanks.

1. $.21 = \dfrac{\boxed{}}{10} + \dfrac{\boxed{}}{100}$ 2. $.351 = \dfrac{\boxed{}}{10} + \dfrac{\boxed{}}{100} + \dfrac{\boxed{}}{1000}$

3. $.877 = \dfrac{\boxed{}}{10} + \dfrac{\boxed{}}{100} + \dfrac{\boxed{}}{1000}$

Now try the exercises for Topic A on page 106.

B. DECIMAL DIGITS

Digits (0, 1, 2, 3, 4, 5, 6, 7, 8, 9) to the *right* of the *decimal point* are called **decimal digits**. The first decimal digit indicates the number of *tenths*; the second decimal digit, the number of *hundredths*; the third decimal digit, the number of *thousandths*; the fourth decimal digit the number of *ten-thousandths*, and so on.

For the decimal

$$.5\ 8\ 3$$

tenths
hundredths
thousandths

5 is the *tenths* digit
8 is the *hundredths* digit
3 is the *thousandths* digit

Example 3 ▶ For the decimal .8084

   i. What is the tenths digit?
  ii. What is the hundredths digit?
 iii. What is the thousandths digit?
  iv. What is the ten-thousandths digit?

Solution.

$$.8\ 0\ 8\ 4$$

tenths
hundredths
thousandths
ten-thousandths

i. 8   *ii.* 0   *iii.* 8   *iv.* 4   ◀

Note that

$$\frac{3}{10} = .3 \qquad \text{whereas} \qquad \frac{3}{100} = \frac{03}{100} = .03$$

tenths
hundredths

Also

$$\frac{57}{1000} = \frac{057}{1000} = .057$$

tenths
hundredths
thousandths

and

$$\frac{7}{1000} = \frac{007}{1000} = .007$$

tenths
hundredths
thousandths

Note that the number of zeros in the denominator is also the number of decimal digits. Thus $\frac{7}{1000}$ equals .007, with *three* decimal digits. Finally, observe that

$$.7 = .70 = .700$$

because

$$\frac{7}{10} = \frac{70}{100} = \frac{700}{1000}$$

In general, you can add 0's to the right of the right-most decimal digit without changing the value. In other words, the new decimal is equivalent to the given one. **Thus**

$$.43 = .430 = .4300$$

and

$$.505 = .5050 = .505\,00$$

PRACTICE EXERCISES FOR TOPIC B

4. For the decimal .5105

   i. What is the tenths digit?
   ii. What is the hundredths digit?
   iii. What is the thousandths digit?
   iv. What is the ten-thousandths digit?

5. For the decimal .036, what is the tenths digit?

Now try the exercises for Topic B on page 106.

C. WORDS AND DECIMAL NOTATION

Example 4 ▶   Write "three thousand five hundred eight ten-thousandths" in decimal notation.

Solution.

Example 5▶ Write *i.* seven tenths   *ii.* seven hundredths   *iii.* seven thousandths
in decimal notation.

Solution.

i.   .7
    ↑
   tenths

ii.   .0 7
    ↑ ↑
   tenths │
     hundredths

iii.   .0 0 7
    ↑ ↑ ↑
   tenths │ │
    hundredths│
     thousandths   ◀

Example 6▶ Write in words.

i. .6   *ii.* .39

Solution.

i. six tenths   *ii.* thirty-nine hundredths   ◀

PRACTICE EXERCISES FOR TOPIC C

In Exercises 6–8, write in decimal notation.

6. six hundred twenty-nine thousandths
7. five tenths
8. five thousandths

In Exercises 9 and 10, write in words.

9. .4   10. .17

Now try the exercises for Topic C on page 107.

D. DECIMALS AND MIXED NUMBERS

A mixed number, such as $42\frac{9}{10}$, can be expressed as a "mixed decimal" by first writing the *whole part, 42, to the left of the decimal point. The fractional part, $\frac{9}{10}$, is then expressed to the right of the decimal point.* Thus

$$\frac{9}{10} = .9$$

and

$$42\frac{9}{10} = 42 + \frac{9}{10}$$
$$= 42 + .9 \qquad \textit{Omit the "+".}$$
$$= 42.9$$

Read: "forty-two and nine tenths" or "forty-two point nine."

When expressing 466 in words we say

four hundred sixty-six

and *not* "four hundred and sixty-six." The word "and" is used when a number has both a whole part and a decimal part, as in Example 7.

Example 7▶ Write "four hundred and sixty-six thousandths" in decimal notation.

Solution.

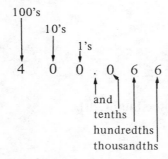

Note the difference between 10's (tens) and ten*ths*, between 100's (hundreds) and hundred*ths*, and so on.

PRACTICE EXERCISES FOR TOPIC D

Express in decimal notation.

11. $5\frac{1}{10}$ 12. $63\frac{9}{100}$

13. twenty-five and seventeen hundredths

Now try the exercises for Topic D on page 107.

E. FRACTIONS TO DECIMALS

To change a fraction to a decimal, divide the numerator by the denominator.

Example 8▶ Change $\frac{5}{8}$ to a decimal.

Solution. Divide the numerator, 5, by the denominator, 8. First write 5 as 5.000. (Add as many 0's as necessary to the right of the decimal point.)

$$
8\overline{)5.000} \quad \overset{.625}{} \longleftarrow
$$

.625 ⟵ Line up the decimal points, and proceed as if dividing
8⟌5.000   whole numbers.

Thus

$$
\frac{5}{8} = .625 \qquad\qquad\qquad\qquad \blacktriangleleft
$$

PRACTICE EXERCISES FOR TOPIC E

14. Change $\frac{3}{4}$ to a decimal.   15. Change $\frac{7}{20}$ to a decimal.

Now try the exercises for Topic E on page 108.

F. DECIMALS TO FRACTIONS

To change from a decimal to a fraction, rewrite the decimal as a fraction whose denominator is a power of 10. Then reduce to lowest terms.

Example 9 ▶   Change .075 to a fraction in lowest terms.

Solution.

$$
.075 = \frac{75}{1000} \qquad \text{Divide numerator and denominator by 25.}
$$
$$
= \frac{3}{40} \qquad\qquad\qquad\qquad\qquad \blacktriangleleft
$$

PRACTICE EXERCISES FOR TOPIC F

16. Change .24 to a fraction in lowest terms.
17. Change .004 to a fraction in lowest terms.

Now try the exercises for Topic F on page 108.

G. INFINITE REPEATING DECIMALS

The fraction $\frac{1}{3}$ can be expressed as an **infinite repeating decimal.** Divide the numerator by the denominator:

$$
3\overline{)1.000\,000} \quad \overset{.333\,333}{}
$$

Thus

$$
\frac{1}{3} = .333\,333 \ldots \qquad \textit{Here the digit 3 repeats, as indicated}
$$
$$
\textit{by the 3 dots.}
$$

The fraction $\frac{2}{11}$ can be expressed as an infinite repeating decimal in which *two digits*, 1 and 8, *repeat*:

$$11\overline{\smash{)}2.00\,00\,00}^{\displaystyle\,.18\,18\,18}$$

Thus

$$\frac{2}{11} = .18\,18\,18\,.\,.\,.\qquad \textit{Again, the 3 dots indicate this repetition.}$$

Every fraction can be written as either a (regular) decimal or as an infinite repeating decimal.

Example 10 ▶ Change $\frac{5}{12}$ to an infinite repeating decimal.

Solution. Divide the numerator by the denominator.

$$12\overline{\smash{)}5.00\,000\,000}^{\displaystyle\,.41\,666\,666}$$

Here, beginning with the third decimal digit, the 6's repeat. Thus

$$\frac{5}{12} = .41\,666\,666\,.\,.\,.$$

◀

PRACTICE EXERCISES FOR TOPIC G

Change each fraction to an infinite repeating decimal.

18. $\frac{4}{9}$ 19. $\frac{5}{6}$

Now try the exercises for Topic G on page 108.

EXERCISES

A. *Fill in the blanks.*

1. $.7 = \dfrac{\Box}{10}$ 2. $.31 = \dfrac{\Box}{10} + \dfrac{\Box}{100}$ 3. $.83 = \dfrac{\Box}{10} + \dfrac{\Box}{100}$

4. $.537 = \dfrac{\Box}{10} + \dfrac{\Box}{100} + \dfrac{\Box}{1000}$ 5. $.661 = \dfrac{\Box}{10} + \dfrac{\Box}{100} + \dfrac{\Box}{1000}$

6. $.8029 = \dfrac{\Box}{10} + \dfrac{\Box}{100} + \dfrac{\Box}{1000} + \dfrac{\Box}{10,000}$

B. 7. What is the tenths digit of .53?   *Answer:*

8. What is the hundredths digit of .054?   *Answer:*

9. What is the thousandths digit of .371?   *Answer:*

*In Exercises 10–12,   i. What is the tenths digit?   ii. What is the hundredths digit?
iii. What is the thousandths digit?*

10. .792   *Answer: i.*   *ii.*   *iii.*

11. .805   *Answer: i.*   *ii.*   *iii.*

12. .0535   *Answer: i.*   *ii.*   *iii.*

*In Exercises 13 and 14,   i. What is the hundredths digit?   ii. What is the thousandths
digit?   iii. What is the ten-thousandths digit?*

13. .8195   *Answer: i.*   *ii.*   *iii.*

14. .717 72   *Answer: i.*   *ii.*   *iii.*

C. *In Exercises 15–22, write in decimal notation.*

15. twenty-nine hundredths   *Answer:*

16. five hundred eighty-one thousandths   *Answer:*

17. six tenths   *Answer:*

18. six hundredths   *Answer:*

19. forty-three hundredths   *Answer:*

20. forty-three thousandths   *Answer:*

21. fifteen thousandths   *Answer:*

22. fifteen ten thousandths   *Answer:*

In Exercises 23–28, write in words.

23. .9   *Answer:*

24. .53   *Answer:*

25. .401   *Answer:*

26. .07   *Answer:*

27. .007   *Answer:*

28. .0007   *Answer:*

D. *Express in decimal notation.*

29. $4\frac{3}{10} =$ **30.** $1\frac{15}{100} =$ **31.** $17\frac{3}{100} =$

32. $9\frac{1}{1000} =$ **33.** $100\frac{1}{100} =$ **34.** $10\frac{1}{10,000} =$

35. six and three tenths   *Answer:*

36. ten and four hundredths   *Answer:*

37. one hundred and five thousandths   *Answer:*

38. forty-nine and sixty-five ten-thousandths   *Answer:*

E. *Change each fraction to a decimal.*

39. $\frac{4}{5} =$ 40. $\frac{3}{4} =$ 41. $\frac{5}{2} =$ 42. $\frac{3}{8} =$

43. $\frac{7}{20} =$ 44. $\frac{1}{25} =$ 45. $\frac{7}{25} =$ 46. $\frac{3}{50} =$

47. $\frac{1}{200} =$ 48. $\frac{21}{40} =$

F. *Change each decimal to a fraction in lowest terms.*

49. .6 =   50. .39 =   51. .24 =   52. .35 =

53. .02 =   54. .08 =   55. .625 =   56. .035 =

57. 1.2 =   58. 1.25 =

G. *Change each fraction to an infinite repeating decimal.*

59. $\frac{2}{3} =$ 60. $\frac{5}{9} =$ 61. $\frac{7}{11} =$ 62. $\frac{1}{6} =$

63. $\frac{7}{12} =$ 64. $\frac{1}{7} =$ 65. $\frac{7}{6} =$ 66. $\frac{6}{13} =$

67. $\frac{1}{30} =$ 68. $\frac{1}{110} =$

12. Size of Decimals

Deciding which of two decimals is the smaller is considerably easier than the comparable problem for fractions.

A. COMPARING TENTHS

Clearly,

$$.7 < .9$$

because

$$\frac{7}{10} < \frac{9}{10}$$

Example 1 ▶ Which of the following numbers is the smallest?

(A) .4   (B) .5   (C) .2   (D) .8   (E) .7

Solution. Each of these numbers has one decimal digit, and thus represents *tenths*. Among these, the smallest is .2 $\left(\text{or } \frac{2}{10}\right)$. Thus the correct choice is (C).   ◀

PRACTICE EXERCISES FOR TOPIC A

1. Which of the following numbers is the smallest?

(A) .6   (B) .3   (C) .1   (D) .8   (E) .4

Now try the exercises for Topic A on page 113.

B. COMPARING HUNDREDTHS, THOUSANDTHS

Clearly,

$$.38 < .41$$

because

$$\frac{38}{100} < \frac{41}{100}$$

Thus to compare .$\boxed{3}$8 with .$\boxed{4}$1, first compare their tenths digits. Because $3 < 4$, it follows that

$$.38 < .41$$

Furthermore,

$$.43 < .47$$

because

$$\frac{43}{100} < \frac{47}{100}$$

Here both decimals, .43 and .47, have the same tenths digit, 4. Thus compare their hundredths digits. Because $3 < 7$, it follows that .4$\boxed{3}$ $<$ .4$\boxed{7}$.

To compare .6 with .64, observe that

.6 = .60

and

.6⟦0⟧ < .6⟦4⟧

Similarly, to compare .83 with .829, observe that

.83 = .830

and

.8⟦2⟧9 < .8⟦3⟧0

To compare the size of two decimals:

1. First compare their tenths digits. If they differ, the *smaller* decimal has the *smaller* tenths digit.
2. If the tenths digits are equal, compare the hundredths digits. If they differ, the *smaller* decimal has the *smaller* hundredths digit.
3. If the hundredths digits are equal, compare the thousandths digits, and so on.

Thus

.⟦5⟧6 < .⟦6⟧3

because 5 < 6;

.7⟦6⟧4 < .7⟦8⟧3

because 6 < 8; and

.59⟦4⟧3 < .59⟦6⟧2

because 4 < 6.

Example 2▶ Which of these numbers is the smallest?

(A) .27   (B) .29   (C) .31   (D) .25   (E) .35

Solution. First compare the tenths digits. Among these numbers, (A), (B), and (D) have the smaller tenths digit, 2. Now compare the hundredths digits of (A), (B), and (D). The smallest hundredths digit among these is 5, so that .25 is the smallest of the given numbers. Thus the correct choice is (D), or .25. ◀

Example 3 ▶   Which of these numbers is the smallest?

(A) .614   (B) .579   (C) .597   (D) .609   (E) .578

Solution. First compare the tenths digits, then the hundredths digits, then the thousandths digits. Choices (B), (C), and (E) each have the smallest tenths digit, 5. *Among these*, (B) and (E) each have the smallest hundredths digit, 7. Between (B) and (E), choice (E) has the smaller thousandths digit, 8. Thus .578 is the smallest of these numbers. The correct choice is (E).   ◀

Example 4 ▶   Which of these numbers is the smallest?

(A) .54   (B) .541   (C) .504   (D) .5   (E) .55

Solution. Add 0's at the right, so that each of these has three decimal digits. Thus compare:

(A) .540   (B) .541   (C) .504   (D) .500   (E) .550

All have the same tenths digit, 5. Choices (C) and (D) each have the smallest hundredths digit, 0. Finally, between (C) and (D), choice (D) has the smaller thousandths digit, 0. Thus the smallest number is .500, or .5, so that the correct choice is (D).   ◀

PRACTICE EXERCISES FOR TOPIC B

Which is the smallest number?

2. (A) .34   (B) .38   (C) .35   (D) .39   (E) .32
3. (A) .54   (B) .46   (C) .49   (D) .55   (E) .52
4. (A) .313   (B) .133   (C) .331   (D) .311   (E) .131
5. (A) .47   (B) .407   (C) .471   (D) .4701   (E) .41

Now try the exercises for Topic B on page 113.

C. MIXED DECIMALS

Clearly,

$$4.3 < 5.2$$

whereas

$$3.14 < 3.24$$

and

$$3.87 < 3.89$$

> *To compare "mixed decimals":*
>
> First compare their whole parts—then, if necessary, their tenths digits, then their hundredths digits, and so on. Thus
>
> 1. if the whole parts differ, the *smaller* number has the *smaller* whole part.
>
> 2. If the whole parts are equal but the tenths digits differ, the *smaller* number has the *smaller* tenths digit.
>
> (And so on).

Example 5 ▶   Which is the smallest number?

(A) 2.03   (B) .032   (C) .023   (D) 0.23   (E) 0.03

Solution. In (B) and (C) the whole part is understood to be 0. Thus choices (B), (C), (D), and (E) each have the smallest whole part, 0. Among these, (B), (C), and (E) each have the smallest tenths digit, 0. And among these, (C) has the smallest hundredths digit, 2. The smallest of these numbers is thus .023, which is choice (C).   ◀

PRACTICE EXERCISES FOR TOPIC C

Which is the smallest number?

6.   (A) 3.4   (B) 3.2   (C) 2.9   (D) 2.7   (E) 3.1

7.   (A) 1.01   (B) 1.10   (C) 1.09   (D) 1.101   (E) 1.110

8.   (A) 10.37   (B) 10.369   (C) 10.396   (D) 10.366   (E) 10.375

Now try the exercises for Topic C on page 114.

D. LARGEST DECIMAL

Example 6 ▶   Which is the largest number?

(A) .048   (B) .084   (C) .804   (D) .840   (E) .480

Solution. The *largest* tenths digit is 8, in choices (C) and (D). Between these two, the larger hundredths digit is 4, in choice (D). Thus, the largest number is .840, that is, choice (D).   ◀

PRACTICE EXERCISES FOR TOPIC D

Which is the largest number?

9.   (A) .419   (B) .238   (C) .491   (D) .444   (E) .489

10. (A) .999   (B) 1.01   (C) 1.10   (D) 1.09   (E) 1.099

Now try the exercises for Topic D on page 114.

EXERCISES

A. *In each exercise, which number is the smallest? Draw a circle around the correct letter.*

1. (A) .7   (B) .4   (C) .3   (D) .9   (E) .1
2. (A) .6   (B) .5   (C) .4   (D) .3   (E) .2

B. *In each exercise, which number is the smallest? Draw a circle around the correct letter.*

3. (A) .15   (B) .19   (C) .12   (D) .11   (E) .13
4. (A) .68   (B) .66   (C) .67   (D) .69   (E) .65
5. (A) .42   (B) .41   (C) .48   (D) .38   (E) .39
6. (A) .72   (B) .27   (C) .29   (D) .92   (E) .91
7. (A) .09   (B) .30   (C) .29   (D) .30   (E) .21
8. (A) .61   (B) .51   (C) .41   (D) .31   (E) .81
9. (A) .41   (B) .53   (C) .39   (D) .43   (E) .51
10. (A) .08   (B) .09   (C) .11   (D) .18   (E) .10
11. (A) .451   (B) .458   (C) .455   (D) .459   (E) .453
12. (A) .308   (B) .302   (C) .306   (D) .304   (E) .307
13. (A) .753   (B) .575   (C) .759   (D) .751   (E) .579
14. (A) .039   (B) .093   (C) .309   (D) .390   (E) .309
15. (A) .111   (B) .101   (C) .110   (D) .100   (E) .102
16. (A) .083   (B) .008   (C) .038   (D) .003   (E) .033
17. (A) .148   (B) .139   (C) .149   (D) .137   (E) .136
18. (A) .707   (B) .770   (C) .777   (D) .077   (E) .070
19. (A) .707   (B) .703   (C) .77   (D) .737   (E) .730
20. (A) .04   (B) .039   (C) .03   (D) .304   (E) .43
21. (A) .614   (B) .6014   (C) .6104   (D) .61   (E) .601
22. (A) .899   (B) .89   (C) .9   (D) .891   (E) .8901
23. (A) .704   (B) .740   (C) .709   (D) .741   (E) .7
24. (A) .321   (B) .3201   (C) .3210   (D) .33   (E) .3202
25. (A) .195   (B) .159   (C) .19   (D) .155   (E) .1501
26. (A) .1401   (B) .1104   (C) .1004   (D) .101   (E) .11

C. *In Exercises 27–38, which number is the smallest? Draw a circle around the correct letter.*

27. (A) 1.07   (B) 2.07   (C) 3.07   (D) .07   (E) 1.09
28. (A) 4.2   (B) 2.4   (C) 2.24   (D) 4.02   (E) 2.04
29. (A) 12.9   (B) 10.9   (C) 10.09   (D) 12.009   (E) 10.009
30. (A) 6.2   (B) 2.6   (C) 2.606   (D) 2.06   (E) 2.066
31. (A) 3.4   (B) 3.389   (C) 3.408   (D) 3.409   (E) 3.4089
32. (A) 1.85   (B) 1.08   (C) 0.85   (D) .088   (E) .009
33. (A) .004   (B) .04   (C) 1.004   (D) .045   (E) .4
34. (A) .6   (B) .06   (C) .0606   (D) .066   (E) 6.006
35. (A) 3.079   (B) .0379   (C) .0039   (D) .0309   (E) 3.009
36. (A) .0030   (B) .03   (C) .029   (D) .031   (E) .0309
37. (A) 2.02   (B) 2.20   (C) .022   (D) .0202   (E) 2.022
38. (A) .848   (B) .884   (C) .088   (D) .0884   (E) .0808

In Exercises 39–42, which is the smallest amount of money? Draw a circle around the correct letter.

39. (A) $.47   (B) $.39   (C) $.41   (D) $.49   (E) $.42
40. (A) $1.06   (B) $1.09   (C) $1.00   (D) $.99   (E) $1.01
41. (A) $10.08   (B) $10.11   (C) $10.01   (D) $11.00   (E) $9.98
42. (A) $100.09 (B) $99.98   (C) $98.99   (D) $101.01 (E) $99.09

D. *In Exercises 43–50, which number is the largest? Draw a circle around the correct letter.*

43. (A) .73   (B) .78   (C) .87   (D) .83   (E) .79
44. (A) .305   (B) .503   (C) .055   (D) .053   (E) .505
45. (A) 1.2   (B) 1.21   (C) 1.02   (D) 2.01   (E) 2.11
46. (A) .022   (B) .032   (C) .201   (D) .2001   (E) .2031
47. (A) 6.1   (B) 4.9   (C) 5.2   (D) 5.9   (E) 6.01
48. (A) 4.08   (B) .048   (C) .008   (D) .084   (E) 8.04
49. (A) 6.01   (B) .061   (C) .6011   (D) 6.009   (E) 6.0101
50. (A) .099   (B) .0909   (C) .0991   (D) .0919   (E) .0990

In Exercises 51–54, which is the largest amount of money? Draw a circle around the correct letter.

51. (A) $.88   (B) $.91   (C) $.79   (D) $.89   (E) $.90
52. (A) $5.04   (B) $4.05   (C) $4.45   (D) $5.15   (E) $5.20
53. (A) $12.00   (B) $11.89   (C) $11.98   (D) $11.99   (E) $12.01
54. (A) $4000.01 (B) $400.91 (C) $400.99 (D) $3999.99 (E) $4000.10

13. Adding and Subtracting Decimals

In everyday life we must frequently add or subtract decimals. For example, several items may be purchased in a department store, say for $13.98, $5.95, and $20.58. Here we add decimals to find the total price. A child's temperature may rise from 98.6° (Fahrenheit) to 102.4°. Subtraction of decimals indicates the rise in temperature.

A. ADDING DECIMALS

Decimals are added or subtracted in columns. Thus

$$\begin{array}{r} .31 \\ + .48 \\ \hline .79 \end{array} \qquad \text{because} \qquad \frac{31}{100} + \frac{48}{100} = \frac{79}{100}$$

> *To add or subtract decimals:*
>
> 1. Line up the decimal points.
> 2. Add or subtract in columns.

Example 1 ▶   Add. $10.45 + 11.04 + 8.82$

Solution.

$$\begin{array}{r} 10.45 \\ 11.04 \\ 8.82 \\ \hline 30.31 \end{array}$$

The sum is 30.31.   ◀

You may want to add 0's to the right of the last decimal digit, although this is not necessary.

Example 2 ▶   Add.   14.42 + 10.044 + 8

Solution.

$$
\begin{array}{r}
14.\ 42\ \boxed{0} \\
10.\ 04\ 4 \\
\underline{8.\ \boxed{00\ 0}} \\
32.\ 46\ 4
\end{array}
$$

The sum is 32.464.   ◀

PRACTICE EXERCISES FOR TOPIC A

Add.

1. .6 + .5 + .7 =   2. 8.6 + 3.8 + 4.9 =

3. 10.05 + 9.95 + 12.27 =

Now try the exercises for Topic A on page 118.

B. SUBTRACTING DECIMALS

Example 3 ▶   Subtract.   49.3 − 4.84

Solution.   First line up the decimal points. Change 49.3 to 49.3 $\boxed{0}$. Then subtract in columns.

$$
\begin{array}{cc}
\begin{array}{r}
{\scriptstyle 8\ \ 12\ 10} \\
49.\ \cancel{3}\ \cancel{0} \\
-\ \ 4.\ 8\ \ 4 \\
\hline
44.\ 4\ \ 6
\end{array}
& \text{or} \quad
\begin{array}{r}
{\scriptstyle 13\ 10} \\
49.\ \cancel{3}\ \cancel{0} \\
{\scriptstyle 5\ \ 9} \\
-\ \cancel{4.8}\ \ 4 \\
\hline
44.\ 4\ \ 6
\end{array}
\end{array}
$$

The difference is 44.46.   ◀

PRACTICE EXERCISES FOR TOPIC B

Subtract.

4. .8 − .5 =   5. 19.7 − 12.9 =   6. 45.5 − 37.53 =

Now try the exercises for Topic B on page 118.

C. FURTHER EXAMPLES

Example 4 ▶ $14.2 - 2.04 =$

Solution.

$$
\begin{array}{r}
\overset{1\ 10}{14.2\ \cancel{0}} \\
-\ \ 2.0\ 4 \\
\hline
12.1\ \ 6
\end{array}
\quad \text{or} \quad
\begin{array}{r}
14.2\overset{10}{\cancel{0}} \\
\overset{1}{\ 2.\cancel{0}4} \\
-\ \ 2.\cancel{0}4 \\
\hline
12.1\ 6
\end{array}
$$
◀

PRACTICE EXERCISES FOR TOPIC C

7. $8.1 + 17.04 + 10 =$ 8. $20.1 - 15.98 =$

Now try the exercises for Topic C on page 118.

D. APPLICATIONS

Example 5 ▶   At a store a woman buys a dress for $37.95, a pair of gloves for $13.98, a pair of shoes for $48.25, and a kerchief for $8.50. How much does she spend on these items?

Solution.   Add the cost of the various items purchased.

$$
\begin{array}{r}
\$\ 37.95 \\
13.98 \\
48.25 \\
\underline{8.50} \\
\$108.68
\end{array}
$$

She spends a total of $108.68 on her various purchases.   ◀

Example 6 ▶   Room temperature drops from 72.4° (Fahrenheit) to 69.8°. What is the drop in temperature?

Solution.   Subtract.

$$
\begin{array}{r}
72.4 \\
-\ 69.8 \\
\hline
2.6
\end{array}
$$

The drop in temperature is 2.6°.   ◀

PRACTICE EXERCISES FOR TOPIC D

9. A worker earns $53.45 on Monday, $49.75 on Tuesday, and $58.90 on Wednesday. What are the total earnings for these three days?

10. A saleswoman earns $120.35 in commission, but must pay her assistant $45.50 of this money. How much does she get to keep?

Now try the exercises for Topic D on this page.

EXERCISES

A. *Add.*

1. .4 + .3 + .2 =
2. .8 + .5 + .7 =
3. 7.2 + 3.8 + 9.5 =
4. 6.4 + 3.6 + 8.4 =
5. 6.82 + 4.85 + 3.06 =
6. 10.41 + 8.09 + 7.96 =
7. 7.71 + 12.9 + 6 =
8. 8.03 + 3.1 + 16 =
9. 19.92 + 9.908 + 10 =
10. 27.2 + 2.04 + 12 =
11. 1.04 + 104 + 10.4 =
12. 6.08 + 60.8 + 68 =
13. 12.01 + 1.201 + 120 =
14. 5.99 + 15.09 + 1.909 =
15. 109.9 + 10.99 + 1.099 =
16. 19.09 + 1.099 + 109 =
17. 16.7 + 21.09 + 19.12 + 7.84 =
18. 53.01 + 27.92 + 3.093 + 16.019 =

B. *Subtract.*

19. .9 – .3 =
20. 1.2 – .5 =
21. 12.7 – 11.3 =
22. 16.4 – 8.4 =
23. 42.3 – 31.4 =
24. 13.7 – 7.9 =
25. 12.1 – 6.06 =
26. 10.03 – 8.09 =
27. 43.8 – 2.07 =
28. 4.06 – .093 =
29. 1.092 – .909 =
30. 52.05 – 8.559 =
31. 25.8 – 1.29 =
32. 41.7 – 14.03 =
33. 54.6 – 9.83 =
34. 101.4 – 1.014 =

C. *Add or subtract, as indicated.*

35. 2.05 + 5.2 + 4.52 =
36. 7.83 + 3.87 + 5.087 =
37. 6.03 + 16.3 + 11 =
38. 15.83 + 5.805 + 12 =
39. 10.09 + 9.021 + 8 =
40. 94.01 + 4.094 + 19 =
41. 9.093 + 90.39 + 18 =
42. 9.08 – 6.12 =
43. 10 – 4.09 =
44. 24.1 – 4.02 =
45. 19.03 – 2.89 =
46. 39.6 – 9.88 =
47. 17.8 – 2.036 =
48. 41.9 – 3.28 =

D. *In Exercises 49–54, add or subtract, as indicated.*

49. $9.85
    7.56
    3.95

50. $58.40
    27.82
    63.98

51. $521.45
    389.50
    435.55
    295.47

52. $5841.01
    3892.72
    2989.83
    5192.58

53. $592.25
    –389.95

54. $1045.40
    –989.95

55. At a supermarket a man buys juice for $1.29, cheese for $2.18, and milk for $.78. What is the total that he spends for these items? *Answer:*

56. A woman writes out checks for $49.90, $82.15, $53.37, and $15.19. What is the total amount of these checks? *Answer:*

57. Tom buys a phonograph record for $6.85. He pays for it with a ten-dollar bill. How much change does he receive? *Answer:*

58. At noon the temperature is 57.2° (Fahrenheit). In the next hour the temperature rises 1.8°. What is the temperature at 1:00 that afternoon? *Answer:*

59. One day a salesman makes trips of 12.6 miles, 5.9 miles, 13.8 miles, 7.7 miles, and 18.4 miles. How many miles does he travel that day? *Answer:*

60. A patient's temperature drops from 103.6° (Fahrenheit) to 101.9°. What is the drop in temperature? *Answer:*

14. Multiplying and Dividing Decimals

Suppose that a pound of cherries costs $1.28 and a scale shows that there are 2.5 pounds of them in a bag. To find the cost of these cherries, you multiply decimals. On the other hand, if a hunk of cheese weighing 3.8 pounds costs $11.21, to find the cost *per* pound, you divide decimals.

A. MULTIPLYING BY A POWER OF 10

First consider a decimal multiplied by 10.

$$.3 \times 10 = \frac{3}{\cancel{10}} \times \frac{\overset{1}{\cancel{10}}}{1} = \frac{3}{1} = 3 \qquad [\ 3.]$$

1. *To multiply a decimal by* 10, *move the decimal point* 1 *digit to the right.*
2. *To multiply by* 10^n, *move the decimal point n digits to the right.*

Example 1 ▶   Multiply.

    i. .174 × 10   *ii.* .174 × 100   *iii.* .174 × 1000

Solution.

    i. .174 × 10 = 1.74   [1.74]

    ii. 100 = 10^2. Thus,

        .174 × 100 = 17.4   [17.4]

    iii. 1000 = 10^3. Thus,

        .174 × 1000 = 174   [174.]

PRACTICE EXERCISES FOR TOPIC A

Multiply.

1. .236 × 10 =   2. 27.53 × 100 =   3. .0028 × 1000 =

Now try the exercises for Topic A on page 125.

B. MULTIPLYING DECIMALS

Here is how decimals are multiplied.

$$.7 \quad × \quad .11 \quad = \frac{7}{10} × \frac{11}{100} = \frac{77}{1000} = .\boxed{0}77$$

    1 decimal   2 decimal   3 decimal
      digit   digits   digits

Note that you must insert a 0 *immediately to the right of the decimal point* in order to obtain 3 decimal digits in the product.

> When one or both factors of a product are decimals, count the total number of decimal digits in the factors in order to place the decimal point in the product.

Example 2 ▶   Multiply.   18 × .25

Solution.

```
        18  ◄────────  0 decimal digits
      × .25  ◄──────── + 2 decimal digits
        90
        3 6
       4.50  ◄────────  2 decimal digits
```

The final 0 counts as a decimal digit, even though it can be dropped. Thus

    18 × .25 = 4.50 = 4.5

Example 3 ▶   Multiply.   .38 × 6.2

Solution.

```
    .3 8  ←————— 2 decimal digits
  × 6.2  ←————— + 1 decimal digit
    7 6
  2 28
  2.35 6  ←————— 3 decimal digits
```
◀

PRACTICE EXERCISES FOR TOPIC B

Multiply.

4. 16 × 2.4 =   5. .38 × .2 =   6. 1.07 × .03 =

Now try the exercises for Topic B on page 125.

C. POWERS OF DECIMALS

When finding a power of a decimal, multiply the number of decimal digits in the base by the exponent. This indicates the number of decimal digits in the resulting power.

Example 4 ▶   Find: $(.2)^2$

Solution.

$$2^2 = 4$$

Now place the decimal point.

$$(.2)^2 = (.2)(.2)$$

1 decimal digit

$$= .04$$

2 decimal digits
◀

Example 5 ▶   Find: $(.01)^3$

Solution.

$$(.01)^3 = (.01)(.01)(.01)$$

2 decimal digits

$$= .000\,001$$

6 decimal digits
◀

PRACTICE EXERCISES FOR TOPIC C

Find each power.

7. $(.5)^2 =$ 8. $(.3)^3 =$ 9. $(.1)^4 =$

Now try the exercises for Topic C on page 125.

D. COST PROBLEMS

Example 6▶   Find the cost of 17 gallons of gas at $1.28 per gallon.

Solution. Each gallon of gas costs the same amount. Thus you *multiply* the number of gallons, 17, by the fixed cost per gallon, $1.28.

$$
\begin{array}{r}
\$\ 1.28 \longleftarrow \text{2 decimal digits}\\
\times\ \ \ 17 \longleftarrow +\ 0 \text{ decimal digits}\\
\hline
8\ 96\\
12\ 8\\
\hline
\$21.76 \longleftarrow \text{2 decimal digits}
\end{array}
$$

The cost of 17 gallons is $21.76.   ◀

PRACTICE EXERCISES FOR TOPIC D

10. Find the cost of eight rolls of tape that sell for $1.45 apiece.
     Answer:

Now try the exercises for Topic D on page 125.

E. DIVIDING BY A POWER OF 10

When multiplying by a power of 10, the decimal point moves to the right. Now observe that when *dividing* by a power of 10, the decimal point moves to the *left.*

$$.3 \div 10 = \frac{3}{10} \times \frac{1}{10} = \frac{3}{100} = .03 \qquad [.0\,3]$$

1. *To divide a decimal by* 10, *move the decimal point* 1 *digit to the left.*
2. *To divide by* 10^n, *move the decimal point n digits to the left.*

Example 7 ▶   Divide.

    i. $3.92 \div 10$ *ii.* $3.92 \div 100$ *iii.* $3.92 \div 1000$

Solution.

    i. $3.92 \div 10 = .392$ $[.3\,92]$

    ii. $3.92 \div 100 = .0392$ $[.03\,92]$

    iii. $3.92 \div 1000 = .003\,92$ $[.003\,92]$ ◀

PRACTICE EXERCISES FOR TOPIC E

Divide.

11. $5.85 \div 10 =$ 12. $2.736 \div 100 =$ 13. $72.14 \div 1000 =$

Now try the exercises for Topic E on page 126.

F. DIVIDING DECIMALS

When dividing by *ten*ths, first multiply the dividend and divisor by 10, so that the divisor will then be an integer. For example,

$$\frac{24}{.6} = \frac{24 \times 10}{.6 \times 10} = \frac{240}{6} = 40$$

When dividing by *hundred*ths, first multiply the dividend and divisor by 100, so that the divisor will then be an integer. Thus

$$\frac{45}{.03} = \frac{45 \times 100}{.03 \times 100} = \frac{4500}{3} = 1500$$

Example 8 ▶   Divide 139.4 by .17.

Solution.

$$\frac{139.4}{.17} = \frac{139.4 \times 100}{.17 \times 100} = \frac{13,940}{17}$$

Note that to multiply

139.4 × 100

move the decimal point 2 digits to the right. In order to do so, insert a 0 to the right of the 4.

139.4 × 100 = 13,94⬚0 .

Now divide 13,940 by 17.

```
        820
17) 13 940
    13 6
       34
       34
        0
```

Thus the quotient is 820.   ◀

PRACTICE EXERCISES FOR TOPIC F

Divide.

14. $\dfrac{48}{.6} =$ 15. $\dfrac{6.95}{.05} =$ 16. $\dfrac{169.6}{.032} =$

Now try the exercises for Topic F on page 126.

G. COST PROBLEMS

Example 9 ▶   Cans of soda cost $.40 each. How many can be bought for $12?

> *Solution.* Divide the total cost, $12, by the cost per can, $.40. Note that
> .40 = .4.
>
> *Multiply numerator and denominator by* 10.
>
> $$\frac{\$\,12}{\$.40} = \frac{12 \times 10}{.4 \times 10} = \frac{120}{4} = 30$$
>
> Thus 30 cans of soda at $.40 each can be bought for $12.   ◀

PRACTICE EXERCISES FOR TOPIC G

17. If cold cuts sell for $2.50 per pound, how many pounds can be purchased
     for $18.50?
     Answer:

Now try the exercises for Topic G on page 126.

H. FURTHER APPLICATIONS

> It is important to recognize whether to multiply or divide in a given example.

Example 10 ▶   48 bagels at $.24 per bagel cost a total of

(A) $.20   (B) $2.00   (C) $200   (D) $1152   (E) $11.52

> *Solution.* You are given the cost *per* bagel, $.24, and want to find the cost of
> 48 bagels. Thus *multiply* $.24 by 48.

$$
\begin{array}{r}
\$\quad.24 \\
\times\ 48 \\
\hline
1\,92 \\
9\,6\ \ \ \\
\hline
\$11.52
\end{array}
$$

$ .24 ⟵——— 2 decimal digits
× 48 ⟵——— + 0 decimal digits
$11.52 ⟵——— 2 decimal digits

> The cost is $11.52. The correct choice is (E).   ◀

Example 11 ▶   A slice of cheese is .15 inch thick. How many slices of this cheese are in a
package that is 1.8 inches thick?

(A) 3   (B) 10   (C) 12   (D) 16   (E) 120

> *Solution.* *Divide* the total thickness of the package by the thickness of each
> individual slice.
>
> $$\frac{1.8}{.15} = \frac{1.8 \times 100}{.15 \times 100} = \frac{180}{15} = 12$$
>
> There are 12 slices altogether, choice (C).   ◀

PRACTICE EXERCISES FOR TOPIC H

18. A cardboard backing is .05 inch thick. How high is a stack of 60 of these backings?

   (A) 1.2 inches   (B) 12 inches   (C) 3 inches

   (D) 30 inches   (E) 300 inches

19. At $7.50 per hour, how long would it take to earn $90?

   (A) 6.75 hours   (B) 67.5 hours   (C) 675 hours

   (D) 12 hours   (E) 120 hours

Now try the exercises for Topic H on page 126.

EXERCISES

A. *Multiply.*

   1. $.192 \times 10 =$ 2. $18.4 \times 100 =$ 3. $5.04 \times 1000 =$

   4. $.065 \times 10 =$ 5. $.0052 \times 100 =$ 6. $.000\,85 \times 10\,000 =$

   7. $4.092 \times 10 =$ 8. $15.004 \times 1000 =$

B. *Multiply.*

   9. $.3 \times .5 =$ 10. $.4 \times .2 =$ 11. $.14 \times .8 =$

   12. $.12 \times .3 =$ 13. $.16 \times .5 =$ 14. $.48 \times .07 =$

   15. $.92 \times .29 =$ 16. $22 \times .5 =$ 17. $48 \times .05 =$

   18. $36 \times 1.2 =$ 19. $280 \times .41 =$ 20. $38.9 \times 1.02 =$

   21. $503 \times 2.15 =$ 22. $.395 \times 92.1 =$

C. *Find each power.*

   23. $(.3)^2 =$ 24. $(.4)^2 =$ 25. $(.02)^2 =$ 26. $(.005)^2 =$

   27. $(.1)^3 =$ 28. $(.3)^3 =$ 29. $(.03)^3 =$ 30. $(.1)^4 =$

D. 31. Find the cost of 12 ice cream cones at $.65 per cone.   *Answer:*

   32. Find the cost of 8 melons at $1.15 per melon.   *Answer:*

   33. A book salesman sells 6 copies of a book priced at $10.25 per copy. What is the total sale price?   *Answer:*

   34. A can of coffee sells for $2.95. What is the price of 7 cans of this coffee? *Answer:*

   35. Apples sell for $.68 per pound. What is the price of 5 pounds of apples? *Answer:*

   36. How much do 6 quarts of milk cost at $.71 per quart?   *Answer:*

37. Each page of a book is .003 inch thick. How thick is a 500-page book?
    Answer:

38. A roll of wallpaper costs $9.75. If a room requires 26 rolls of this wallpaper, what is the cost of wallpaper for this room?   *Answer:*

39. What is the cost of 47 gallons of paint at $8.95 per gallon?   *Answer:*

40. A woman earns $6.25 per hour of work. How much does she earn for 36 hours of work?   *Answer:*

E. *Divide.*

41. $.54 \div 10 =$ 42. $3.5 \div 100 =$ 43. $.372 \div 1000 =$

44. $4.08 \div 10 =$ 45. $.053 \div 1000 =$ 46. $.0042 \div 100 =$

47. $55.08 \div 10 =$ 48. $.053 \div 10,000 =$

F. *Divide.*

49. $25 \div .5 =$ 50. $120 \div .6 =$ 51. $32 \div .04 =$

52. $910 \div .07 =$ 53. $4800 \div .12 =$ 54. $121 \div .11 =$

55. $2.6 \div .13 =$ 56. $840 \div .12 =$ 57. $13.5 \div 4.5 =$

58. $14.4 \div .48 =$ 59. $5.2 \div .13 =$ 60. $10.24 \div .032 =$

G. 61. Pencils cost $.07 each. How many can be bought for $1.26?   *Answer:*

62. Oranges cost $.30 each. How many can be bought for $5.10?   *Answer:*

63. Erasers sell for $.60 each. How many can you buy for $14.40?   *Answer:*

64. Grapes sell for $1.60 per pound. How many pounds can you buy for $11.20?
    Answer:

65. Sugar sells for $.75 per pound. How many pounds can you buy for $11.25?
    Answer:

66. A man buys a car for $10,560. If he pays $440 in equal monthly installments, for how many months will the payments last?   *Answer:*

67. A book sells for $15. One week, total sales for the book are $1080. How many copies of the book are sold that week?   *Answer:*

68. A woman pays $21.60 for a tank of gas. If gas sells for $1.20 per gallon, how many gallons does she buy?   *Answer:*

H. *In Exercises 69–74, draw a circle around the correct letter.*

69. 32 cans of peas at $.45 per can cost

    (A) $14.40   (B) $13.40   (C) $2.88   (D) $129.60   (E) $24.40

70. If bars of soap cost $.85 apiece, how many can you buy for $16.15?

    (A) 17   (B) 18   (C) 19   (D) 20   (E) 21

71. If a motorist uses 22 gallons of gasoline at $1.35 per gallon, what is the total price she pays for gasoline?

    (A) $27.90   (B) $29.70   (C) $30.70   (D) $39.70   (E) $27.70

72. If each booklet is .18 inch thick and a pile of these booklets is 4.68 inches high, how many booklets are there in the pile?

    (A) 260   (B) 2.60   (C) 216   (D) 16   (E) 26

73. Each page of a book is .002 inch thick. Excluding the cover, how thick is a 400-page book?

    (A) .8 inch   (B) 8 inches   (C) .2 inch   (D) 2 inches   (E) 20 inches

74. If pencils sell for $.09 each, then 90 of these pencils cost

    (A) $1.00   (B) $10.00   (C) $8.10   (D) $81.00   (E) $810.00

75. Suppose that a pound of cherries costs $1.28 and a scale shows that there are 2.5 pounds of them in a bag. Find the cost of these cherries.   *Answer:*

76. If a hunk of cheese weighing 3.8 pounds costs $11.21, find the cost per pound. *Answer:*

---

15. Cost and Profit

We continue our study of cost and profit when several items are involved.

A. COST

Example 1 ▶ Find the total cost of 5 pounds of potatoes at $.30 per pound and 3 pounds of onions at $.35 per pound.

    (A) $1.50   (B) $1.05   (C) $2.55   (D) $25.55   (E) $.45

Solution. In each case, multiply the cost per pound by the number of pounds.

Potatoes: $.30 × 5 = $1.50

Onions: $.35 × 3 = $1.05

Now add the cost of potatoes and the cost of onions to find the total cost.

$$
\begin{array}{r}
\$1.50 \\
+ \;\$1.05 \\
\hline
\$2.55
\end{array}
$$

The total cost is $2.55. The correct choice is (C).   ◀

Example 2 ▶ A parking lot charges $3 for the first hour and $2 for each additional hour. What is the cost of parking a car there for 6 hours?

Solution. The cost for the first hour is $3. The car is parked there for 6 hours in all, so that there are 6 − 1, or 5, additional hours. For each additional hour the cost is $2. The *total* cost is then

$$\$3 + (6 - 1) \times \$2 = \$3 + (5 \times \$2)$$
$$= \$3 + \$10$$
$$= \$13$$ ◀

PRACTICE EXERCISES FOR TOPIC A

1. If apples cost 35¢ each and oranges cost 30¢ each, find the cost of 4 apples and 3 oranges.

    (A) $2.30   (B) $2.40   (C) $2.50   (D) $3.00   (E) $3.20

2. A car rental agency charges $55 for the first day's rental and $40 for each additional day. How much does it cost to rent a car from the agency for 6 days?

    (A) $240   (B) $330   (C) $200   (D) $275   (E) $255

Now try the exercises for Topic A on page 130.

B. PROFIT

When an item is sold, the **profit** made equals the **total sales** minus the **costs**.

$$\boxed{\text{Profit} = \text{Total Sales} - \text{Costs}}$$

Example 3 ▶ A grocer buys a carton of 18 boxes of cornflakes for $10.50. He sells each box for 85¢. Find his profit for selling all 18 boxes.

    (A) $4.80   (B) $10.50   (C) $15.30   (D) $3.60   (E) $3.80

Solution. First, to find the *total sales*, multiply 18, the number of boxes sold, by 85¢, the price per box. Note that

$$85¢ = \$.85$$

```
$   .85  ◄——— Price per item
  × 18   ◄——— Number of items
  ────
  6 80
  8 5
 ──────
 $15.30  ◄——— Total Sales
```

Next, the *cost* to the grocer is $10.50. To find his *profit*, use

$$
\begin{aligned}
\text{Profit} &= \text{Total Sales} - \text{Costs} \\
&= \$15.30 - \$10.50
\end{aligned}
$$

```
  $15.30  ◄——— Total Sale
- $10.50  ◄——— Costs
 ───────
  $ 4.80  ◄——— Profit
```

His profit is $4.80, choice (A).   ◄

Example 4 ▶ A drama group sells 584 tickets to a play at $5 each. The group pays $750 to rent an auditorium and $325 in other expenses. Find the profit.

Solution. To find the *total sales*, multiply 584, the number of tickets sold, by $5, the price per ticket.

$$584 \times \$5 = \$2920 \,\text{◄——— Total Sales}$$

To find the *costs*, add $750, for renting the auditorium, and $325, for other expenses.

```
  $ 750
+ $ 325
 ──────
  $1075  ◄——— Costs
```

$$
\begin{aligned}
\text{Profit} &= \text{Total Sales} - \text{Costs} \\
&= \$2920 - \$1075
\end{aligned}
$$

```
  $2920  ◄——— Total Sales
- $1075  ◄——— Costs
 ──────
  $1845  ◄——— Profit
```

The profit is $1845.   ◄

PRACTICE EXERCISES FOR TOPIC B

3. A merchant buys a box of 24 candy bars for $3.20. If he sells each candy bar for 30 cents, what is his profit?

4. It costs a bakery 38 cents to bake a loaf of bread. If 40 loaves are sold for the regular price of 85 cents per loaf, and 20 loaves are sold at a "day-old" rate of 45 cents per loaf, the total profit is

(A) $34   (B) $9   (C) $43   (D) $20.20   (E) $21.20

Now try the exercises for Topic B on page 131.

EXERCISES

A. *In Exercises 1–6, draw a circle around the correct letter.*

1. The cost to a store of 92 notebooks at 48¢ per notebook is

(A) $33.16   (B) $34.16   (C) $43.16   (D) $44.06   (E) $44.16

2. Pens cost 25¢ each and erasers cost 35¢ each. The cost of 36 pens and 18 erasers is

(A) $15   (B) $15.30   (C) $6.30   (D) $32.40   (E) $17.10

3. A bookstore orders 27 copies of a history book at $8 per copy and 40 copies of an atlas at $9.50 per copy. The total cost to the bookstore is

(A) $596   (B) $380   (C) $216   (D) $1172.50   (E) $117.25

4. A grocer orders 12 cartons of detergent at $27 per carton, 9 cartons of tissues at $11 per carton, and 15 cartons of paper towels at $13 per carton. The total cost to the grocer is

(A) $423   (B) $618   (C) $294   (D) $324   (E) $519

5. A moving company charges $95 for the first hour of work and $75 for each additional hour. The cost of an 8-hour job is

(A) $835   (B) $845   (C) $695   (D) $620   (E) $740

6. To mail a (first-class) letter in 1984 costs 20¢ for the first ounce and 17¢ for each additional ounce. The cost of mailing a 10-ounce letter is

(A) $2.00   (B) $1.70   (C) $1.73   (D) $1.90   (E) $1.97

7. To rent an auditorium costs $250 per hour for the first two hours and $150 for each additional hour. The cost to rent that auditorium for 5 hours is _____

8. A duplicating center charges 8 cents per copy for the first 5 copies of the same page, 6 cents per copy for the next 10 copies, and 5 cents per copy for each additional copy. The cost of making 100 copies of a page is _____

9. In Exercise 8, the cost of making 50 copies of one page and 20 copies of another page is _____

10. In Exercise 8, the cost of making 10 copies of one page, 25 copies of a second page, and 45 copies of a third page is _____

B. *In Exercises 11–16, draw a circle around the correct letter.*

11. A grocer buys a carton of 24 boxes of crackers for $9.60. She sells each box for 55 cents. Her profit for selling all 24 boxes is

(A) $22.80   (B) $13.20   (C) $3.60   (D) $3.20   (E) $4.60

12. A store buys a box of 32 scarves for $189. It sells each scarf for $9. The profit for selling all of the scarves is

(A) $278   (B) $288   (C) $199   (D) $99   (E) $109

13. A man bought 6 melons at $1.35 each. How much change did he receive from a $10 bill?

(A) $8.00   (B) $8.10   (C) $8.65   (D) $1.90   (E) $2.90

14. A woman bought 22 cans of soup at 65¢ each. How much change did she receive from a $20 bill?

(A) $14.30   (B) $6.70   (C) $5.70   (D) $.57   (E) $7.00

15. Tickets to a concert cost $9 each. If 880 tickets are sold and the expenses of running the concert are $6400, the profit is

(A) $7920   (B) $14,320   (C) $1520   (D) $5520   (E) $820

16. A druggist buys 48 tubes of toothpaste at $.53 each. She sells them for $.79 each. Her total profit is

(A) $.26   (B) $12.48   (C) $124.80   (D) $25.44   (E) $37.92

17. A department store pays $1450 for a shipment of 35 coats. It sells them all for $79 apiece. The profit is _____

18. At a football game 512 tickets are sold for $6.50 each. It costs $2200 to rent the stadium. If other expenses amount to $750, the profit is _____

19. A theater charges $6 for an orchestra seat and $4 for a balcony seat. It sells 330 orchestra seats and 252 balcony seats. If expenses are $2400, the theater's profit is _____

20. A supermarket buys two dozen loaves of bread for 60¢ per loaf. It sells 19 of these loaves for 87¢ each. If the other loaves get stale and are not sold, the profit is _____

21. A store buys 60 umbrellas for $4.50 each. It sells 35 of the umbrellas for $8.00 each and it sells the remaining umbrellas at a reduced price of $5.95 each. The store's profit is _____

22. 550 people pay $12 each to see a play. The theater company pays the performers a total of $2400. It pays $1750 to rent the theater and $1225 in other expenses. Its profit is _____

16. Rounding

For many purposes, it is not necessary to be precise when describing a number. For example, you may hear that 25,000 fans attended a baseball game, when actually, there were 25,237. Here the number of fans was "rounded" to the nearest thousand.

A. ROUNDING TO THE NEAREST THOUSAND OR MILLION

The number 6492 lies between 6000 and 7000. Observe that 6492 is closer to 6000 than it is to 7000. In fact,

$$
\begin{array}{ccc}
6492 & \text{whereas} & 7000 \\
-\ 6000 & & -\ 6492 \\
\hline
492 & < & 508
\end{array}
$$

Thus 6492 is closer to 6000 than to 7000. We will round 6492 to 6000. On the other hand, 6504 is closer to 7000 than it is to 6000:

$$
\begin{array}{ccc}
6504 & \text{whereas} & 7000 \\
-\ 6000 & & -\ 6504 \\
\hline
504 & > & 496
\end{array}
$$

Thus we will round 6504 to 7000. The key to rounding a number to the nearest 1000 is the 100's digit.

> To round to the nearest 1000:
>
> 1. If the 100's digit is 5 or more, increase the 1000's digit by 1; otherwise, retain the 1000's digit.
> 2. Replace the last 3 digits by 0's (as in 1000).

Example 1 ▶   Round to the nearest 1000.

    i. 7362   *ii.* 8594   *iii.* 49,630

Solution.

  i. The 100's digit of 7362 is 3, which is less than 5. Thus round to 7000.

  ii. The 100's digit of 8594 is 5. Increase the 1000's digit, 8, by 1. Round to 9000.

  iii. The 100's digit of 49,630 is 6 and is therefore more than 5. Increase the 1000's digit from 9 to 10. To do so, you must also consider the 10,000's digit, and increase 49 to 50. Thus

       49,630 rounds to 50,000   ◀

By *convention*, a number such as 6500, which lies midway between 6000 and 7000,

$$
\begin{array}{r}
6500 \\
-\ 6000 \\
\hline
500
\end{array}
\quad = \quad
\begin{array}{r}
7000 \\
-\ 6500 \\
\hline
500
\end{array}
$$

is usually rounded to the higher of the two possibilities. Thus 6500 rounded to the nearest thousand equals 7000.

> To round to the nearest 1,000,000:
>
> 1. If the 100,000's digit is 5 or more, increase the 1,000,000's digit by 1; otherwise, retain the 100,000's digit.
> 2. Replace the last 6 digits by 0's (as in 1,000,000).

Example 2 ▶   Round to the nearest 1,000,000.

    i. 4,052,631   *ii.* 15,936,218

Solution.

  i. The 100,000's digit of 4,052,631 is 0, which is less than 5. Round to 4,000,000.

  ii. The 100,000's digit of 15,936,218 is 9, which is more than 5. Round to 16,000,000.   ◀

These rules can be modified for rounding to the nearest 100, to the nearest 100,000, to the nearest 10,000,000, etc.

Example 3 ▶   Round 126,342,171 to the nearest

i. 100,000   *ii.* 10,000,000

Solution.

i. The 10,000's digit of 126,342,171 is 4. Round to 126,300,000.

ii. The 1,000,000's digit of 126,342,171 is 6, which is more than 5. Increase the 10,000,000's digit from 2 to 3. Round to 130,000,000.   ◀

PRACTICE EXERCISES FOR TOPIC A

In Exercises 1 and 2, round to the nearest 1000.

1. 8174   2. 12,684

In Exercises 3 and 4, round to the nearest 1,000,000.

3. 75,555,555   4. 49,459,392

5. Round 7,391,504 to the nearest 100,000.

Now try the exercises for Topic A on page 138.

B. ROUNDING TO THE NEAREST TENTH, HUNDREDTH, OR THOUSANDTH

Clearly,

.14 is closer to .1 than to .2

$$\begin{array}{lcl} .14 & \text{whereas} & .20 \\ \underline{-\ .10} & & \underline{-\ .14} \\ .04 & < & .06 \end{array}$$

However,

.16 is closer to .2 than to .1

$$\begin{array}{lcl} .16 & \text{whereas} & .20 \\ \underline{-\ .10} & & \underline{-\ .16} \\ .06 & > & .04 \end{array}$$

Thus

.14 rounded to the nearest tenth is .1

but

.16 rounded to the nearest tenth is .2

By convention, .15, which lies midway between .10 and .20,

$$\begin{array}{lcl} .15 & & .20 \\ \underline{-\ .10} & & \underline{-\ .15} \\ .05 & = & .05 \end{array}$$

is rounded to .2.

> To round to the nearest *tenth*:
> 1. If the *hundredths* digit is 4 *or less*, simply drop all digits to the right of the tenths digit.
> 2. If the *hundredths* digit is 5 *or more*, add 1 to the tenths digit and drop all digits to the right.

Example 4 ▶ Round each of the following to the nearest tenth.

  i. .43   *ii.* .47   *iii.* .4509   *iv.* .96

Solution.

  i. .43 rounds to .4

  ii. .47 rounds to .5

  iii. .4509 rounds to .5

  iv. Add 1 to the tenths digit by rounding

         1.0
         .9̶6̶ to 1.0

  (Note that .10 = .1, so that .96 *does not* round to .10.)   ◀

To round to the nearest *hundredth*, consider whether the *thousandths* digit is 5 *or more*. To round to the nearest *thousandth*, consider the *ten thousandths* digit.

Example 5 ▶ Round to the nearest hundredth.

  i. .784   *ii.* .0961

Solution.

  i. .784 rounds to .78

          .10
  ii. .0̶9̶6̶1̶ rounds to .10   ◀

Example 6 ▶ Round to the nearest thousandth.

  i. .362 29   *ii.* .7358   *iii.* .8996

Solution.

  i. .362 29 rounds to .362

  ii. .7358 rounds to .736

          .900
  iii. .8̶9̶9̶6̶ rounds to .900   ◀

PRACTICE EXERCISES FOR TOPIC B

In Exercises 6 and 7, round to the nearest tenth.

6. .86   7. .607

In Exercises 8 and 9, round to the nearest hundredth.

8. .664   9. .7091

In Exercises 10 and 11, round to the nearest thousandth.

10. .8135   11. .491 49

Now try the exercises for Topic B on page 139.

C. ROUNDING TO THE NEAREST WHOLE NUMBER

To round to the nearest whole number:

1. If the *tenths* digit is 4 *or less*, simply drop all decimal digits.
2. If the *tenths* digit is 5 *or more*, add 1 to the 1's digit and drop all decimal digits.

Example 7 ▶ Round to the nearest whole number.

    i. 7.4   *ii.* 18.932   *iii.* .8

Solution.

  i. 7.4 rounds to 7

 ii. 18.932 rounds to 19

iii. .8 = 0.8 . Therefore, .8 rounds to 1.0, or 1   ◀

PRACTICE EXERCISES FOR TOPIC C

Round to the nearest whole number.

12. 3.14   13. 99.919

Now try the exercises for Topic C on page 139.

D. ESTIMATING

Rounding enables you to *estimate* (or *approximate)* a sum, difference, product, or quotient quickly. This is particularly important when checking for gross errors.

Example 8 ▶ Approximate the sum

> 12 234
> 8 916
> 6 102
> 5 518

by first rounding each number to the nearest 1000, and then adding.

Solution. Round, and then add.

> 12 000
> 9 000
> 6 000
> 6 000
> ——————
> 33 000

The sum is approximately 33,000. The actual sum is 32,770.

> 12 234
> 8 916
> 6 102
> 5 518
> ——————
> 32 770   ◀

Example 9 ▶ Estimate the product

> 3.8 × 6.2

by first rounding each factor to the nearest whole number.

Solution.

> 3.8 rounds to 4
>
> 6.2 rounds to 6
>
> 4 × 6 = 24

Thus, the product is *approximately* 24.   ◀

Example 10 ▶ Divide.

$$\frac{.8}{.3}$$

Approximate the quotient to the nearest hundredth.

Solution.

> 2.666 666 ◀— 6's keep repeating.
> 3.⌐8.000 000

To the nearest hundredth, the quotient is 2.67 .   ◀

PRACTICE EXERCISES FOR TOPIC D

14. Approximate the sum

$$
\begin{array}{r}
18\ 419\ 327 \\
12\ 501\ 064 \\
11\ 103\ 841 \\
9\ 712\ 349 \\
\underline{5\ 532\ 849} \\
\end{array}
$$

by first rounding each number to the nearest 1,000,000, and then adding.

15. Estimate the product

$$12.2 \times 9.8 \times 5.01$$

by first rounding each factor to the nearest whole number.

In Exercises 16 and 17, approximate each quotient to the nearest hundredth.

16. $\dfrac{.7}{.3}$ 17. $\dfrac{.091}{.8}$

Now try the exercises for Topic D on page 139.

EXERCISES

A. *In Exercises 1–5, round to the nearest 1000.*

   1. 8496 *Answer:*   2. 6692 *Answer:*   3. 7541 *Answer:*

   4. 19,999 *Answer:*   5. 6500 *Answer:*

In Exercises 6–10, round to the nearest 1,000,000.

   6. 6,328,596 *Answer:*   7. 20,499,532 *Answer:*

   8. 15,503,006 *Answer:*   9. 19,932,147 *Answer:*

10. 121,543,202 *Answer:*

11. Round 536,215 to the nearest *i.* 100   *ii.* 1000   *iii.* 100,000.
    Answer: i.   *ii.*   *iii.*

12. Round 88,599,036 to the nearest *i.* 100,000 *ii.* 1,000,000 *iii.* 10,000,000.
    Answer: i.   *ii.*   *iii.*

In Exercises 13–16, round the populations of the following cities to the nearest 100,000.

13. Minneapolis   370,951   *Answer:*

14. San Francisco   678,974   *Answer:*

15. Chicago   3,005,072   *Answer:*

16. Philadelphia   1,688,210   *Answer:*

In Exercises 17–20, round the asking prices of the following houses to the nearest $1000.

17. Split-level $ 78,898   *Answer:*

18. Brownstone $165,500   *Answer:*

19. Ranch-style $ 89,500   *Answer:*

20. Mobile Home $ 29,999   *Answer:*

B. *In Exercises 21–25, round to the nearest tenth.*

21. .14 *Answer:*   22. .77 *Answer:*   23. .354 *Answer:*

24. 1.88 *Answer:*   25. 99.96 *Answer:*

In Exercises 26–30, round to the nearest hundredth.

26. .743 *Answer:*   27. .936 *Answer:*   28. 3.505 *Answer:*

29. 4.895 *Answer:*   30. 99.995 *Answer:*

In Exercises 31-34, round to the nearest thousandth.

31. .6721 *Answer:*   32. .1349 *Answer:*

33. .5999 *Answer:*   34. 5.0095 *Answer:*

In Exercises 35–38, round to the nearest i. tenth   ii. hundredth   iii. thousandth.

35. .4321   *Answer: i.*   *ii.*   *iii.*

36. .6758   *Answer: i.*   *ii.*   *iii.*

37. .4365   *Answer: i.*   *ii.*   *iii.*

38. 1.1919   *Answer: i.*   *ii.*   *iii.*

39. A stop-watch clocks a sprinter at 9.89 seconds. What is the time measured to the nearest tenth of a second? *Answer:*

40. A wooden beam is 2.366 inches thick. How thick is it to the nearest hundredth of an inch? *Answer:*

C. *In Exercises 41–46, round to the nearest integer.*

41. 4.1   *Answer:*   42. 1.6   *Answer:*   43. 27.49   *Answer:*

44. 31.501 *Answer:*   45. .6   *Answer:*   46. 99.9   *Answer:*

47. A necktie sells for $7.95. What is the price to the nearest dollar? *Answer:*

48. A car's speedometer indicates 4876.8 miles driven. To the nearest mile, how many miles are indicated? *Answer:*

D. *In Exercises 49 and 50, approximate each sum by first rounding each number to the nearest 1000, and then adding.*

49. 53 821   *Answer:*   50. 4004   *Answer:*
    47 143   3982
    52 495   5943
    50 501   6099
    36 182   7509

In Exercises 51–52, approximate each sum by first rounding each number to the nearest 1,000,000, and then adding.

51. 6 132 529   *Answer:*
8 429 599
7 006 381
9 582 111

52. 16 592 401   *Answer:*
18 389 526
21 099 599
14 593 999

In Exercises 53 and 54:
  i. *Approximate each sum by first rounding each number to the nearest 1000, and then adding.*
  ii. *Find the actual sum by adding the numbers as given.*
  iii. *Round the sum in part ii. to the nearest 1000.*
  iv. *Do parts i. and iii. agree?*

53. 6314   *Answer: i.*   *ii.*   *iii.*   *iv.*
2129

54. 5321   *Answer: i.*   *ii.*   *iii.*   *iv.*
3216

55. Approximate the difference

17 814 599
− 12 671 385

by first rounding each number to the nearest 100,000, and then subtracting.
Answer:

In Exercises 56–59, estimate the product by first rounding each factor to the nearest whole number.

56. 7.04 × 12.36   *Answer:*

57. 7.8 × 9.9   *Answer:*

58. 8.9 × 9.1 × 11.9   *Answer:*

59. 10,001.2 × 999.7   *Answer:*

In Exercises 60–62, estimate each quotient by first rounding to the nearest whole number.

60. $\dfrac{18.3}{6.02}$ *Answer:*

61. $\dfrac{19.95}{4.98}$ *Answer:*

62. $\dfrac{35.89 \times 63.91}{143.92}$ *Answer:*

63. A car uses 16.2 gallons of gas to drive 335.6 miles. Estimate the number of miles per gallon it gets on this trip by first rounding each factor to the nearest whole number.   *Answer:*

64. One kilogram is approximately 2.2 pounds. Estimate the number of pounds in 15.9 kilograms by first rounding each factor to the nearest whole number.   *Answer:*

In Exercises 65–68, approximate each quotient to the nearest hundredth.

65. $\frac{.4}{.3}$ *Answer:*   **66.** $\frac{.1}{7}$ *Answer:*

67. $.79\overline{)8.00}$ *Answer:*   **68.** $.093\overline{)5.13}$ *Answer:*

Review Exercises for Unit III

1. Fill in the blank. $.429 = \dfrac{\square}{10} + \dfrac{\square}{100} + \dfrac{\square}{1000}$

2. Consider the decimal .0413 .
   i. What is the tenths digit?   *Answer:*
   ii. What is the hundredths digit?   *Answer:*
   iii. What is the thousandths digit?   *Answer:*

3. Write "forty-five thousandths" in decimal notation.   *Answer:*

4. Write .046 in words.   *Answer:*

5. Express $12\frac{3}{100}$ in decimal notation.   *Answer:*

6. Change $\frac{3}{25}$ to a decimal.   *Answer:*

7. Change .375 to a fraction in lowest terms.   *Answer:*

8. Change $\frac{7}{30}$ to an infinite repeating decimal.   *Answer:*

In Exercises 9–11, which number is the smallest? Draw a circle around the correct letter.

9. (A) .407   (B) .417   (C) .471   (D) .410   (E) .401

10. (A) .03   (B) .30   (C) .033   (D) .303   (E) .330

11. (A) 1.01   (B) 11.0   (C) 1.11   (D) 1.001   (E) 1.101

12. Which is the smallest amount of money? Draw a circle around the correct letter.
   (A) $49.98   (B) $53.01   (C) $48.99   (D) $49.01   (E) $49.89

In Exercises 13 and 14, which number is the largest? Draw a circle around the correct letter.

13. (A) .662   (B) .626   (C) .622   (D) .266   (E) .2666

14. (A) 4.010   (B) 4.101   (C) 4.110   (D) 4.011   (E) 4.1011

In Exercises 15–18, add or subtract, as indicated.

15. 50.4 + 15.99 + 5.89 =   **16.** 1608 – 6.99 =

17. 9.895 + 19.27 + 3.9953 + 6.021 =   **18.** $431.75
                                                    386.09
                                                   291.98

19. A man buys a shirt for $17.95, a pair of gloves for $14.50, and a scarf for $11.85. How much does he spend on his wardrobe?   *Answer:*

20. If the distance between two towns is 49.6 miles and a car has been driven 28.9 miles from one town in the direction of the other, how far is the car from the second town?   *Answer:*

In Exercises 21–24, multiply.

21. .184 × 10 =   22. 7.1721 × 1000 =   23. .2 × .5 =

24. .052 × 1.25 =

25. Find $(.04)^3$.   *Answer:*

26. What is the cost of 6 quarts of paint at $7.95 per quart?   *Answer:*

In Exercises 27–30, divide.

27. 5.05 ÷ 10 =   28. .038 ÷ 100 =   29. 4.2 ÷ .14 =   30. .084 ÷ 1.2 =

In Exercises 31–34, draw a circle around the correct letter.

31. If tickets to a concert cost $6.50 each and 5 tickets are purchased, the cost is
    (A) $32.50   (B) $30.50   (C) $325   (D) $1.50   (E) $1.30

32. A car that sells for $12,000 is paid for in equal monthly installments of $800 each. For how many months will the payments last?
    (A) 12   (B) 15   (C) 18   (D) 32   (E) 96

33. If pens cost 39¢ each and pencils cost 12¢ each, the cost of 15 pens and 20 pencils is
    (A) $4.68   (B) $5.85   (C) $7.80   (D) $7.25   (E) $8.25

34. A store buys 20 boxes of stationery for $6.22 each. Two of the boxes become damaged, and are sold for $4.95 each. The rest are sold for $12.95 each. The store's profit is
    (A) $118.60   (B) $124.40   (C) $243.00   (D) $233.10   (E) $134.60

35. A man buys 3 boxes of razor blades for $2.45 each. If he pays for them with a $20 bill, how much change does he receive?   *Answer:*

36. At a baseball game 12,400 grand stand seats are sold for $5.50 each and 9800 bleacher seats are sold for $4.00 each. What is the total sales?   *Answer:*

37. Round 17,449 to the nearest 1000.   *Answer:*

38. Round 4,850,494 to the nearest   *i.* 10,000   *ii.* 100,000   *iii.* 1,000,000.
    Answer: i.   *ii.*   *iii.*

39. Round .793 to the nearest hundredth.   *Answer:*

40. Round .5094 to the nearest   *i.* tenth   *ii.* hundredth   *iii.* thousandth.
    Answer: i.   *ii.*   *iii.*

41. Approximate the sum

$$
\begin{array}{r}
42\ 084\ 173 \\
53\ 946\ 985 \\
59\ 095\ 993 \\
\underline{29\ 009\ 819}
\end{array}
$$

by first rounding each number to the nearest 1,000,000 and then adding. *Answer:*

42. Estimate the product

$$9.8 \times 4.4 \times 10.6$$

by first rounding each factor to the nearest whole number.   *Answer:*

43. Estimate the quotient

$$\frac{53.98}{6.43}$$

by first rounding to the nearest whole number.   *Answer:*

Review Exercises on Units I and II

1. Express 27,085 in verbal form.

2. Multiply 408 by 1200.

3. $2^3 \times 3^2 =$

4. Which of the following are primes?   *i.* 17   *ii.* 27   *iii.* 37   *iv.* 47

5. Write 5600 as the product of primes.

6. Find *lcm* (4, 5, 12).

7. Find *gcd* (28, 48).

8. Which fraction is the smallest? Draw a circle around the correct letter.

(A) $\frac{3}{10}$ (B) $\frac{1}{2}$ (C) $\frac{1}{5}$ (D) $\frac{3}{5}$ (E) $\frac{1}{4}$

9. $\frac{1}{4} + \frac{1}{2} =$ 10. $\frac{9}{10} - \frac{1}{5} =$ 11. $\frac{2}{5} \times \frac{10}{11} =$

12. $\frac{5}{8} \div \frac{3}{4} =$ 13. $1\frac{1}{2} + 2\frac{1}{4} =$ 14. $2\frac{1}{8} \times 4\frac{1}{2} =$

15. If a pound and a quarter of sugar has been used from a 5-pound package, how much sugar remains?

Practice Exam on Unit III

1. Write "ninety-seven thousandths" in decimal notation.   *Answer:*

2. Change $\frac{7}{40}$ to a decimal.   *Answer:*

3. Change 1.04 to a fraction in lowest terms.   *Answer:*

4. Which number is the smallest?

   (A) .049   (B) .094   (C) .0499   (D) .0491   (E) .4001

In Problems 5–10, add, subtract, multiply, or divide, as indicated.

5. $27.2 + 36.9 + 15.91 + 21.99 =$ 6. $13.08 - 9.98 =$

7. $.045 \times 100 =$ 8. $4.17 \div 1000 =$

9. $.042 \times 2.1 =$ 10. $.505 \div .005 =$

11. Orchestra seats for a concert cost $7.50 each, whereas balcony seats cost $6.00 each. If 800 orchestra seats and 550 balcony seats are sold, what are the total sales? *Answer:*

12. If a worker earns $4.50 per hour, how long does it take him to earn $900?

    (A) 20 hours   (B) 25 hours   (C) 38 hours   (D) 200 hours   (E) 3800 hours

13. Round 474,099 to the nearest 1000.   *Answer:*

14. Round .0909 to the nearest   *i.* tenth   *ii.* hundredth   *iii.* thousandth.

    Answer: i.   *ii.*   *iii.*

15. Approximate the quotient

    $$\frac{.012}{.42}$$

    to the nearest hundredth.   *Answer:*

UNIT IV PERCENTAGE

17. Fractions, Decimals, and Percent

Percentages, like fractions and decimals, are commonly used to express parts of whole objects: A student receives a grade of 89 percent on an English exam. A soap is advertised as being $99 \frac{44}{100}$ percent pure. What exactly is the relationship between fractions, decimals, and percent? How do you express percentages, fractions, and decimals in terms of one another?

A. PERCENT TO FRACTIONS

Percent means *hundredths.* For example,

$$57\% = \frac{57}{100}$$

To express a percent as a fraction:

1. Remove the percent symbol, %.
2. Divide the resulting number by 100.
3. Reduce to lowest terms, if necessary.

Example 1 ▶   What is 85% expressed as a fraction?

Solution.

$$85\% = \frac{85}{100} \qquad \text{Divide numerator and denominator by 5.}$$

$$= \frac{17}{20} \qquad\qquad\qquad\qquad\qquad\qquad ◀$$

Example 2 ▶   What is 120% expressed as a fraction?

Solution.

$$120\% = \frac{120}{100} \qquad \text{Divide numerator and denominator by 20.}$$

$$= \frac{6}{5} \qquad\qquad\qquad\qquad\qquad\qquad ◀$$

PRACTICE EXERCISES FOR TOPIC A

Express as a fraction.

1. 60%   2. 84%   3. 125%

Now try the exercises for Topic A on page 150.

B. PERCENT TO DECIMALS

> To convert a percent to a decimal:
>
> 1. If there is no decimal point, insert one to the left of the percent symbol.
> 2. Remove the percent symbol.
> 3. Move the decimal point two digits to the left.

For example,

$$39\% = 39.\% = .39$$

and

$$8\% = 8.\% = .\boxed{0}\,8$$

Note that when converting 8% to .08, you must insert a $\boxed{0}$ to the left of the 8, so that you can move the decimal point two digits to the left.

Example 3 ▶   Express 160% as a decimal.

Solution.

$$160\% = 160.\% = 1.60 = 1.6$$

Note that the 0 at the far right in 1.60 can be dropped. Thus

$$160\% = 1.6$$ ◀

Example 4 ▶ Express $12\frac{1}{2}\%$ as a decimal.

Solution. Because $\frac{1}{2} = .5$, it follows that

$$12\frac{1}{2}\% = 12.5\% = .125$$ ◀

PRACTICE EXERCISES FOR TOPIC B

Express as a decimal.

4. 67%   5. 180%   6. $10\frac{1}{4}\%$

Now try the exercises for Topic B on page 151.

C. DECIMALS TO PERCENT

Because *percent* means hundredths, it follows that

$$.27 = 27\%$$

> To express a decimal as a percent:
> 1. If there are less than two decimal digits, add 0's at the right.
> 2. Move the decimal point two digits to the right.
> 3. Insert the percent symbol at the right.

Thus

$$.27 = 27.\% \quad \text{or} \quad 27\%$$
$$.3 = .30 = 30.\% \quad \text{or} \quad 30\%$$

Add a zero
at the right.

$$4 = 4.00 = 400.\% \quad \text{or} \quad 400\%$$

Add two zeros
at the right.

Example 5 ▶ Express 2.05 as a percent.

Solution.

$$2.05 = 205.\% = 205\%$$ ◀

Example 6 ▶ Express .005 as a percent.

Solution.

$$.005 = \underset{\displaystyle \curvearrowright}{00.5}\% = .5\%, \text{ or } \frac{1}{2}\% \qquad ◀$$

PRACTICE EXERCISES FOR TOPIC C

Express as a percent.

7. .82   8. .6   9. 2.9   10. 1.003

Now try the exercises for Topic C on page 151.

D. FRACTIONS TO PERCENT

To express $\frac{3}{4}$ as a percent, divide the numerator, 3, by the denominator, 4.

$$4\overline{)3.00}^{\,.75}$$

Thus

$$\frac{3}{4} = .75 = \underset{\displaystyle \curvearrowright}{75.}\% \quad \text{or} \quad 75\%$$

> To express a fraction as a percent:
>
> 1. First divide the numerator by the denominator to express the fraction as a decimal.
> 2. Then express this decimal as a percent, as indicated in Topic C.

Example 7 ▶ Express $\frac{7}{8}$ as a percent.

Solution.

$$8\overline{)7.000}^{\,.875}$$

Thus

$$\frac{7}{8} = .875 = \underset{\displaystyle \curvearrowright}{87.5}\% \qquad ◀$$

Example 8 ▶   Express $\frac{4}{3}$ as a percent.

Solution.

$$3\overline{)4.000\,000} \atop 1.333\,333\,\ldots$$

Thus $\frac{4}{3}$ can be expressed as the repeating decimal 1.333 333 . . . . In other words,

$$\frac{4}{3} = 1.333\,333\,\ldots$$

But if we write out the division process as in long division, and stop after the second decimal digit, we obtain

$$\begin{array}{r} 1.33 \\ 3\overline{)4.00} \\ \underline{3} \\ 10 \\ \underline{9} \\ 10 \\ \underline{9} \\ 1 \end{array} \leftarrow \text{remainder}$$

Here the remainder is 1, so that $\frac{4}{3} = 1.33\frac{1}{3}$. The more usual practice is to write

$$\frac{4}{3} = 1.33\frac{1}{3} = 133\frac{1}{3}\%$$ ◀

PRACTICE EXERCISES FOR TOPIC D

Express as a percent.

11. $\frac{3}{10}$ 12. $\frac{8}{5}$ 13. $\frac{5}{6}$

Now try the exercises for Topic D on page 151.

E. MIXED NUMBERS TO PERCENT

Here is how the mixed number $2\frac{1}{2}$ is expressed as a percent.

$$\begin{aligned} 2\frac{1}{2} &= 2 + \frac{1}{2} \\ &= 2 + .5 \\ &= 2.50 \\ &= 250\% \end{aligned}$$

> To express a mixed number as a percent:
>
> 1. Express as the sum of the whole part and the fractional part.
> 2. Express as a decimal.
> 3. Then express as a percent, as indicated in Topic C.

Example 9 ▶   Express $1\frac{1}{4}$ as a percent.

Solution.

$$1\frac{1}{4} = 1 + \frac{1}{4}$$
$$= 1 + .25$$
$$= 1.25$$
$$= 125\%$$

◀

PRACTICE EXERCISES FOR TOPIC E

Express as a percent.

14. $4\frac{1}{5}$ 15. $1\frac{3}{4}$

Now try the exercises for Topic E on page 151.

EXERCISES

A. *In Exercises 1–10, express each percent as a fraction in lowest terms.*

1. 50% =   2. 40% =   3. 75% =   4. 35% =   5. 22% =

6. 36% =   7. 150% =   8. 175% =   9. 48% =   10. 64% =

In Exercises 11 and 12, draw a circle around the correct letter.

11. What is 15% expressed as a fraction?

    (A) 15   (B) 1500   (C) $\frac{3}{20}$ (D) $\frac{3}{10}$ (E) $1\frac{3}{20}$

12. What is 44% expressed as a fraction?

    (A) $\frac{4}{9}$ (B) $\frac{4}{11}$ (C) $\frac{44}{99}$ (D) $\frac{11}{20}$ (E) $\frac{11}{25}$

13. A theatre is 80% full. Express this percentage as a fraction in lowest terms.
    Answer:

14. If 62.5% of the students at a college graduate, write this percentage as a fraction in lowest terms.   *Answer:*

B. *In Exercises 15–26, express each percent as a decimal.*

15. 53% =   16. 14% =   17. 40% =   18. 90% =

19. 2% =   20. 150% =   21. 225% =   22. 22.5% =

23. 19.05% =   24. $40\frac{1}{2}$% =   25. $10\frac{1}{4}$% =   26. $100\frac{1}{10}$% =

In Exercises 27 and 28, draw a circle around the correct letter.

27. 2.5% =

    (A) 2.5   (B) 25   (C) 250   (D) .25   (E) .025

28. $99\frac{44}{100}$% =

    (A) 99.44   (B) 994,400   (C) .9944   (D) 9.944   (E) 99,440

29. Dry air contains 78% nitrogen. Express this percentage as a decimal. *Answer:*

30. A breakfast cereal supplies 82.7% of the minimum daily requirement of niacin. Express this percentage as a decimal. *Answer:*

C. *Express each decimal as a percent.*

31. .43 =   32. .95 =   33. 1.45 =   34. .2 =

35. .02 =   36. 2.2 =   37. .002 =   38. 3.15 =

39. 10 =   40. 1.007 =   41. 2.55 =   42. 2.055 =

D. *Express each fraction as a percent.*

43. $\frac{53}{100}$ =   44. $\frac{7}{10}$ =   45. $\frac{4}{5}$ =   46. $\frac{5}{4}$ =

47. $\frac{9}{2}$ =   48. $\frac{5}{8}$ =   49. $\frac{1}{12}$ =   50. $\frac{17}{20}$ =

51. $\frac{1}{3}$ =   52. $\frac{1}{9}$ =   53. $\frac{7}{6}$ =   54. $\frac{2}{11}$ =

E. *Express each mixed number as a percent.*

55. $1\frac{1}{2}$ =   56. $3\frac{1}{4}$ =   57. $4\frac{1}{10}$ =   58. $2\frac{1}{5}$ =

59. $12\frac{1}{2}$ =   60. $1\frac{1}{8}$ =   61. $1\frac{1}{20}$ =   62. $5\frac{7}{12}$ =

63. $1\frac{1}{3}$ =   64. $1\frac{1}{9}$ =

18. Percentage Problems

Questions involving percentage arise frequently in real-life situations. For example, you may be told that a student answered 90% of the 40 problems on her math quiz correctly. How many of these problems did she actually answer correctly? Or you may know that 30% of a worker's salary goes into taxes. If the monthly taxes amount to $620, how can you then find the monthly salary? Before considering such practical applications, let us consider several basic types of numerical problems that involve percentage.

A. 20% OF 80

Suppose you are asked

> What is 20% of 80?

Here, the word "of" indicates multiplication, so that what you want to find is

> 20% × 80

Change 20% to a decimal

> 20% = .20 = .2

and then calculate

> .2 × 80 = 16.0 = 16

Thus 20% of 80 is 16.

Example 1 ▶   What is 60% of 75?

(A) 60   (B) 50   (C) 45   (D) 4.5   (E) 125

Solution. You want

$$60\% \times 75$$

Change 60% to a decimal

$$60\% = .60 = .6$$

and then calculate

$$.6 \times 75 = 45.0 = 45$$

Thus 60% of 75 is 45. The correct choice is (C).   ◄

PRACTICE EXERCISES FOR TOPIC A

1. What is 25% of 60?
   (A) 10   (B) 15   (C) 20   (D) 25   (E) 12.5
2. What is 45% of 150?
3. What is 7% of 15?

Now try the exercises for Topic A on page 157.

B. EQUATIONS

In order to solve some percentage problems, it is best to set up an *equation.* Before considering an equation with percent, here is a simple illustration. Suppose you are told:

Four times a number is equal to 12

and are asked to find this number. To do so, let *n* stand for this *unknown* number. Then translate the problem as follows:

Four times a number is equal to 12

4   ×   *n*   =   12

This last statement is known as an *equation.* An **equation** is a statement of *equality.*

An equation is like a balanced scale. In order to preserve the balance, if you change one side, you must do the same thing to the other side. (See the figure to the right.)

Because 4 *multiplies n* on the left side, in order to isolate the unknown, *divide* the left side by 4. But now the scale is unbalanced. (See the figure to the right.)

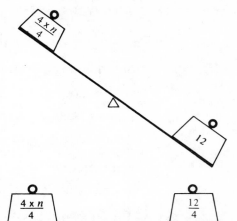

What you do to the left side, also do to the right side. Thus divide both sides by 4 to preserve the balance. (See the figure to the right.)

$$4 \times n = 12 \qquad \text{Divide both sides by 4.}$$

$$\frac{4 \times n}{4} = \frac{12}{4}$$

$$\frac{\overset{1}{\cancel{4}} \times n}{\underset{1}{\cancel{4}}} = \frac{\overset{3}{\cancel{12}}}{\underset{1}{\cancel{4}}}$$

$$n = 3$$

Thus the number *n* is now known to be 3. (Note that $4 \times 3 = 12$.)

Example 2 ▶ Solve the equation

$$5 \times n = 60$$

Solution. Because 5 multiplies *n* on the left side, to isolate *n*, *divide* both sides by 5.

$$\frac{5 \times n}{5} = \frac{60}{5}$$

$$\frac{\overset{1}{\cancel{5}} \times n}{\underset{1}{\cancel{5}}} = \frac{\overset{12}{\cancel{60}}}{\underset{1}{\cancel{5}}}$$

$$n = 12$$

Thus the solution is 12.   ◀

Now here is an equation involving a percentage.

Example 3 ▶ Solve the equation

$$50\% \times n = 7$$

Solution. Change 50% to .5.

$$.5 \times n = 7$$

To eliminate decimals, multiply both sides by 10.

$$10 \times .5 \times n = 10 \times 7$$
$$5 \times n = 70$$

Divide both sides by 5.

$$\frac{\overset{1}{\cancel{5}} \times n}{\underset{1}{\cancel{5}}} = \frac{\overset{14}{\cancel{70}}}{\underset{1}{\cancel{5}}}$$

$$n = 14$$

Thus the solution is 14.   ◄

PRACTICE EXERCISES FOR TOPIC B

Solve each equation for *n.*

4. $5 \times n = 45$ 5. $8 \times n = 44$ 6. $20\% \times n = 5$ 7. $90\% \times n = 18$

Now try the exercises for Topic B on page 158.

C. 20% OF WHAT NUMBER?

Example 4 ▶ If 30 is 20% of a certain number, find that number.

Solution. Reword the question as follows.

30 is equal to 20% of *what* number?

Now let *n* stand for the *unknown* number. Recall that "of" often indicates multiplication. Translate the problem as follows:

30 is equal to 20% of what number?

$$30 \quad = \quad 20\% \times \quad n$$

$$30 = .2 \times n \qquad \text{Multiply both sides by 10 to eliminate decimals.}$$
$$10 \times 30 = 10 \times .2 \times n$$
$$300 = 2 \times n \qquad \text{Divide both sides by 2 to isolate the } n.$$
$$\frac{300}{2} = \frac{2 \times n}{2}$$
$$150 = n$$

Thus 30 is 20% of 150.   ◄

PRACTICE EXERCISES FOR TOPIC C

8. If 40 is 50% of a number, what is the number?

    (A) 60   (B) 80   (C) 90   (D) 100   (E) 200

9. If 90 is 60% of a certain number, find that number.

10. If 50 is 200% of a number, what is the number?

Now try the exercises for Topic C on page 158.

D. WHAT PERCENT?

Example 5 ▶ What percent of 40 is 30?

    (A) 75%   (B) 25%   (C) 30%   (D) 40%   (E) 133.3%

Solution. To see what is involved here, reword the question.

What percent is 30 of 40?

You are really asked here to write the fraction $\frac{30}{40}$ as a percent. Thus

$$\frac{30}{40} = \frac{3}{4}$$

Divide 3 by 4

$$4\overline{)3.00}^{\,.75}$$

to obtain

$$\frac{3}{4} = .75 = 75\%$$

Thus 30 is 75% of 40. The correct choice is (A).   ◀

PRACTICE EXERCISES FOR TOPIC D

11. What percent of 50 is 20?   12. What percent of 20 is 50?

13. What percent of 32 is 12?

Now try the exercises for Topic D on page 159.

E. APPLIED PROBLEMS

Problems involving percentages arise in many applications. Often you must first reword the problem.

Example 6 ▶ Sue correctly answered 90% of the 40 problems on her math quiz. How many problems did she answer correctly?

    (A) 4   (B) 9   (C) 35   (D) 36   (E) 38

Solution. Reword the problem.

What is 90% of 40?

Here again, the word "of" indicates multiplication. Thus find

.9 × 40 = 36.0 = 36

She answered 36 problems correctly. The correct choice is (D).   ◀

Example 7 ▶ 15% of Bob's salary goes into taxes. If his taxes are $30.75 each week, find his weekly salary (before taxes).

Solution. 15% of Bob's salary "equals" taxes. Write in $30.75 for his taxes, and find his salary, s.

15% of Bob's salary equals $30.75.

15% × s =   30.75

.15 × s = 30.75

100 × .15 × s = 100 × 30.75

15 × s = 3075   Divide both sides by 15.

$$\frac{15 \times s}{15} = \frac{3075}{15}$$

s = 205

His weekly salary is $205.   ◀

PRACTICE EXERCISES FOR TOPIC E

14. A jug of punch contains 40% grape juice. If the jug contains 15 gallons of punch, how much of this is grape juice?

(A) 4 gallons   (B) 5 gallons   (C) 6 gallons

(D) 8 gallons   (E) 10 gallons

15. 60% of the students at a school are women. If there are 330 women at the school, how many students are there altogether?

Now try the exercises for Topic E on page 159.

EXERCISES

A. *In Exercises 1–6, draw a circle around the correct letter.*

1. What is 50% of 66?

(A) 50   (B) 33   (C) 132   (D) 13.2   (E) 99

  2. What is 40% of 8?

     (A) 32   (B) 3.2   (C) 320   (D) 4   (E) 12

  3. What is 75% of 60?

     (A) 15   (B) 42.5   (C) 425   (D) 45   (E) 48

  4. What is 70% of 200?

     (A) 70   (B) 14   (C) 140   (D) 1400   (E) 60

  5. What is 30% of 90?

     (A) 30   (B) 33   (C) 270   (D) 27   (E) 120

  6. What is 80% of 120?

     (A) 100   (B) 96   (C) 90   (D) 108   (E) 24

  7. What is 20% of 1440?   *Answer:*   8. What is 60% of 55?   *Answer:*

  9. What is 35% of 80?   *Answer:*   10. What is 65% of 90?   *Answer:*

11. What is 85% of 80?   *Answer:*   12. What is 17% of 60?   *Answer:*

B. *Solve each equation for n.*

  13. $2 \times n = 20$ *Answer:*   14. $3 \times n = 36$ *Answer:*

  15. $4 \times n = 28$ *Answer:*   16. $10 \times n = 130$ *Answer:*

  17. $7 \times n = 56$ *Answer:*   18. $5 \times n = 75$ *Answer:*

  19. $20 \times n = 140$ *Answer:*   20. $8 \times n = 60$ *Answer:*

  21. $50\% \times n = 9$ *Answer:*   22. $40\% \times n = 6$ *Answer:*

  23. $30\% \times n = 27$ *Answer:*   24. $80\% \times n = 120$ *Answer:*

  25. $75\% \times n = 15$ *Answer:*   26. $60\% \times n = 30$ *Answer:*

  27. $90\% \times n = 270$ *Answer:*   28. $35\% \times n = 105$ *Answer:*

C. *In Exercises 29–34, draw a circle around the correct letter.*

29. 20% of what number is 12?

     (A) 240   (B) 24   (C) 204   (D) 60   (E) 14.4

30. 25% of what number is 8?

     (A) 2   (B) 10   (C) 200   (D) 32   (E) 33

31. 60% of what number is 24?

     (A) 14.4   (B) 144   (C) 40   (D) 84   (E) 15

32. 75% of what number is 84?

     (A) 112   (B) 125   (C) 21   (D) 63   (E) 105

33. If 36 is 25% of a certain number, find that number.

     (A) 9   (B) 27   (C) 61   (D) 48   (E) 144

34. If 40% of a certain number is 20, find that number.

     (A) 8   (B) 80   (C) 800   (D) 50   (E) 60

35. If 90% of a certain number is 90, find that number.   *Answer:*

36. If 30% of a certain number is 18, find that number.   *Answer:*

37. If 25% of a number is 120, what is the number?   *Answer:*

38. If 75% of a number is 72, what is the number?   *Answer:*

39. If 45% of a number is 4.5, what is the number?   *Answer:*

40. If 22% of a number is 33, what is the number?   *Answer:*

D. *In Exercises 41–44, draw a circle around the correct letter.*

41. What percent of 50 is 25?

   (A) 50%   (B) 25%   (C) 2%   (D) 20%   (E) 200%

42. What percent of 48 is 36?

   (A) $33\frac{1}{3}$%   (B) 133%   (C) 25%   (D) 75%   (E) 125%

43. What percent of 90 is 72?

   (A) 72%   (B) 90%   (C) 20%   (D) 80%   (E) 120%

44. What percent of 40 is 12?

   (A) 30%   (B) 20%   (C) 12%   (D) 48%   (E) 52%

45. What percent of 80 is 72?   *Answer:*

46. What percent of 125 is 75?   *Answer:*

47. What percent of 36 is 3.6?   *Answer:*

48. What percent of 24 is 36?   *Answer:*

E. *In Exercises 49–54, draw a circle around the correct letter.*

49. A student answers 82% of the questions on an exam correctly. If there are 50 questions on the exam, how many does she answer correctly?

   (A) 50   (B) 40   (C) 41   (D) 42   (E) 32

50. Harry answers 36 out of 60 questions correctly. What percent of his answers are correct?

   (A) 36%   (B) 50%   (C) 54%   (D) 60%   (E) 72%

51. A pitcher throws 50 called strikes out of a total of 125 pitches in a game. What percent of his pitches are called strikes?

   (A) 25%   (B) 40%   (C) 50%   (D) 85%   (E) 250%

52. An alloy contains 60% copper. How much copper is there in 400 tons of the alloy?

   (A) 24 tons   (B) 240 tons   (C) 640 tons   (D) 320 tons   (E) $666\frac{2}{3}$ tons

53. 20% of the people in a village are retired. If there are 120 retired people there, what is the population of the village?

   (A) 24   (B) 2400   (C) 24,000   (D) 600   (E) 144

54. A family pays $550 a month for rent. This amounts to 25% of its monthly income. What is the family's monthly income?

    (A) $112.50   (B) $662.50   (C) $2000   (D) $2200   (E) $1375

55. A basketball player made 80% of her foul shot attempts. If she made 16 foul shots, how many did she attempt?   *Answer:*

56. Al pays 14% of his weekly salary in taxes. If his taxes are $22.40, his weekly salary (before taxes) is _____

57. An airplane pilot earns $38,000 per year. 22% of his salary is withheld for taxes. How much money is withheld?   *Answer:*

58. A man pays a sales tax of $13.05 on a suit. The sales tax rate is 9%. What is the price of the suit (before the tax)?   *Answer:*

59. At a certain college, 35% of the students are enrolled in at least one mathematics course. If 2310 students are enrolled in at least one math course, how many students are there at the college?   *Answer:*

60. If the sales tax on a $4.50 item is $.36, what is the sales tax rate?   *Answer:*

19. Taxes, Sales, and Percent Increase or Decrease

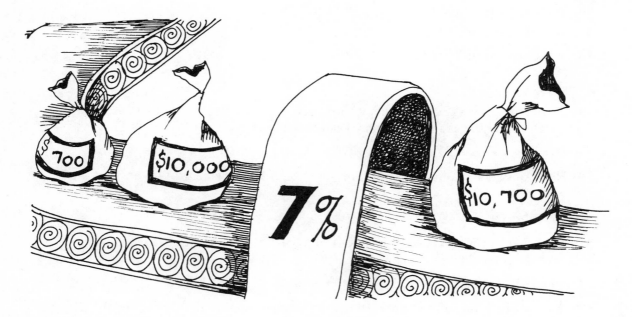

Many practical applications involve percent increase or decrease. Problems concerning percent increase or decrease arise when computing sales tax, or when considering price fluctuations or population shifts. For example, by what *percentage* has the price of milk increased over the past year? What is the percent decrease in the urban population of Massachusetts?

A. PERCENT INCREASE AND SALES TAX

A part-time student earns $10,000 a year. He receives a 7% *increase* in salary. The increase amounts to

$$\$10,000 \times .07 = \$700$$

This *increase* is in *addition* to *all* of (100% of) his former salary. In other words, he receives 100% of his former salary *plus* an additional 7%. Thus his new salary is 100% + 7%, or 107%, of his former salary. To determine his new salary, multiply $10,000 by 107%.

$$\$10,000 \times 107\% = \$10,000 \times 1.07$$
$$= \$10,700$$

His new salary is $10,700.

Similarly, for an increase of 8%, multiply the former salary by 108% or 1.08. For an increase of 10%, multiply by 110% or 1.1 .

Example 1 ▶ A jacket sells for $80 plus a 6% sales tax. What is the total price?

(A) $86   (B) $4.80   (C) $75.20   (D) $84.80   (E) $140

Solution. The 6% sales tax is *in addition to* the price of the jacket. Thus 100% of the original price is increased by 6%. To find the *total* price, multiply the price of the jacket, $80, by 1.06.

$$
\begin{array}{r}
1.06 \quad \longleftarrow \quad \text{2 decimal digits} \\
\times \ \$80 \quad \longleftarrow + \text{0 decimal digits} \\
\hline
\$84.80 \quad \longleftarrow \quad \text{2 decimal digits}
\end{array}
$$

The total price is $84.80, choice (D).   ◀

Example 2 ▶ The price of coffee, which was originally $2.80 per pound, is increased by 5%. What is the new price?

Solution. An *increase* of 5% means you must multiply the original price by 105% or 1.05 .

$$
\begin{array}{r}
\$2.80 \quad \longleftarrow \quad \text{2 decimal digits} \\
\times \ 1.05 \quad \longleftarrow + \text{2 decimal digits} \\
\hline
14\ 00 \\
2\ 80\ 0 \\
\hline
\$2.94\ 00 \quad \longleftarrow \quad \text{4 decimal digits}
\end{array}
$$

Because 2.9400 = 2.94, the new price is $2.94.   ◀

Example 3 ▶ The population of a city, which was originally 200,000, is now 280,000. What percent increase is there in the population?

Solution. The increase in population is given by

$$
\begin{array}{l}
280{,}000 \ \cdots \text{ new population} \\
- \ 200{,}000 \ \cdots \text{ original population} \\
\hline
80{,}000 \ \cdots \text{ increase in population}
\end{array}
$$

To determine the *percent* increase, divide the increase in population by the original population. Thus

$$\frac{80,000}{200,000} = \frac{8}{20} = \frac{2}{5} = 40\%$$

The percent increase is 40%.   ◄

PRACTICE EXERCISES FOR TOPIC A

1. A hat sells for $12.00 plus an 8% sales tax. What is the total price?
   (A) $12.08   (B) $12.80   (C) $12.96   (D) $13.08   (E) $96.00

2. A town of 50,000 people increases in population by 4%. What is the new population of this town?

3. A 50% increase in the price of bread will increase a 64-cent loaf to
   (A) 32 cents   (B) 80 cents   (C) 88 cents   (D) 96 cents   (E) $1.14

4. A college enrollment goes up from 9000 students to 12,000 students. What percent increase does this represent?

5. A book whose list price is $16.00 sells for $17.12, including sales tax. What percent of the list price is the sales tax?

6. A bank officer who earns $50,000 one year is raised to $56,000 the next year. What percent increase is this in her salary?

Now try the exercises for Topic A on page 164.

B. PERCENT DECREASE AND REDUCTIONS

During hard times, a man who was earning $10,000 a year is asked to take a 5% *reduction* in his salary. The *reduction* (or *decrease*) amounts to

$10,000 × .05   or $500

This decrease is *subtracted from* 100% of his former salary. Thus he receives 100% of his former salary *minus* the 5% reduction. His new salary is therefore only 100% – 5%, or 95%, of his former salary. To determine his new salary, multiply $10,000 by 95%.

$$\$10,000 \times 95\% = \$10,000 \times .95$$
$$= \$9500$$

His new salary is $9500.

Example 4▶   A town with 24,000 people loses 6% of its population. What is the new population?

Solution.

$$100\% - 6\% = 94\% = .94$$

$$
\begin{array}{r}
240\ 00 \\
\times\ .94 \\
\hline
960\ 00 \\
21\ 600\ 0 \\
\hline
22560.00
\end{array}
\quad
\begin{array}{l}
\longleftarrow\quad 0\ \text{decimal digits} \\
\longleftarrow\ +\ 2\ \text{decimal digits} \\
\\
\\
\longleftarrow\quad 2\ \text{decimal digits}
\end{array}
$$

The new population is 22,560.   ◀

Example 5▶   At a sale, a $28 sweater is reduced by 20%. Its sale price is

(A) $48   (B) $8   (C) $33.60   (D) $23.40   (E) $22.40

Solution.

$$100\% - 20\% = 80\% = .80$$

$$\$28 \times .80 = \$22.40$$

The sale price is $22.40, choice (E).   ◀

Example 6▶   A scarf that originally sold for $15 was reduced to $12. What percent reduction was this?

Solution.   The reduction in price is given by

$$
\begin{array}{r}
\$15 \cdots \text{original price} \\
-\ \$12 \cdots \text{new price} \\
\hline
\$\ 3 \cdots \text{reduction in price}
\end{array}
$$

To determine the *percent* reduction, divide the reduction in price by the original price. Thus

$$\frac{\$3}{\$15} = \frac{3}{15} = \frac{1}{5} = 20\%$$

The percent reduction was 20%.   ◀

PRACTICE EXERCISES FOR TOPIC B

7. A baseball that sold for $7.50 was reduced by 10%. What was the new price?

(A) $.75   (B) $6.00   (C) $6.50   (D) $6.75   (E) $7.00

8. A state with 4 million people loses $12\frac{1}{2}\%$ of its population. What is its new population?

9. A man who earns $40,000 per year receives $34,000 after taxes. What percent of his salary does he pay in taxes?

(A) 6%   (B) 12%   (C) 15%   (D) 18%   (E) 20%

10. A publisher sells 50,000 copies of a textbook one year. The next year there is a 25% reduction in sales of this text. How many copies are sold?

Now try the exercises for Topic B on page 165.

EXERCISES

A. *In Exercises 1–6, draw a circle around the correct letter.*

1. Bob's weekly paycheck, which used to be $200, is increased by 12%. His new paycheck is

(A) $212   (B) $224   (C) $2400   (D) $188   (E) $200.12

2. Jane's monthly paycheck, which used to be $1800, is increased by 8%. Her new paycheck is

(A) $1944   (B) $1656   (C) $144   (D) $1808   (E) $1814.44

3. A coat sells for $140 plus a 7% sales tax. What is the total price?

(A) $140.98   (B) $1498   (C) $149.80   (D) $147   (E) $980

4. A tie sells for $7.50 plus a 6% sales tax. What is the total price?

(A) $7.56   (B) $8.10   (C) $7.95   (D) $8.00   (E) $7.92

5. The price of gasoline, which was $1.25 per gallon, is increased by 4%. The new price per gallon is

(A) $1.29   (B) $6   (C) $1.21   (D) $1.30   (E) $1.20

6. A business that makes a profit of $180,000 one year increases its profit by 25% the next year. Its profit is then

(A) $45,000   (B) $25,000   (C) $205,000   (D) $225,000   (E) $135,000

7. A man who earns $46,000 per year receives a 4% increase in salary. What is his new salary?   *Answer:*

8. The population of a city increases by 5%. If the population was formerly 4,400,000, its new population is _____

9. A pair of gloves sells for $16 plus sales tax. If the total price is $16.96, what percent sales tax is there?   *Answer:*

10. A woman who earned $42,000 last year earns $56,000 this year. What percent increase is there in her salary?   *Answer:*

11. A store increases the price of its chocolates by 12%. If the old price was $4.50 per pound, the new price per pound is _____

12. Season's ticket sales for Laker games increased by 10% this season. If last season the Lakers sold 4230 season's tickets, how many do they sell this season?
Answer:

13. A saleswoman who earns $26,600 per year receives a 6% increase in salary. Her new salary is _____

14. The price of milk, which was $2.20 per gallon, is increased by 5%. The new price is _____

15. A salesman who sells $400,000 of life insurance one year sells $600,000 the next year. What percent increase is this in sales?   *Answer:*

16. The price of a long-distance telephone call increases from $1.50 per minute to $2.00 per minute. By what percent does the price increase?   *Answer:*

B. *In Exercises 17–22, draw a circle around the correct letter.*

17. A man who was earning $20,000 per year receives a 2% reduction in his salary. His new salary is
(A) $18,000   (B) $19,800   (C) $16,000   (D) $19,840   (E) $19,600

18. A town with 10,000 people loses 8% of its population. Its new population is
(A) 9200   (B) 2000   (C) 8000   (D) 9920   (E) 9992

19. A sweater that sold for $20 was reduced by 15%. Its new price is
(A) $18.50   (B) $5.00   (C) $15.00   (D) $17.00   (E) $17.50

20. A coat that sold for $199 was reduced by 10%. Its new price is
(A) $180.10   (B) $179.10   (C) $19.90   (D) $189   (E) $179

21. An alarm clock that sold for $29.60 was reduced by 25%. Its new price is
(A) $22.10   (B) $27.10   (C) $22.20   (D) $22.00   (E) $27.00

22. A business that makes a profit of $80,000 one year decreased its profit by 30% the next year. Its new profit is
(A) $50,000   (B) $56,000   (C) $24,000   (D) $104,000   (E) $79,760

23. A town with 180,000 people loses 3% of its population. What is its new population?
Answer:

24. A mine produces 450 million tons of ore one year. The next year there is a 6% decrease in production. How many tons of ore are produced the next year?
Answer:

25. A book is reduced in price from $12.80 to $9.60. What percent reduction is this?
Answer:

26. The population of a country decreases from 48,000,000 to 44,000,000. What percent decrease is there in population?   *Answer:*

27. A worker who earns $16,000 has 12% deducted from his salary for taxes and union dues. What is his take-home pay?   *Answer:*

28. A suit that sold for $97.50 was reduced by 8%. Its new price is _____

29. A basketball team that wins 60 games one year wins only 48 games the next year. By what percent does the number of games won decrease?   *Answer:*

30. A pair of shoes that originally sold for $120 was reduced by 20%. A month later the new price was reduced by another 10%. After this second reduction, what did the pair of shoes sell for?   *Answer:*

31. A woman whose annual salary is $45,000 pays 20% of it in taxes. What is her salary after taxes?   *Answer:*

32. If a man who earns a weekly salary of $320 takes home $240, and the reduction is due to taxes, what percent of his salary goes into taxes?   *Answer:*

Review Exercises for Unit IV

In Exercises 1–3, express each percent as a fraction in lowest terms.

1. 80% =

2. 68% =

3. 175% =

In Exercises 4–6, express each percent as a decimal.

4. 70% =

5. 2.5% =

6. $12\frac{1}{4}\%$ =

7. A football stadium is 45% full. Express this percentage
   i. as a fraction in lowest terms   *Answer:*
   ii. as a decimal   *Answer:*

In Exercises 8–11, express each decimal as a percent.

8. .4 =

9. .04 =

10. .004 =

11. 1.025 =

In Exercises 12–15, express each fraction as a percent.

12. $\frac{9}{10}$ =

13. $\frac{3}{20}$ =

14. $\frac{8}{5}$ =

15. $\frac{2}{3}$ =

In Exercises 16 and 17, express each mixed number as a percent.

16. $2\frac{1}{4} =$ **17.** $3\frac{5}{12} =$

In Exercises 18 and 19, draw a circle around the correct letter.

18. What is 20% of 70?

(A) 14   (B) 28   (C) 3.5   (D) 140   (E) 1400

19. What is 75% of 120?

(A) 9   (B) 75   (C) 80   (D) 84   (E) 90

20. What is 15% of 60?   *Answer:*   **21.** What is 40% of 1250?   *Answer:*

In Exercises 22–25, solve each equation for n.

22. $5 \times n = 35$ *Answer:*   **23.** $10 \times n = 45$ *Answer:*

24. $30\% \times n = 15$ *Answer:*   **25.** $90\% \times n = 36$ *Answer:*

In Exercises 26 and 27, draw a circle around the correct letter.

26. 40% of what number is 32?

(A) 8   (B) 12   (C) 12.8   (D) 64   (E) 80

27. If 39 is 75% of a certain number, find that number.

(A) 12   (B) 36   (C) 52   (D) 55   (E) 60

28. If 9% of a number is 36, find that number.   *Answer:*

29. If 33% of a number is 66, find that number.   *Answer:*

In Exercises 30 and 31, draw a circle around the correct letter.

30. What percent of 60 is 45?

(A) 15%   (B) 25%   (C) 45%   (D) 75%   (E) $133\frac{1}{3}\%$

31. What percent of 40 is 28?

(A) 12%   (B) 16%   (C) 60%   (D) 70%   (E) 75%

32. What percent of 25 is 8?   *Answer:*

33. What percent of 49 is .49?   *Answer:*

In Exercises 34–37, draw a circle around the correct letter.

34. Edith answers 54 out of 60 questions correctly. What percent of her answers are correct?

(A) 80%   (B) 85%   (C) 90%   (D) 94%   (E) 95%

35. A family pays $75 a week for food. This amounts to 30% of its weekly income. What is the family's weekly income?

    (A) $200   (B) $225   (C) $250   (D) $275   (E) $300

36. A pair of shoes sells for $56 plus a 5% sales tax. What is the total price?

    (A) $57.60   (B) $58.00   (C) $58.50   (D) $58.80   (E) $61.00

37. A woman who earns $52,000 per year receives a 2% increase in salary. Her new salary is

    (A) $52,200   (B) $52,010.40   (C) $52,104   (D) $53,040   (E) $54,000

38. A salesman who sells 180 automobiles one year, sells 25% more the next year. How many does he sell the next year?   *Answer:*

39. The population of a city increases by 12%. If the population was formerly 250,000, the new population is _____

In Exercises 40 and 41, draw a circle around the correct letter.

40. A hat that sold for $30 was reduced by 25%. Its new price is

    (A) $20   (B) $21   (C) $22.50   (D) $24   (E) $25

41. A doctor who earns $180,000 one year pays 40% in taxes. His earnings that year after taxes are

    (A) $108,000   (B) $112,000   (C) $115,000   (D) $120,000   (E) $72,000

42. A country with 16,000,000 people experiences a 4% decline in population. How many people are there then in the country?   *Answer:*

43. If 650,000 fans attend the Redskins' home games one year and there is a decrease in attendance of 10% the next year, how many fans attend Redskins' home games the next year?   *Answer:*

44. A man earns $40,000 a year. By what percent must his salary increase for him to earn $50,000 the next year.   *Answer:*

45. A man earns $50,000 a year. By what percent must his salary decrease for him to earn $40,000 the next year?   *Answer:*

46. A dress that originally sold for $60 was reduced by 10%. A week later the dress was reduced by another 10%. What was the price after this second reduction?
    Answer:

47. A woman who earns $48,000 per year has 25% of her salary deducted for taxes and 10% of her salary deducted for benefits. What is her take-home pay?   *Answer:*

Review Exercises on Units I–III

1. Express 47,002 in words.   *Answer:*

2. 5805 × 24 =

3. Write 14,400 as the product of primes.   *Answer:*

4. Find *lcm* (54, 81).   *Answer:*   5. Find *gcd* (84, 96).   *Answer:*

6. $\dfrac{7}{10} \times \dfrac{3}{14} =$ 7. $\dfrac{12}{25} \div \dfrac{3}{5} =$

8. $1\dfrac{3}{4} + 2\dfrac{5}{8} =$ 9. $3\dfrac{1}{2} \times 5\dfrac{1}{4} =$

10. Express $\dfrac{27}{1000}$ as a decimal.   *Answer:*

11. Express $\dfrac{3}{50}$ as a decimal.   *Answer:*

12. Express .025 as a fraction in lowest terms.   *Answer:*

13. .028 × .004 =   14. $\dfrac{1.2}{.003} =$

15. If each book is .35 inch thick, how high is a stack of 60 such books?   *Answer:*

16. Pens cost $.35 each. How many pens can be bought for $14.00?   *Answer:*

Practice Exam on Unit IV

1. Express 52% as a fraction in lowest terms.   *Answer:*

2. Express 5% as a decimal.   *Answer:*

3. Express $\dfrac{7}{8}$ as a percent.   *Answer:*

4. Express .004 as a percent.   *Answer:*

5. Express $2\frac{1}{5}$ as a percent.   *Answer:*

6. What is 25% of 68?   *Answer:*

7. Solve for n: $80\% \times n = 40$ *Answer:*

In Problems 8–10, draw a circle around the correct letter.

8. If 40% of a number is 200, the number is

   (A) 80   (B) 140   (C) 250   (D) 500   (E) 800

9. A student answered 54 out of 60 questions correctly. What percent of her answers were correct?

   (A) 88%   (B) 90%   (C) 92%   (D) 94%   (E) 96%

10. The price of a magazine subscription, which was $24.00 per year, was increased by 20%. The new price is

    (A) $28.80   (B) $29.00   (C) $29.20   (D) $30.00   (E) $44.00

11. The population of a state decreased by 5%. If the population was formerly 8,000,000, the new population is _____

12. A man who earns $30,000 per year pays 15% in taxes. His take-home pay after taxes is _____

13. A sales tax on a $16 hat is 8%. The total price of the hat is _____

You are now given two sample midterm exams for practice. Complete solutions to these practice exam questions are given, beginning on page 370.

Practice Midterm Exam A

When several choices are given, draw a circle around the correct letter.

1.   Two hundred twenty thousand five hundred is written

   (A) 220,000.05   (B) 225,000   (C) 220,500

   (D) 220,005   (E) 225,000.05

2. $390 - 72 =$

   (A) 218   (B) 318   (C) 328   (D) 462   (E) 228

3. $5616 \div 18 =$

   (A) 31   (B) 31.2   (C) 312   (D) 3120   (E) 321

4. $3^4 =$

5. $4^2 \times 5 =$

6.   Express 1440 as the product of primes. *Answer:*

7.   Which of these fractions is the smallest?

   (A) $\dfrac{3}{4}$ (B) $\dfrac{3}{5}$ (C) $\dfrac{4}{5}$ (D) $\dfrac{4}{7}$ (E) $\dfrac{5}{7}$

8. $\dfrac{2}{5} + \dfrac{1}{9} =$ 9. $4\dfrac{1}{5} - 2\dfrac{1}{2} =$ 10. $3\dfrac{1}{4} \div 1\dfrac{3}{4} =$

11. $\left(\dfrac{2}{5}\right)^2 =$

12.   Which of these numbers is the smallest?

   (A) .071   (B) .008   (C) .069   (D) .0079   (E) .017

13. $27.1 - 12.45 =$

14.   Change $\dfrac{10}{13}$ to a decimal rounded to the nearest hundredth.

   (A) .76   (B) .77   (C) .769   (D) .8   (E) .80

15.   If a pound of coffee costs \$2.95, how much do 12 pounds of coffee cost? *Answer:*

16.   What is 60% of 75? *Answer:*

17.   If 60 is 75% of a certain number, find that number. *Answer:*

18.   Change $\dfrac{7}{12}$ to an infinite repeating decimal. *Answer:*

19. $(.02)^3 =$

20.   A suit sells for \$140 plus a 4% sales tax. What is the total price? *Answer:*

Practice Midterm Exam B

When several choices are given, draw a circle around the correct letter.

1.   Five thousand and two tenths is written

   (A) 5210   (B) 5002.10   (C) 5000.2   (D) 5000.02   (E) 5002.1

2. $1855 \div 35 =$

3. $2^6 =$

4.   Express 13,200 as the product of primes. *Answer:*

5.   Find the least common multiple of 8 and 28. *Answer:*

6. $\dfrac{1}{10} + \dfrac{3}{5} =$ 7. $4\dfrac{7}{8} - 1\dfrac{1}{4} =$

8.   Which of the following fractions is the *smallest*?

   (A) $\dfrac{1}{4}$ (B) $\dfrac{2}{5}$ (C) $\dfrac{3}{10}$ (D) $\dfrac{5}{12}$ (E) $\dfrac{3}{13}$

9.   Change $\dfrac{3}{7}$ to an infinite repeating decimal. *Answer:*

10.   Which of the following numbers is the *smallest*?

   (A) .045   (B) .0054   (C) .4005   (D) .0045   (E) .004 49

11. $10 \div 3\dfrac{1}{4} =$

12. $4.5 \times .48 =$

   (A) 216   (B) 21.6   (C) 2.16   (D) .216   (E) .0216

13. $(.12)^2 =$

14.   If pencils sell for $.12 each, how many can you buy for $18?

   (A) 216   (B) 15   (C) 150   (D) 1500   (E) 2160

15. $27.9 - 4.95 =$

16.   Express 88% as a fraction.

(A) $\frac{8}{10}$ (B) $\frac{12}{25}$ (C) $\frac{22}{25}$ (D) $1\frac{8}{10}$ (E) 88

17.   30% of the students at River Junction College are freshmen. If there are 900 freshmen at this college, how many students are there at the college? *Answer:*

18.   If 70% of a number is 7, find this number. *Answer:*

19.   At a sale, a $16 sweater is reduced by 25%. Its sale price is

(A) $4   (B) $20   (C) $12   (D) $15.75   (E) $15.60

20.   A storage company charges $100 to store a full-sized piano for the first month and $40 for each additional month. How much does it cost to store this piano for six months?

20. Measure

It is important to know how to work with units, such as hours and minutes or pounds and ounces.

A. TIME

> 1 hour = 60 minutes
> 1 minute = 60 seconds

To add or subtract time units, you may have to change seconds to minutes, minutes to hours, or hours to minutes.

Example 1 ▶ 　 4 hours 33 minutes
　　　　　　　 + 6 hours 29 minutes

Solution. Begin by adding the smaller units.

　　　　 33 minutes
　　　 + 29 minutes
　　　　 62 minutes

Because 62 minutes is more than 1 hour, change to hours and minutes:

$$62 \text{ minutes} = 60 \text{ minutes} + 2 \text{ minutes}$$
$$= 1 \text{ hour } 2 \text{ minutes}$$

Thus

```
    4 hours  33 minutes
 +  6 hours  29 minutes
   ─────────────────────
   10 hours  62 minutes
 +  1 hour    2 minutes
   ─────────────────────
   11 hours   2 minutes
```

◄

Example 2 ►   8 minutes   30 seconds
      – 5 minutes   40 seconds

Solution. Begin at the right with the smaller units. Because you are subtracting 40 seconds from 30 seconds, borrow 1 minute, or 60 seconds. Thus change

      8 minutes   to   7 minutes 60 seconds

so that

      8 minutes 30 seconds   becomes   7 minutes 90 seconds

```
     7 minutes    90 seconds
     8 minutes    30 seconds
 -   5 minutes    40 seconds
    ────────────────────────
     2 minutes    50 seconds
```

Alternatively, you could add 1 minute (or 60 seconds) to *both* given time measures. Thus

      8 minutes 30 seconds   is increased to   8 minutes 90 seconds

and

      5 minutes 40 seconds   is increased to   6 minutes 40 seconds

```
                  90 seconds
     8 minutes    30 seconds
     6 minutes
 -   5 minutes    40 seconds
    ────────────────────────
     2 minutes    50 seconds
```

◄

Example 3 ►   A baseball game started at 2:30 P.M. and ended at 5:15 P.M. How long did it last?

Solution. Write

      2:30 P.M.   as   2 hours   30 minutes
      5:15 P.M.   as   5 hours   15 minutes

Subtract 2 hours 30 minutes from 5 hours 15 minutes.

```
     4 hours    75 minutes
     5 hours    15 minutes
 -   2 hours    30 minutes
    ──────────────────────
     2 hours    45 minutes
```

The game lasted 2 hours and 45 minutes.

◄

Sometimes units of time must be multiplied or divided by a number, as illustrated in Example 4.

Example 4 ▶ A two-hour television show is divided into six equal parts. How long is each part?

Solution. Change hours to minutes. Thus

2 hours = 2 × 60 minutes = 120 minutes

Now divide by 6.

$$\frac{120 \text{ minutes}}{6} = 20 \text{ minutes}$$

Each part is 20 minutes long.   ◀

PRACTICE EXERCISES FOR TOPIC A

1.   2 hours 35 minutes
     + 1 hour   45 minutes

2.   10 minutes 20 seconds
     − 7 minutes 45 seconds

3. A movie began at 8:45 P.M. and ended at 11:00 P.M. How long did it last?

4. How many 6-minute periods are there in an hour and a half?

Now try the exercises for Topic A on page 181.

B. WEIGHT

There are two widely-used systems of weight (and length)—the American system and the metric system. We begin by considering the American system.

> 1 pound = 16 ounces

Example 5 ▶ Add 9 pounds 12 ounces and 6 pounds 8 ounces.

Solution.

```
      9 pounds  12 ounces
  +   6 pounds   8 ounces
     15 pounds  20 ounces
  +   1 pound    4 ounces
     16 pounds   4 ounces
```

Note that you must convert 20 ounces to 1 pound 4 ounces.   ◀

In the *metric system*, the basic unit of weight is called a **gram**. There are about 454 grams to a pound. A paper clip weighs about 1 gram.

```
┌─────────────────────────────────┐
│  1 kilogram = 1000 grams        │
│  1 gram =    100 centigrams     │
└─────────────────────────────────┘
```

Example 6▶   Subtract 3 kilograms 650 grams from 5 kilograms 500 grams.

Solution. Borrow 1 kilogram, or 1000 grams.

$$
\begin{array}{ll}
4 \text{ kilograms} & 1500 \text{ grams} \\
\cancel{5 \text{ kilograms}} & \cancel{500 \text{ grams}} \\
-\ 3 \text{ kilograms} & 650 \text{ grams} \\
\hline
1 \text{ kilogram} & 850 \text{ grams}
\end{array}
$$

◀

Example 7▶   A company ships 20 packages, each of which weighs 2 pounds 2 ounces. What is the total weight of the shipment?

Solution.

$$
\begin{array}{ll}
2 \text{ pounds} & \cancel{2 \text{ ounces}} \\
\times\ 20 & \\
\hline
40 \text{ pounds} & \cancel{40 \text{ ounces}} \\
2 \text{ pounds} & 8 \text{ ounces} \\
\hline
42 \text{ pounds} & 8 \text{ ounces}
\end{array}
$$

Here note that

$$40 \text{ ounces} = 32 \text{ ounces} + 8 \text{ ounces}$$
$$= 2 \text{ pounds } 8 \text{ ounces}$$

Thus the total weight is 42 pounds 8 ounces, or $42\frac{1}{2}$ pounds.   ◀

PRACTICE EXERCISES FOR TOPIC B

5.   6 pounds 11 ounces
    + 3 pounds   9 ounces

6.   20 pounds   5 ounces
    − 10 pounds 10 ounces

7.   2 kilograms 750 grams
    + 3 kilograms 250 grams

8. A 4-kilogram load of sand is divided into five equal piles. How many grams does each pile weigh?

Now try the exercises for Topic B on page 182.

C. LENGTH

In the American system,

$$1 \text{ foot} = 12 \text{ inches}$$

Example 8 ▶
```
  4 feet   9 inches
- 1 foot  11 inches
```

Solution. Borrow 1 foot, or 12 inches.

```
  3 feet    21 inches
  4 feet     9 inches
- 1 foot    11 inches
  2 feet    10 inches
```
◀

In the metric system, the basic unit of length is called a **meter**. One meter is approximately 39.37 inches.

$$1 \text{ kilometer} = 1000 \text{ meters}$$
$$1 \text{ meter} = 100 \text{ centimeters}$$

Example 9 ▶
```
    8 meters   45 centimeters
  + 6 meters   95 centimeters
```

Solution.

```
    8 meters    45 centimeters
  + 6 meters    95 centimeters
   14 meters   140 centimeters
  + 1 meter     40 centimeters
   15 meters    40 centimeters
```
◀

PRACTICE EXERCISES FOR TOPIC C

9.  ```
      5 feet  10 inches
    + 3 feet   8 inches
    ```

10. ```
      6 meters  20 centimeters
    - 2 meters  30 centimeters
    ```

11. A 3 foot-4 inch rope is cut into 4 equal parts. How long is each part?

Now try the exercises for Topic C on page 183.

D. AMERICAN AND METRIC SYSTEMS

To convert units of weight from one system to the other, the following approximations are useful.

$$\begin{array}{l} 1 \text{ pound} \approx 454 \text{ grams} \\ 1 \text{ kilogram} \approx 2.2 \text{ pounds} \end{array}$$

Read: 1 pound is *approximately* equal to 454 grams.

Example 10 ▶ Convert 20 pounds to

    i. grams   *ii.* kilograms   *iii.* centigrams

Solution. Use: 1 pound ≈ 454 grams

    i.
$$\begin{array}{r} 454 \text{ grams} \\ \times\ 20 \\ \hline 9080 \text{ grams} \end{array}$$

Thus 20 pounds equal approximately 9080 grams.

    ii.   1000 grams = 1 kilogram

Thus *divide* 9080 grams by 1000 to find the number of kilograms in 20 pounds. Move the decimal point in 9080. three places to the left.

$$\frac{9080 \text{ grams}}{1000} = 9.08\!\!\not{0} \text{ kilograms}$$

Thus 20 pounds equal approximately 9.08 kilograms.

    iii.   1 gram = 100 centigrams

Therefore *multiply* 9080 grams by 100 to find the number of centigrams in 20 pounds. Move the decimal point in 9080. two places to the right to obtain 9080 00.

$$\begin{array}{r} 9080 \text{ grams} \\ \times\ 100 \\ \hline 908{,}000 \text{ centigrams} \end{array}$$

Thus 20 pounds are equivalent to approximately 908,000 centigrams. ◀

The most important conversions of units of length between the two systems are given by

$$\begin{array}{ll} 1 \text{ inch} & \approx 2.54 \text{ centimeters} \\ 1 \text{ kilometer} & \approx\ .62 \text{ mile} \\ 1 \text{ mile} & \approx 1.6 \text{ kilometers} \end{array}$$

Example 11 ▶ Convert:

     i. 1 foot 3 inches to centimeters.
    ii. 30 kilometers to miles.

Solution.

    i. 1 foot 3 inches = 15 inches

$$\begin{array}{r} 2.54 \text{ centimeters} \\ \times\ 15 \\ \hline 12\ 70 \\ 25\ 4 \\ \hline 38.1\cancel{0} \text{ centimeters} \end{array}$$

Thus 1 foot 3 inches equals approximately 38.1 centimeters.

  ii.

$$\begin{array}{r} .62 \text{ miles} \\ \times\ 30 \\ \hline 18.6\cancel{0} \text{ miles} \end{array}$$

Thus 30 kilometers equal approximately 18.6 miles. ◀

PRACTICE EXERCISES FOR TOPIC D

Convert. Express your answer to the nearest tenth of a unit.

12. 6 pounds to grams   13. 3.2 pounds to centigrams

14. 15 kilograms to pounds   15. 2 feet to centimeters

16. 12 miles to kilometers

Now try the exercises for Topic D on page 184.

EXERCISES

A. *Add or subtract.*

1.   5 hours 24 minutes
     + 3 hours 28 minutes

2.   6 hours 12 minutes
     + 3 hours 48 minutes

3.   4 hours 20 minutes
     + 3 hours 50 minutes

4.   8 minutes 21 seconds
     + 7 minutes 49 seconds

5.   8 hours 25 minutes
     - 4 hours   5 minutes

6.   6 hours 15 minutes
     - 4 hours 30 minutes

7.   7 hours 40 minutes
     - 2 hours 48 minutes

8.   15 hours 13 minutes
     - 6 hours 40 minutes

9. 10 hours 15 minutes
    – 6 hours 36 minutes

10. 9 minutes 40 seconds
     – 5 minutes 55 seconds

11. 13 minutes 28 seconds
     – 8 minutes 37 seconds

12. 11 hours 5 minutes
     – 6 hours 6 minutes

13. A show started at 8:40 P.M. and ended at 11:05 P.M. How long did it last?
    Answer:

14. Diane left her mother's house at 3:35 P.M. and arrived home at 5:20 P.M. How long did she travel?   *Answer:*

15. A movie started at 7:40 P.M. and ended at 9:25 P.M. How long did it last?
    Answer:

16. A flight left at 8:42 A.M. and arrived at 11:35 A.M. How long did it last?
    Answer:

17. A lecture began at 8:05 P.M. and ended at 9:50 P.M. How long did it last?
    Answer:

18. A meeting began at 11:10 A.M. and ended at 1:05 P.M. How long did it last?
    Answer:

19. Bob has four 50-minute classes in a row, beginning at 10:30 A.M. What time does he finish?   *Answer:*

20. Half-way through a three-and-a-half-hour movie there is an intermission. How long after the movie begins does the intermission begin?   *Answer:*

21. It takes Bill 6 minutes to complete one telephone sales call. How many of these sales calls can Bill make in 2 hours?   *Answer:*

22. Brenda needs 15 minutes to solve each of her math problems. How long should it take her to solve 8 math problems?   *Answer:*

B. *Add.*

23. 6 pounds 10 ounces
     + 5 pounds   6 ounces

24. 7 pounds 13 ounces
     + 8 pounds 11 ounces

25. 10 kilograms 120 grams
     + 7 kilograms 380 grams

26. 5 kilograms 600 grams
     + 4 kilograms 700 grams

27. 8 grams 75 centigrams
     + 4 grams 52 centigrams

28. 12 grams 96 centigrams
     + 14 grams 89 centigrams

Subtract.

29. 5 pounds 12 ounces
     – 2 pounds 10 ounces

30. 9 pounds 6 ounces
     – 3 pounds 9 ounces

31. 7 pounds
    – 4 pounds 12 ounces

32. 9 pounds   7 ounces
    – 6 pounds 15 ounces

33. 12 kilograms 390 grams
    – 7 kilograms 165 grams

34. 8 kilograms   45 grams
    – 4 kilograms 132 grams

35. 10 grams 17 centigrams
    – 6 grams 62 centigrams

36. 9 grams 12 centigrams
    –   95 centigrams

37. Add 5 pounds 6 ounces and 9 pounds 11 ounces.   *Answer:*

38. Add 17 pounds 12 ounces and 8 pounds 13 ounces.   *Answer:*

39. Subtract 5 pounds 10 ounces from 10 pounds 5 ounces.   *Answer:*

40. Subtract 11 pounds 12 ounces from 16 pounds.   *Answer:*

41. Subtract 3 kilograms 500 grams from 10 kilograms.   *Answer:*

42. Subtract 7 grams 12 centigrams from 9 grams.   *Answer:*

43. What is the total weight of 50 3-ounce packages?   *Answer:*

44. A 5-kilogram package of sugar is divided into 20 equal portions. How many centigrams does each portion weigh?   *Answer:*

C. *Add.*

45. 5 feet 6 inches
    + 3 feet 6 inches

46. 6 feet   5 inches
    + 3 feet 11 inches

47. 6 kilometers 450 meters
    + 3 kilometers 550 meters

48. 8 meters 25 centimeters
    + 6 meters 85 centimeters

Subtract.

49. 7 feet 4 inches
    – 3 feet 5 inches

50. 8 feet 2 inches
    – 2 feet 8 inches

51. 9 feet
    – 4 feet 9 inches

52. 12 kilometers 10 meters
    – 10 kilometers 12 meters

53. 3 meters 47 centimeters
    – 2 meters 58 centimeters

54. 15 meters 45 centimeters
    – 5 meters 85 centimeters

55. Each step he takes moves Bill 2 feet 2 inches forward. How many steps must he take to advance 26 feet?   *Answer:*

56. A football team must gain 3 yards in 4 plays to score a touchdown. How many inches must be gained per play? (Use: 1 yard = 3 feet)   *Answer:*

57. A 2 foot-6 inch wire is cut into three equal parts. How long is each part? *Answer:*

D. *Convert. Express your answer to the nearest tenth of a unit.*

58. 15 pounds to grams   *Answer:*   59. 25 kilograms to pounds   *Answer:*

60. 60 pounds to kilograms   *Answer:*   61. 9 inches to centimeters   *Answer:*

62. 2 feet 4 inches to centimeters   *Answer:*

63. 55 miles to kilometers   *Answer:*   64. 200 kilometers to miles   *Answer:*

65. If you are 58 kilometers from Paris, what is your distance from Paris in miles? *Answer:*

66. If a Russian basketball player is 203.2 centimeters tall, what is his height in feet and inches?   *Answer:*

21. Rectangles and Squares

Arithmetic methods are used to solve problems concerning the area or the perimeter of a rectangular region. Sometimes these problems also involve a cost factor (Topics C and E).

A. AREA OF A RECTANGLE

Area involves two "dimensions" and is measured in terms of *square* inches, *square* feet, and so on. We write

in.^{2}	for	in. $\times$ in.	[square inches]
ft.^{2}	for	ft. $\times$ ft.	[square feet]

In the metric system, we write

cm.^{2}	for	cm. $\times$ cm.	[square centimeters]
m.^{2}	for	m. $\times$ m.	[square meters]

DEFINITION

> **A rectangle** is a closed 4-sided figure whose sides meet at right angles.

The area of a rectangle equals its length times its width.

Area of Rectangle = length $\times$ width

width

length

For example, if the length is 5 inches and the width is 2 inches, then

Area = length $\times$ width

= 5 in. $\times$ 2 in.

= 10 in.^{2}

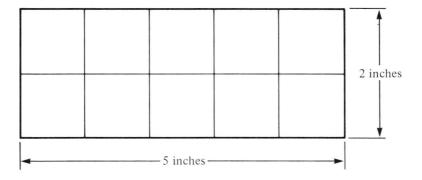

2 inches

5 inches

In the preceding figure, each box represents one square inch. Altogether, there are 10 square inches. Thus

$$5 \text{ in.} \times 2 \text{ in.} = 10 \text{ in.}^2$$

Example 1 ▶ Find the area of a rectangular parking lot that is 100 meters long and 60 meters wide.

(A) 160 m.   (B) 320 m.   (C) 600 m.²   (D) 6000 m.²   (E) 13,600 m.²

Solution.

$$\text{Area} = \text{length} \times \text{width}$$
$$= 100 \text{ m.} \times 60 \text{ m.}$$
$$= 6000 \text{ m.}^2 \quad (square\ meters)$$

The area is 6000 m.². The correct choice is (D).   ◀

In the case of a vertical rectangular region, such as a wall or a television screen, the area is often given in terms of *height* and width.

> Area = height × width

Example 2 ▶ Find the area of a rectangular movie screen that is 20 feet high and 30 feet wide.

(A) 600 ft.²   (B) 6000 ft.²   (C) 50 ft.   (D) 100 ft.   (E) 1300 ft.²

Solution.

$$\text{Area} = \text{height} \times \text{width}$$
$$= 20 \text{ ft.} \times 30 \text{ ft.}$$
$$= 600 \text{ ft.}^2$$

Choice (A) is correct.   ◀

Some problems concern *solid* rectangular figures, such as boxes, and you must select the area of one of the faces.

Example 3 ▶ A rectangular dresser is 28 inches long, 20 inches wide, and 40 inches high. Find the area of its base.

Solution. The base is the *bottom* of the dresser. The height is of no concern in this problem. Thus

$$\text{Area} = \text{length} \times \text{width}$$
$$= 28 \text{ in.} \times 20 \text{ in.}$$
$$= 560 \text{ in.}^2$$

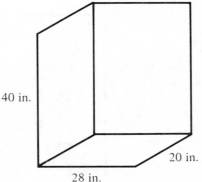

40 in.

28 in.

20 in.   ◀

Sometimes you are given the area of a rectangle as well as *either* the length or width. You are then asked to find the other dimension.

Example 4 ▶ The area of a rectangular tabletop is 35 square feet. If the length is 7 feet, find the width.

Solution.

Area = length × width

35 ft.² = 7 ft. × width

Divide both sides by 7 ft. and note that

$$\frac{\text{ft.}^2}{\text{ft.}} = \frac{\overset{1}{\cancel{\text{ft.}}} \times \text{ft.}}{\underset{1}{\cancel{\text{ft.}}}} = \text{ft.}$$

$$\frac{35 \text{ ft.}^2}{7 \text{ ft.}} = \text{width}$$

$$5 \text{ ft.} = \text{width} \qquad\qquad ◀$$

PRACTICE EXERCISES FOR TOPIC A

1. Find the area of a page that is 10 inches by 7 inches.

2. A trunk is 4 feet long, 2 feet wide, and 3 feet high. Find the area of its base.

3. The area of a vertical wall is 120 square feet. If the width of the wall is 15 feet, find the height.

Now try the exercises for Topic A on page 191.

B. AREA OF A SQUARE

When the length and width of a rectangle are equal, the rectangle is called a **square**. The accompanying figure shows a square of side length 4 inches.

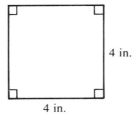

4 in.

4 in.

Because

length = width = 4 inches

the area of this square is

4 in. × 4 in., or 16 in.²

In general, the area of a square equals the square of its side (length).

$$\boxed{\text{Area of a square} = \text{side}^2}$$

Example 5 ▶ Find the area of a square garden whose side length is 90 feet.

Solution.

$$
\begin{aligned}
\text{Area} &= \text{side}^2 \\
&= (90 \text{ ft.})^2 \\
&= 8100 \text{ ft.}^2
\end{aligned}
$$
◄

PRACTICE EXERCISES FOR TOPIC B

4. Find the area of a square handkerchief whose side length is 11 inches.

Now try the exercises for Topic B on page 192.

C. AREA AND COST

Example 6 ▶ What is the cost of the material for a tablecloth that is to be 8 feet by 6 feet, if the material costs $4 per square foot?

Solution. First find the area of the tablecloth.

$$
\begin{aligned}
\text{Area} &= \text{length} \times \text{width} \\
&= 8 \text{ ft.} \times 6 \text{ ft.} \\
&= 48 \text{ ft.}^2
\end{aligned}
$$

To find the cost, now multiply 48 square feet by the cost of $4 per square foot. Note that $4 *per* square foot can be expressed by the fraction $\dfrac{\$4}{\text{ft.}^2}$.

$$\text{Cost} = 48 \; \overset{1}{\cancel{\text{ft.}^2}} \times \frac{\$4}{\cancel{\text{ft.}^2}} = \$192$$
◄

PRACTICE EXERCISES FOR TOPIC C

5. Find the cost of the material for a rectangular bedspread that is 12 feet by 9 feet if the material costs $6 per square foot.   *Answer:*

6. It costs 15 cents per square yard to paint a wall. How much does it cost to paint a hallway wall that is 25 yards by 3 yards?

    (A) $112.50 (B) $50.00   (C) $5.00   (D) $11.25 (E) $75.00

Now try the exercises for Topic C on page 193.

D. PERIMETER

DEFINITION

> The **perimeter** of a closed figure, such as a rectangle, is the sum of the lengths of its sides.

In other words, the perimeter of a closed figure is the length of its boundary.

Perimeter of rectangle = length + width + length + width

Thus,

> Perimeter of rectangle = 2 × length + 2 × width

In the accompanying figure, the length is 12 inches and the width is 4 inches.

$$\begin{aligned}
\text{Perimeter} &= 2 \times \text{length} + 2 \times \text{width} \\
&= 2 \times 12 \text{ in.} + 2 \times 4 \text{ in.} \\
&= 24 \text{ in.} + 8 \text{ in.} \\
&= 32 \text{ in.}
\end{aligned}$$

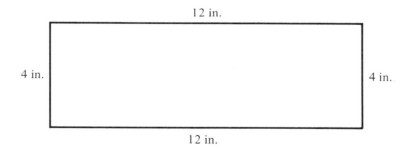

12 in.

4 in.   4 in.

12 in.

Note that perimeter measures boundary length (and *not* area). Units of perimeter are inches, feet, and centimeters (and *not* square inches, and so on).

Example 7 ▶ Find the perimeter of a rectangle that is 9 feet long and 8 feet wide.

(A) 17 ft.   (B) 72 ft.   (C) 34 ft.   (D) 72 ft.²   (E) 144 ft.

Solution.

$$\begin{aligned}
\text{Perimeter} &= 2 \times \text{length} + 2 \times \text{width} \\
&= 2 \times 9 \text{ ft.} + 2 \times 8 \text{ ft.} \\
&= 18 \text{ ft.} + 16 \text{ ft.} \\
&= 34 \text{ ft.}
\end{aligned}$$

The correct choice is (C).   ◀

Example 8 ▶   The perimeter of a rectangle is 30 yards and the width is 5 yards. Find
i. the length, *ii.* the area.

Solution.

i.   Perimeter = 2 × length + 2 × width

30 yd. = 2 × length + 2 × 5 yd.

= 2 × length + 10 yd.

Subtract 10 yd. from both sides.

30 yd. = 2 × length + 10 yd.
– 10 yd.   – 10 yd.

20 yd. = 2 × length

Now divide both sides by 2.

$$\frac{\overset{10}{\cancel{20}}\text{yd.}}{\underset{1}{\cancel{2}}} = \frac{\overset{1}{\cancel{2}} \times \text{length}}{\underset{1}{\cancel{2}}}$$

10 yd. = length

ii.   Area = length × width

= 10 yd. × 5 yd.

= 50 yd.^{2} ◀

In a square, the length equals the width. Consequently,

Perimeter of square = 4 × side

In the figure to the right, the
side length of the square is
5 centimeters, so that the
perimeter is

4 × 5 cm., or 20 cm.

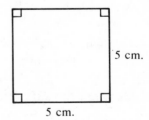

5 cm.

5 cm.

PRACTICE EXERCISES FOR TOPIC D

7. Find the perimeter of a rectangular plot of land that is 100 feet by 60 feet.

8. The perimeter of a rectangle is 800 meters and the length is 250 meters.
Find *i.* the width, *ii.* the area.

9. Find the perimeter of a square of side length 30 centimeters.

Now try the exercises for Topic D on page 193.

E. PERIMETER AND COST

Example 9 ▶   A woman wants to fence in a garden that is 20 feet by 15 feet. If fencing costs $6 per foot, how much will the fencing cost?

Solution. Here perimeter (and *not* area) is involved because the fence goes around the *boundary* of the garden. First find the perimeter.

$$\text{Perimeter} = 2 \times \text{length} + 2 \times \text{width}$$
$$= 2 \times 20 \text{ ft.} + 2 \times 15 \text{ ft.}$$
$$= 40 \text{ ft.} + 30 \text{ ft.}$$
$$= 70 \text{ ft.}$$

Now find the cost. Multiply 70 feet by $6 *per* foot. Note that $6 per foot can be expressed by the fraction $\frac{\$6}{\text{ft.}}$.

$$\text{Cost} = 70 \overset{1}{\cancel{\text{ft.}}} \times \frac{\$6}{\underset{1}{\cancel{\text{ft.}}}}$$
$$= \$420 \qquad \blacktriangleleft$$

PRACTICE EXERCISES FOR TOPIC E

10. A rectangular field that is 300 feet by 250 feet is enclosed by wiring that costs $.40 per foot. How much does the wiring for the field cost?

Now try the exercises for Topic E on page 194.

EXERCISES

A. *When several choices are given, draw a circle around the correct letter.*

1. Find the area of a rug that is 6 feet by 3 feet.
   (A) 9 ft.²   (B) 12 ft.²   (C) 18 ft.²   (D) 18 ft.   (E) 36 ft.²

2. Find the area of a tabletop that is 40 inches long and 20 inches wide.
   (A) 80 in.²   (B) 60 in.²   (C) 120 in.²   (D) 800 in.²   (E) 8000 in.²

3. Find the area of a blotter that is 35 inches long and 20 inches wide.
   (A) 55 in.   (B) 110 in.   (C) 70 in.²   (D) 700 in.²   (E) 7000 in.²

4. Find the area of a floor that is 20 meters by 14 meters.
   (A) 280 m.²   (B) 560 m.²   (C) 68 m.²   (D) 340 m.²   (E) 680 m.²

5. Find the area of a page that is 9 inches by 6 inches.   *Answer:*

6. Find the area of a wall that is 9 meters high and 12 meters wide.   *Answer:*

7. Find the area of a rectangular television screen that is 11 inches high and 20 inches wide.   *Answer:*

8. Find the area of a rectangular mirror that is 7 feet high and 2 feet wide.   *Answer:*

9. Find the area of the base of a rectangular cabinet that is 30 inches long, 10 inches wide, and 14 inches high.   *Answer:*

10. Find the area of the floor of a room that is 16 feet long, 12 feet wide, and 10 feet high.   *Answer:*

11. Find the area of one wall of a room that is 15 feet long, 15 feet wide, and 9 feet high.   *Answer:*

12. Find the area of the ceiling of a room that is 22 feet long, 20 feet wide, and 10 feet high.   *Answer:*

13. Find the length of a rectangle whose width is 9 yards and whose area is 90 square yards.

    (A) 10 yd.   (B) 10 yd.²   (C) 81 yd.   (D) 36 yd.   (E) 810 yd.²

14. The length of a rectangle is 11 feet and the area is 88 square feet. Find the width.

    (A) 8 ft.   (B) 77 ft.   (C) 38.5 ft.   (D) 99 ft.   (E) 968 ft.

15. The area of a rectangular carpet is 24 square feet and the length is 8 feet. Find the width.   *Answer:*

16. The area of a rectangular bedspread is 54 square feet and the width is 6 feet. Find the length.   *Answer:*

B. *In Exercises 17 and 18, draw a circle around the correct letter.*

17. A square has a side length of 7 centimeters. Its area is

    (A) 49 cm.   (B) 49 cm.²   (C) 70 cm.²   (D) 28 cm.   (E) 7 cm.²

18. A square has a side length of 15 feet. Its area is

    (A) 30 ft.   (B) 60 ft.   (C) 60 ft.²   (D) 150 ft.²   (E) 225 ft.²

19. Find the area of a square table top whose side length is 2.5 centimeters.   *Answer:*

20. Find the area of a square floor that measures 14 meters on each side.   *Answer:*

21. One yard equals 3 feet. How many square feet are there in 1 square yard?   *Answer:*

22. How many square inches are there in 1 square foot?   *Answer:*

C. *In Exercises 23–26, draw a circle around the correct letter.*

23. What does it cost to carpet a room that is 15 feet by 10 feet, if the carpeting costs $3 per square foot?

    (A) $150   (B) $45   (C) $30   (D) $450   (E) $75

24. What does it cost to carpet a room that is 7 meters by 5 meters, if the carpeting costs $12 per square meter?

    (A) $144   (B) $420   (C) $35   (D) $300   (E) $600

25. Find the cost of a tablecloth that is 5 yards by 3 yards at 12 dollars per square yard.

    (A) $150   (B) $96   (C) $192   (D) $180   (E) $360

26. It costs 20 cents per square yard to varnish a floor. How much will it cost to varnish a hallway floor that is 40 yards by 10 yards?

    (A) $40   (B) $400   (C) $800   (D) $80   (E) $8000

27. A blanket is to be 7 feet by 4 feet. The material for the blanket costs 3 dollars per square foot. Find the cost of the material for this blanket.   *Answer:*

28. Glass costs $2.50 per square yard. How much will a pane of glass that is 6 feet long and 3 feet wide cost?   *Answer:*

29. A woman buys a piece of material that is 20 feet by 3 feet. If the material costs $3.50 per square foot, how much does she pay?   *Answer:*

30. Carpeting costs $12.50 per square foot. How much will it cost to carpet a floor that is 15 feet by 12 feet?   *Answer:*

31. It costs $2700 to carpet a floor that is 20 feet by 15 feet. How much does this carpeting cost per square foot?   *Answer:*

32. Material for a tablecloth that is to be 6 feet by 4 feet costs $132. How much does the material cost per square foot?   *Answer:*

D. *When several choices are given, draw a circle around the correct letter.*

33. Find the perimeter of a rectangle whose length is 8 centimeters and whose width is 5 centimeters.

    (A) 13 cm.   (B) 26 cm.   (C) 40 cm.   (D) 20 cm.   (E) 80 cm.

34. Find the perimeter of a rectangle that is 3 yards by 2 yards.

    (A) 5 yd.   (B) 6 yd.   (C) 10 yd.   (D) 12 yd.   (E) 24 yd.

35. A woman wants to sew a silk border around a blanket that is 7 feet by 4 feet. How long a piece of silk does she need?

    (A) 11 ft.   (B) 15 ft.   (C) 22 ft.   (D) 28 ft.   (E) 56 ft.

36. How much fencing is needed to fence in a park that is 75 feet long and 35 feet wide?

    (A) 110 ft.   (B) 220 ft.   (C) 2625 ft.   (D) 2625 ft.^{2} (E) 262.5 ft.

37. How long a rope is needed to enclose a rectangular parking lot that is 100 yards by 54 yards?   *Answer:*

38. A rectangular field is 400 yards by 300 yards. How many yards of fencing would it take to enclose the field?   *Answer:*

39. A rectangle has length 5 inches and perimeter 18 inches. Find its width.

   (A) 13 in.   (B) 23 in.   (C) $\frac{18}{5}$ in.   (D) 6.5 in.   (E) 4 in.

40. A rectangle has width 10 feet and perimeter 50 feet. Find its length.

   (A) 5 ft.   (B) 40 ft.   (C) 15 ft.   (D) 10 ft.   (E) 25 ft.

41. A rectangle is 14 inches long. If its perimeter is 40 inches, what is its width?
   Answer:

42. The perimeter of a rectangle is 100 inches. If its width is 10 inches, its length
   is _____

43. The perimeter of a rectangle is 20 inches and its length is 6 inches. Its *area*
   is _____

44. The area of a rectangle is 20 square inches and its length is 5 inches. Find its
   perimeter.   *Answer:*

45. The perimeter of a rectangle is 60 centimeters and its width is 10 centimeters. Find
   its *area.*   *Answer:*

46. The area of a rectangle is 100 square meters and its width is 5 meters. Find its
   perimeter.   *Answer:*

47. Find the perimeter of a square of side length 9 centimeters.   *Answer:*

48. Find the side length of a square whose perimeter measures 28 feet.   *Answer:*

49. Find the area of a square whose perimeter is 20 inches.   *Answer:*

50. Find the area of a square whose perimeter is 44 inches.   *Answer:*

E. *In Exercises 51–54, draw a circle around the correct letter.*

51. Fencing costs $5 per foot. How much does it cost to fence in a rectangular garden
   that is 20 feet by 10 feet?
   (A) $100   (B) $500   (C) $300   (D) $1000   (E) $2000

52. A ribbon is used as the border of a blanket that is 6 feet by 4 feet. If the ribbon
   costs $2 per foot, what is the cost of the border?
   (A) $24   (B) $48   (C) $12   (D) $20   (E) $40

53. A rectangular field is enclosed by wiring that costs $.50 per yard. If the field is 900
   yards by 500 yards, the cost of the wiring is
   (A) $450,000   (B) $225,000   (C) $700   (D) $1400   (E) $2800

54. A fence is needed for a rectangular park that is 50 yards by 40 yards. If the fencing
   costs $4 per yard, the total cost of the fencing is
   (A) $8000   (B) $800   (C) $4000   (D) $360   (E) $720

55. A garden that is 20 feet by 12 feet is fenced in for $1600. What is the cost per foot of the fencing?   *Answer:*

56. A rectangular lot is fenced in for $1000. If the fencing costs $5 per foot and the lot is 40 feet wide, how long is the yard?   *Answer:*

57. Piping that goes around a rectangular rooftop 40 feet long and 30 feet wide costs $3 per foot. What is the cost of the piping?   *Answer:*

58. A rope encloses a football field that is 100 yards long and 53 yards wide. If the rope costs $1.50 per yard, what is the total cost?   *Answer:*

22. Triangles, Circles, and Rectangular Boxes

We continue our study of some of the most important geometric figures. We first consider the area and perimeter of a triangle. (The Pythagorean Theorem for right triangles is treated in Section 33.) For circles we consider the area; the perimeter is now called the *circumference*. We also examine the *volumes* and *surface areas* of solid figures called *rectangular boxes* and *cubes*. Other geometric figures (parallelograms, trapezoids, spheres, cylinders, and cones) will be discussed in Section 28.

A. TRIANGLES

DEFINITION

A **triangle** is a closed 3-sided figure.

Each of the figures below shows a triangle with **sides** a, b, and c and with **vertices** (corner points) A, B, and C. The line segment from any vertex drawn perpendicular to the opposite side is called a **height** of the triangle. The corresponding perpendicular side of the triangle is called a **base**. A base, b, and corresponding height, h, are indicated in each figure. The first triangle is known as a **right triangle**. One of its angles is a right angle.

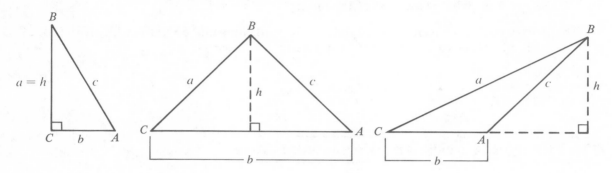

The formula for the area of a triangle can be obtained from the formula for the area of a rectangle. A right triangle can be thought of as "half of a rectangle." Therefore, its area is half that of the corresponding rectangle.

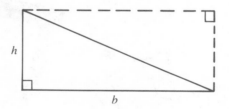

$$\text{Area of triangle} = \frac{1}{2} \text{ Area of rectangle}$$

$$= \frac{1}{2} b \times h$$

$$= \frac{b \times h}{2}$$

Example 1 ▶ Find the area of a right triangle with base 6 inches and height 5 inches.

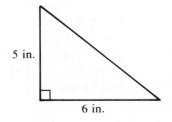

5 in.

6 in.

Solution.

$$\text{Area} = \frac{b \times h}{2}$$

$$= \frac{\overset{3 \text{ in.}}{\cancel{6 \text{ in.}}} \times 5 \text{ in.}}{\underset{1}{\cancel{2}}}$$

$$= 15 \text{ in.}^2$$

◀

For other types of triangles, the formula for area is the same. For example, consider the triangle in the figure below to the left, with base 10 inches and height 5 inches. Its area is the sum of the areas of the two right triangles in the figure below to the right, with bases 4 inches and 6 inches and with common height 5 inches. Note that

$$A_1 = \frac{\overset{2 \text{ in.}}{\cancel{4 \text{ in.}}} \times 5 \text{ in.}}{\underset{1}{\cancel{2}}} = 10 \text{ in.}^2 \text{ and } A_2 = 15 \text{ in.}^2 \quad \text{(as in Example 1).}$$

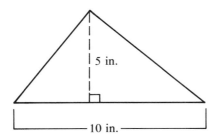
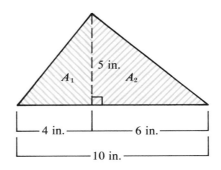

Let A be the area of the triangle.

$$A = A_1 + A_2$$
$$= 10 \text{ in.}^2 + 15 \text{ in.}^2$$
$$= 25 \text{ in.}^2$$

Observe that

$$25 \text{ in.}^2 = \frac{50 \text{ in.}^2}{2}$$
$$= \frac{10 \text{ in.} \times 5 \text{ in.}}{2}$$

that is,

$$\frac{base \times height}{2}$$

To sum up, for *any* triangle,

$$\boxed{\text{Area of triangle} = \frac{1}{2} \text{ base} \times \text{height}}$$

Example 2 ▶ Find the base of a triangle whose height is 8 centimeters and whose area is 12 square centimeters.

Solution.

$$\text{Area} = \frac{\text{base} \times \text{height}}{2}$$

$$12 \text{ cm.}^2 = \frac{\text{base} \times \overset{4 \text{ cm.}}{\cancel{8 \text{ cm.}}}}{\underset{1}{\cancel{2}}}$$

Divide both sides by 4 cm.

$$\frac{12 \text{ cm.}^2}{4 \text{ cm.}} = \text{base}$$

$$3 \text{ cm.} = \text{base} \qquad ◀$$

The *perimeter* of a triangle is the sum of the lengths of its sides. Thus the perimeter of the triangle pictured here is 18 centimeters.

$$\begin{array}{r} 6 \text{ cm.} \\ 4 \text{ cm.} \\ + \ \ 8 \text{ cm.} \\ \hline 18 \text{ cm.} \end{array}$$

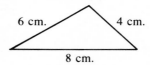

PRACTICE EXERCISES FOR TOPIC A

1. The area of a triangle with base 10 inches and height 4 inches is

   (A) 20 in.²   (B) 40 in.²   (C) 14 in.²   (D) 28 in.²   (E) 28 in.

2. Find the height of a triangle with base 5 centimeters and area 25 square centimeters.

3. Find the perimeter of a triangle with side lengths 3 centimeters, 6 centimeters, and 7 centimeters.

Now try the exercises for Topic A on page 207.

B. CIRCUMFERENCE OF A CIRCLE

DEFINITION

> A **circle** is a closed figure every point of which is at a fixed distance from a given point, known as the **center** of the circle. The fixed distance is called the **radius** of the circle.

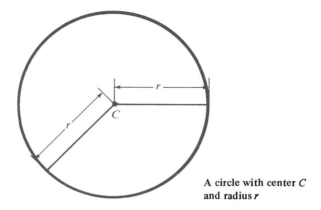

A circle with center C
and radius r

Consider any line through the center of a circle. It cuts the circle at two points, P and Q. The length of the line segment from P to Q is called the **diameter** of the circle. Clearly,

$$\text{diameter} = 2 \times \text{radius}$$

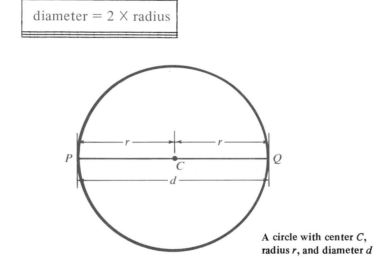

A circle with center C,
radius r, and diameter d

Example 3 ▶ Find the radius of a circle whose diameter is 8 centimeters.

Solution.

$$\text{diameter} = 2 \times \text{radius}$$

$$8 \text{ cm.} = 2 \times \text{radius}$$

Divide both sides by 2.

$$4 \text{ cm.} = \text{radius} \qquad ◀$$

The **circumference** of a circle can be thought of as the distance around the circle. Thus the circumference of a circle plays the same role as the perimeter of a rectangle or triangle. The ancient Greeks discovered that the circumference of any circle is always equal to the diameter times the *irrational* number

π. (An **irrational number** is a number that cannot be written as the quotient of two integers, and cannot be expressed as either an "ordinary" or an infinite *repeating* decimal.) To the nearest hundredth,

$$\pi \approx 3.14 \qquad \textit{Read: } \pi \text{ is approximately equal to 3.14.}$$

Thus

$$\boxed{\text{Circumference} = \pi \times \text{diameter}}$$

Example 4 ▶ Find the circumference of a circle that has diameter 5 inches.
  i. Express your answer in terms of π.
  ii. Approximate to the nearest tenth of an inch. (Use $\pi = 3.14$.)

Solution.

  i.   Circumference $= \pi \times$ diameter
                    $= \pi \times 5$ in., or 5π in.

  ii. $C \approx \underline{5 \times 3.14}$ in.
              15.70

To the nearest tenth of an inch,

Circumference ≈ 15.7 in.   ◀

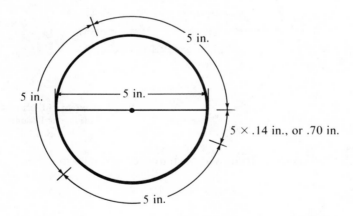

Because

Circumference $= \pi \times$ diameter

and

diameter $= 2 \times$ radius

it follows that

Circumference $= \pi \times (2 \times \text{radius})$

or

$$\boxed{\text{Circumference} = 2\pi \times \text{radius}}$$

In other words, the circumference of a circle is 2π times the radius.

When approximations involving π are called for, if nothing to the contrary is stated, use

$$\pi \approx 3.14$$

Example 5 ▶ Find the circumference of a circle of radius 10 centimeters.
   i. Express your answer in terms of π.
   ii. Approximate to the nearest centimeter.

Solution.

   i.   Circumference $= 2\pi \times$ radius

$= 2\pi \times 10$ cm.

$= 20\pi$ cm.

   ii.   Circumference $\approx \underline{20 \times 3.14}$ cm.

62.80

To the nearest centimeter,

Circumference ≈ 63 cm.   ◀

PRACTICE EXERCISES FOR TOPIC B

4. Find the diameter of a circle with radius 7 centimeters.

5. Find the radius of a circle with diameter 7 centimeters.

6. Find the circumference of a circle with diameter 9 inches. Express your answer in terms of π.

7. Find the circumference of a circle with radius 8 inches. Express your answer to the nearest tenth of an inch.

8. Find the radius of a circle with circumference 6π feet.

Now try the exercises for Topic B on page 207.

C. AREA OF A CIRCLE

The Greeks also discovered the formula for the area of a circle.

$$\boxed{\text{Area of circle} = \pi \times \text{radius}^2}$$

Thus the area is π times the square of the radius, or *approximately* 3.14 times the square of the radius.

Example 6 ▶ Find the area of a circle with radius 10 centimeters.
     i. Express your answer in terms of π.
     ii. Approximate to the nearest centimeter.

Solution.

  i.   Area $= \pi \times$ radius^2
                $= \pi(10 \text{ cm.})^2$
                $= \pi \times 100 \text{ cm.}^2$
                $= 100\pi \text{ cm.}^2$

  ii.   Area $\approx 100 \times 3.14 \text{ cm.}^2$
                $= 314 \text{ cm.}^2$ ◀

Example 7 ▶ Find the area of a circle with circumference 8π inches. Express your answer in terms of π.

Solution. First use

        Circumference $= 2\pi \times$ radius

to find the radius.

        8π in. $= 2\pi \times$ radius

Divide both sides by 2π.

$$\frac{\overset{4 \text{ in.}}{\cancel{8\pi \text{ in.}}}}{\underset{1}{\cancel{2\pi}}} = \text{radius}$$

The radius is thus 4 inches. Now use

        Area $= \pi \times$ radius^2

to find

        Area $= \pi \times (4 \text{ in.})^2$
             $= \pi \times 16 \text{ in.}^2$
             $= 16\pi \text{ in.}^2$ ◀

Example 8 ▶ Find the area of the shaded region in the figure. Express your answer in terms of π.

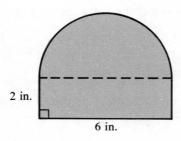

2 in.

6 in.

Solution. The figure consists of a semicircle (half of a circle) on top of a rectangle.

$$\text{Area of rectangle} = 2 \text{ in.} \times 6 \text{ in.}$$
$$= 12 \text{ in.}^2$$

The diameter of the circle is 6 inches, so that the radius is 3 inches.

$$\text{Area of circle} = \pi \times (3 \text{ in.})^2$$
$$= \pi \times 9 \text{ in.}^2$$
$$= 9\pi \text{ in.}^2$$

The area of a semicircle is $\frac{1}{2}$ of the area of the circle, or $\frac{9\pi}{2}$ in.^{2}. The total area is

$$12 \text{ in.}^2 + \frac{9\pi}{2} \text{ in.}^2 = \left(12 + \frac{9\pi}{2}\right) \text{in.}^2 = \left(\frac{24 + 9\pi}{2}\right) \text{in.}^2$$

In the numerator, the sum, $24 + 9\pi$, of the counting number 24 and the irrational number 9π cannot be further simplified. Thus the area of the given region is $\left(\frac{24 + 9\pi}{2}\right)$ in.^{2}. ◄

PRACTICE EXERCISES FOR TOPIC C

In Exercises 9–11, find the area of the circle specified by the given information. Express your answer to the nearest tenth of a unit.

9. radius 2 inches

10. diameter 10 centimeters

11. circumference 8π feet

In Exercise 12, draw a circle around the correct letter.

12. If the circumference of a circle is 5π inches, then the area is

   (A) 10π in.   (B) 25π in.^{2} (C) 2.5π in.^{2}

   (D) 6.25π in.^{2} (E) 6.25 in.^{2}

13. Find the area of the shaded region shown here, which consists of a right triangle together with a semicircle. Express your answer in terms of π.

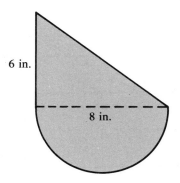

6 in.

8 in.

Now try the exercises for Topic C on page 208.

D. RECTANGULAR BOXES AND CUBES

The **volume** of a solid figure, such as a rectangular box, is a measure of how many (cubic) units it can hold.

As you know, the area of a rectangle is the product of its length and width.

Area of rectangle = length × width

Now consider a box that has all rectangular faces, as shown here. The volume of such a **rectangular box** is the product of its length, width, and height.

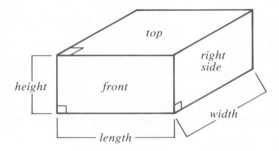

> Volume of rectangular box = length × width × height

The **surface area** of a rectangular box, that is, the sum of the areas of its 6 faces, is obtained by considering the above figure, which shows only 3 of the 6 faces. The combined area of all 6 faces is given by

2 × length × width	(*top* and *bottom*)
+ 2 × length × height	(*front* and *back*)
+ 2 × width × height	(*left* and *right sides*)

Thus

> Surface area of rectangular box =
>   2 × length × width + 2 × length × height
>   + 2 × width × height

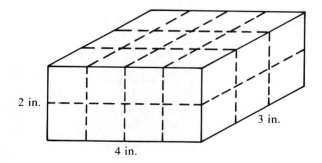

For example, the length of the box at the bottom of page 204 is
4 inches, the width is 3 inches, and the height is 2 inches. The volume is

$$\underbrace{\text{length}}\quad\times\quad\underbrace{\text{width}}\quad\times\quad\underbrace{\text{height}}$$
$$\overline{\text{4 inches}} \times \overline{\text{3 inches}} \times \overline{\text{2 inches}} = \text{24 inches} \times \text{inches} \times \text{inches},$$
$$\text{or } \text{24 cubic inches}$$

The surface area is given by

$$2 \times \text{length} \times \text{width} + 2 \times \text{length} \times \text{height} + 2 \times \text{width} \times \text{height}$$

$$= 2 \times 4 \text{ inches} \times 3 \text{ inches} + 2 \times 4 \text{ inches} \times 2 \text{ inches}$$
$$+ 2 \times 3 \text{ inches} \times 2 \text{ inches}$$

$$= 24 \text{ square inches} + 16 \text{ square inches} + 12 \text{ square inches}$$

$$= 52 \text{ square inches}$$

Note that the unit of volume is cubic inches, whereas the unit of surface area is square inches.

$$\text{inches} \times \text{inches} \times \text{inches} = \text{cubic inches}$$

Write

in.^{3} for   cubic inches
ft.^{3} for   cubic feet
cm.^{3} for   cubic centimeters

Example 9 ▶   Find
  i. the volume and
  ii. the surface area
       of the rectangular box
       shown at the right.

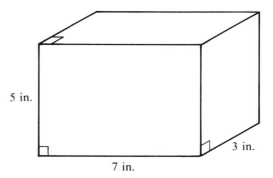

5 in.

7 in.

3 in.

Solution.

  i.   Volume = length × width × height

$$= 7 \text{ in.} \times 3 \text{ in.} \times 5 \text{ in.}$$

$$= 105 \text{ in.}^3$$

  ii.   Surface area = 2 × length × width + 2 × length × height
             + 2 × width × height

$$= 2 \times 7 \text{ in.} \times 3 \text{ in.} + 2 \times 7 \text{ in.} \times 5 \text{ in.} + 2 \times 3 \text{ in.} \times 5 \text{ in.}$$

$$= 42 \text{ in.}^2 + 70 \text{ in.}^2 + 30 \text{ in.}^2$$

$$= 142 \text{ in.}^2$$ ◀

DEFINITION

> A **cube** is a rectangular box in which the length, width, and height are equal.

Because (the length of) any side of a cube equals the length, width, and height, the volume of a cube is given by

side × side × side, or side³

Thus

> Volume of cube = side³

Similarly, the area of each of the six faces of a cube is the same, namely

side × side, or side²

Consequently, because there are 6 such faces,

> Surface area of cube = 6 × side²

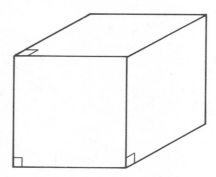

Example 10 ▶ Find

i. the volume and

ii. the surface area of a cube if the length of a side is 3 centimeters.

Solution.

i.   Volume = side³

= (3 cm.)³

= 27 cm.³

ii.   Surface area = 6 × side²

= 6 × (3 cm.)²

= 6 × 9 cm.²

= 54 cm.² ◀

PRACTICE EXERCISES FOR TOPIC D

14. Find *i.* the volume and *ii.* the surface area of a rectangular box of length 3 feet, width 1 foot, and height 4 feet.

15. Find *i.* the volume and *ii.* the surface area of a cube if the length of each side is 4 inches.

Now try the exercises for Topic D on page 209.

EXERCISES

A. *In Exercises 1–6, find the area of each triangle.*

1. base 4 inches, height 6 inches   *Answer:*

2. base 5 centimeters, height 10 centimeters   *Answer:*

3. base 3 feet, height 2 feet   *Answer:*

4. base 7 centimeters, height 4 centimeters   *Answer:*

5. base 9 inches, height 5 inches   *Answer:*

6. base = height = 10 inches   *Answer:*

In Exercises 7 and 8, draw a circle around the correct letter.

7. Find the base of a triangle with area 36 square inches and height 9 inches.
   (A) 4 in.   (B) 2 in.   (C) 8 in.   (D) 324 in.   (E) 162 in.

8. The base of a triangle is 5 inches and the area is 25 square inches. The height is
   (A) 2.5 in.   (B) 5 in.   (C) 10 in.   (D) 20 in.   (E) 125 in.

9. If the height of a triangle is 7 centimeters and the area is 63 square centimeters, find the base.   *Answer:*

10. If the base of a triangle is 1 foot and the area is 1 square foot, find the height.   *Answer:*

In Exercises 11–14, find the perimeter of a triangle whose sides are as indicated.

11. 6 cm., 3 cm., 4 cm.   *Answer:*   12. 5 in., 4 in., 3 in.   *Answer:*

13. 9 ft., 10 ft., 12 ft.   *Answer:*

14. 4.2 cm., 3.8 cm., 6.1 cm.   *Answer:*

B. *In Exercises 15–18, find the diameter of the indicated circle.*

15. radius 3 inches   *Answer:*   16. radius 7.5 inches   *Answer:*

17. circumference 12π inches   *Answer:*

18. circumference 9π meters   *Answer:*

In Exercises 19–22, find the radius of the indicated circle.

19. diameter 4 feet   *Answer:*   **20.** diameter 5 inches   *Answer:*

21. circumference 6π inches   *Answer:*

22. circumference 11π centimeters   *Answer:*

In Exercises 23–26, find the circumference of the given circle. Express your answer to the nearest tenth of a unit.

23. diameter 4 feet   *Answer:*   **24.** diameter 7 centimeters   *Answer:*

25. radius 5 inches   *Answer:*   **26.** radius 2 feet   *Answer:*

In Exercises 27 and 28, draw a circle around the correct letter.

27. If the circumference of a circle is 28π inches, its radius is
   (A) 28 inches   (B) 14π inches   (C) 14 inches   (D) 56 inches   (E) 7 inches

28. If the radius of a circle is 100 meters, to the nearest meter, the circumference is
   (A) 300 m.   (B) 314 m.   (C) 600 m.   (D) 628 m.   (E) 63 m.

C. *Find the area of the indicated circle. Express your answer to the nearest tenth of a unit.*

   29. radius = 7 centimeters   *Answer:*

   30. radius = 5 inches   *Answer:*   **31.** radius = 4 feet   *Answer:*

   32. radius = 6 meters   *Answer:*   **33.** diameter = 8 inches   *Answer:*

   34. diameter = 10 centimeters   *Answer:*

   35. diameter = 1 foot 4 inches   *Answer:*

   36. diameter = 3 meters   *Answer:*

   37. circumference = 8π meters   *Answer:*

   38. circumference = 18π inches   *Answer:*

   39. circumference = 9π centimeters   *Answer:*

   40. circumference = 3π inches   *Answer:*

In Exercises 41 and 42, draw a circle around the correct letter.

   41. The area of a circle with diameter 12 centimeters is
      (A) 6 cm.^{2} (B) 36 cm.^{2} (C) 36π cm.^{2} (D) 72π cm.^{2} (E) 144π cm.^{2}

   42. The area of a circle with circumference 6π inches is
      (A) 3π in.^{2} (B) 6π in.^{2} (C) 9π in.^{2} (D) 12π in.^{2} (E) 36π in.^{2}

In Exercises 43–45, find the area of each shaded region. Express your answer in terms of π.

43.   **44.**   **45.**

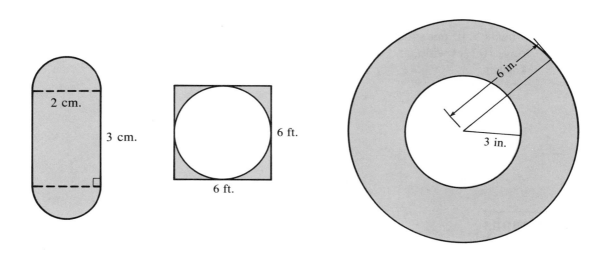

       Answer: *Answer:*   *Answer:*

46. A small (circular) pizza pie has radius 5 inches and costs $2.00. At this rate, how much should a large pizza pie with radius 10 inches cost?   *Answer:*

D. *In Exercises 47–54, for each rectangular box find: i. the volume ii. the surface area.*

47. length = 4 feet, width = 2 feet, height = 1 foot   *Answer: i.   ii.*

48. length = 5 centimeters, width = 2 centimeters, height = 4 centimeters
Answer: i.   ii.

49. length = width = 6 inches, height = 3 inches   *Answer: i.   ii.*

50. length = 10 inches, width = height = 5 inches   *Answer: i.   ii.*

51. length = 4 meters, width = 3 meters, height = 2 meters
Answer: i.   ii.

52. length = 1 foot 6 inches, width = 1 foot, height = 6 inches
Answer: i.   ii.

53. length = width = height = 5 inches   *Answer: i.   ii.*

54. length = width = height = 20 centimeters   *Answer: i.   ii.*

55. How much paper is needed to line the inside of a cigar box that is 10 inches by 6 inches by 3 inches.   *Answer:*

56. How much air is contained in a room that is 15 feet by 12 feet by 9 feet?
Answer:

57. Gas that costs $3.50 per cubic foot is pumped into a cubical container whose side length is 7 feet. What is the cost of filling this container?   *Answer:*

58. It costs $.40 per square foot to paint the walls and ceiling of a room that is 12 feet long, 10 feet wide, and 8 feet high. What is the cost of painting this room?
Answer:

23. Graphs

It is important to know how to interpret a graph that might appear in a newspaper, magazine, or textbook.

A. BAR GRAPHS

In a bar graph the *height* of a column or bar represents a number.

1985 SALES
(Thousands)

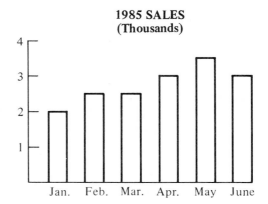

Example 1 ▶ Use the graph at the right to find out the total number of magazines sold by Collegiate Press from April through June 1985.

(A) 3000 (B) 3500
(C) 3.5 (D) 9.5
(E) 9500

Solution. Each bar represents the number of *thousands* of magazines sold each month from January through June, 1985. To find how many thousands were sold, compare the top of the bar with the scale at the left. In April, 3000 copies were sold and in June, 3000 copies were sold. In May, 3500 copies were sold. (The top of the bar for May lies midway between the 3 and the 4 at the left. Thus the bar represents 3500, which is midway between 3000 and 4000. Altogether, there were 9500 copies sold during the months April, May, and June, 1985. The correct choice is (E).   ◀

PRACTICE EXERCISES FOR TOPIC A

These exercises refer to the graph at the right. Draw a circle around the correct letter.

1. Find the total number of magazines sold by the *Weekly Journal* from June through August, 1985.

(A) 5500 (B) 7000
(C) 7500 (D) 8000
(E) 7.5

2. For which two months was there a combined total of 7000 magazines sold?

(A) June and July
(B) July and August
(C) August and September
(D) June and September

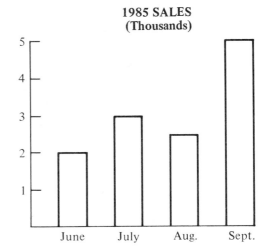

1985 SALES
(Thousands)

(E) July and September

Now try the exercises for Topic A on page 214.

B. CIRCLE GRAPHS

In a circle graph, various items are presented as a *percentage* of the total dollar amount.

Example 2 ▶ In the graph at the right, find the annual expenditure for schools.

(A) 10%   (B) $15,500,000
(C) $1,550,000
(D) $155,000
(E) $155,000,000

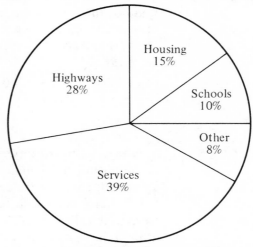

ANNUAL EXPENDITURES: $15,500,000

Solution. The graph indicates that 10% of the total annual expenditures of $15,500,000 is spent on schools. Because 10% = .1,

$$10\% \times \$15,500,000 = .1 \times \$15,500,000$$
$$= \$1,550,000$$

Choice (C) is correct.   ◀

PRACTICE EXERCISES FOR TOPIC B

These exercises refer to the graph at the right. Draw a circle around the correct letter.

3. Find the annual expenditure for housing.

(A) $1,920,000   (B) $1,600,000
(C) $1,880,000   (D) $4,000,000
(E) $19,200,000

4. For which item was there an annual expenditure of $4,000,000?

(A) highways   (B) other
(C) schools   (D) housing
(E) services

ANNUAL EXPENDITURES: $16,000,000

Now try the exercises for Topic B on page 215.

C. LINE GRAPHS

Line graphs are particularly useful when you want to compare two successive periods.

Example 3 The graph at the right shows the number of students enrolled in colleges in New England from 1980 to 1984. What was the increase in enrollment from 1982 to 1983?

(A) 250,000   (B) 500,000
(C) 750,000
(D) 1,750,000
(E) 2,250,000

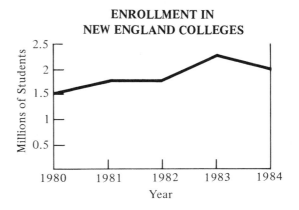

ENROLLMENT IN
NEW ENGLAND COLLEGES

Solution. In 1982 there were 1,750,000 students enrolled. In 1983 there were 2,250,000. To find the *increase* in enrollment, subtract:

```
    2 250 000
  - 1 750 000
      500 000
```

The increase in enrollment was 500,000 [Choice (B)].   ◄

PRACTICE EXERCISES FOR TOPIC C

These exercises refer to the above graph. Draw a circle around the correct letter.

5. Which year had the lowest enrollment?

    (A) 1980   (B) 1981   (C) 1982   (D) 1983   (E) 1984

6. In which year did the enrollment decrease?

    (A) 1980   (B) 1981   (C) 1982   (D) 1983   (E) 1984

Now try the exercises for Topic C on page 216.

EXERCISES

A. *Draw a circle around the correct letter.*

GRADUATES

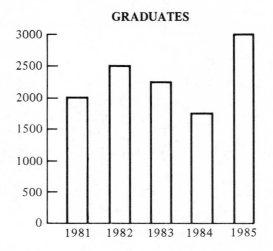

For Exercises 1–5, use the graph at the right, which shows the number of students who graduated from Wilson College each year from 1981 to 1985.

1. The number that graduated in 1982 was

   (A) 2000   (B) 2500
   (C) 2250   (D) 12,500,000
   (E) 250,000

2. The number that graduated in 1984 was closest to
   (A) 1500   (B) 1750   (C) 2000   (D) 2250   (E) 8500

3. The *total* number that graduated in the two years 1981 and 1982 was closest to
   (A) 2000   (B) 2500   (C) 4000   (D) 4500   (E) 5000

4. The *total* number that graduated in the two years 1983 and 1984 was closest to
   (A) 3000   (B) 3500   (C) 4000   (D) 4500   (E) 5000

5. The year in which the *smallest* number of students graduated was
   (A) 1981   (B) 1982   (C) 1983   (D) 1984   (E) 1985

For Exercises 6–10, use the graph at the right, which shows the number of students attending colleges in New York City in 1985, by borough.

STUDENTS IN 1985
(Thousands)

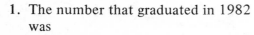

6. The number attending colleges in Queens in 1985 was closest to

   (A) 70   (B) 80
   (C) 60,000   (D) 70,000
   (E) 80,000

7. The *total* number attending colleges in the Bronx and Brooklyn in 1985 was closest to
   (A) 60,000   (B) 80,000   (C) 70,000   (D) 140,000   (E) 100,000

8. The *total* number attending colleges in Manhattan and Queens was closest to
   (A) 70,000   (B) 80,000   (C) 100,000   (D) 170,000   (E) 180,000

9. How many more students attended a college in Manhattan than in Brooklyn?

(A) 100,000   (B) 80,000   (C) 180,000   (D) 10,000   (E) 20,000

10. The *total* number attending colleges in all five boroughs in 1985 was closest to

(A) 340,000   (B) 320,000   (C) 300,000   (D) 35,000   (E) 40,000

For Exercises 11–15, use the graph at the right, which shows the monthly income of the XYZ Corporation in 1984 in millions of dollars.

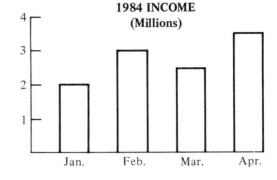

11. The *total* income for January and February was

(A) $2,000,000   (B) $3,000,000
(C) $4,000,000   (D) $5,000,000
(E) $5000

12. The *total* income for February and March was closest to

(A) $5,000,000   (B) $5,500,000   (C) $6,000,000
(D) $6,500,000   (E) $7,000,000

13. The *total* income for February, March, and April was closest to

(A) $8,000,000   (B) $8,500,000   (C) $9,000,000
(D) $9,500,000   (E) $10,000,000

14. How much more income was there in February than in January?

(A) $500,000   (B) $1,000,000   (C) $1,500,000
(D) $2,000,000   (E) $3,000,000

15. How much more income was there in April than in March?

(A) $250,000   (B) $500,000   (C) $750,000
(D) $1,000,000   (E) $6,000,000

B. *Draw a circle around the correct letter.*

For Exercises 16–20, use the graph at the right, which indicates a family's allotment of its monthly income of $2500.

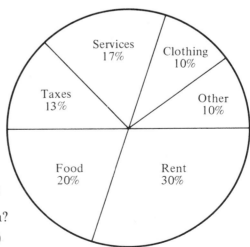

MONTHLY INCOME: $2500

16. How much does the family spend each month for rent?

(A) $750   (B) $75
(C) $7500   (D) $1250   (E) $775

17. How much money goes for taxes each month?

(A) $300   (B) $325   (C) $350
(D) $400   (E) $500

18. Each month the family spends $500 for

(A) clothing   (B) services   (C) taxes   (D) food   (E) rent

19. What is the *total* amount the family spends each month for food and rent?

(A) $500   (B) $750   (C) $1000   (D) $1250   (E) $1500

20. How much does the family spend a *year* for clothing?

(A) $250   (B) $2500   (C) $750   (D) $300   (E) $3000

For Exercises 21–25, use the graph at the right, which indicates the after-tax disbursements of a company during one year.

21. How much did the company spend for supplies?

(A) $250,000   (B) $800,000
(C) $900,000   (D) $2,500,000
(E) $9,000,000

22. What was the company's profit that year?

(A) $252,000   (B) $25,200
(C) $2,520,000   (D) $70,000
(E) $700,000

23. The total spent for rent and supplies was

(A) $684,000   (B) $1,188,000   (C) $1,872,000
(D) $1,584,000   (E) $2,000,000

24. For which item did the company spend $576,000?

(A) supplies   (B) salaries   (C) rent   (D) overhead   (E) profit

25. How much more did the company spend for salaries than for supplies?

(A) $900,000   (B) $1,188,000   (C) $288,000
(D) $252,000   (E) $324,000

C. *Draw a circle around the correct letter.*

For Exercises 26–30, use the graph at the right, which indicates the annual profits of the Honest Lenders Finance Corporation, in thousands of dollars.

26. What was the profit in 1983?

(A) $200,000   (B) $250,000
(C) $250   (D) $400,000
(E) $550,000

CORPORATE PROFITS

Thousands of Dollars

600
500
400
300
200
100

1980   1981   1982   1983   1984

Year

27. What was the increase in profit
    from 1980 to 1981?

    (A) $300,000   (B) $400,000
    (C) $700,000   (D) $350,000
    (E) $100,000

28. What was the increase in profit from 1983 to 1984?
    (A) $50,000   (B) $100,000   (C) $150,000   (D) $300,000   (E) $350,000

29. What was the *decrease* in profit from 1981 to 1982?
    (A) $150,000   (B) $200,000   (C) $250,000   (D) $400,000   (E) $600,000

30. What were the total profits for the years 1981 and 1982?
    (A) $200,000   (B) $300,000   (C) $400,000   (D) $500,000   (E) $600,000

*For Exercises 31–35, use the graph
at the right, which shows the number
of Tokyo subway passengers, in
millions, during one work-week.*

31. The number of passengers on
    Wednesday was closest to

    (A) 2,500,000   (B) 2,750,000
    (C) 3,000,000   (D) 3,250,000
    (E) 3,500,000

32. The smallest number of passengers
    used the subway on

    (A) Monday   (B) Tuesday
    (C) Wednesday   (D) Thursday
    (E) Friday

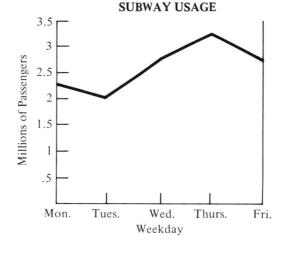

33. The increase from Tuesday to Thursday was
    (A) 750,000   (B) 1,000,000   (C) 1,250,000
    (D) 1,500,000   (E) 3,250,000

34. The decrease from Monday to Tuesday was
    (A) 250,000   (B) 500,000   (C) 1,000,000
    (D) 2,000,000   (E) 2,500,000

35. The decrease from Thursday to Friday was
    (A) 250,000   (B) 500,000   (C) 750,000
    (D) 2,750,000   (E) 3,250,000

Review Exercises for Unit V

When several choices are given, draw a circle around the correct letter.

1.   4 hours 45 minutes
      + 2 hours 35 minutes

2.   9 minutes 28 seconds
      − 6 minutes 39 seconds

3. A party began at 8:45 P.M. and ended at 1:30 A.M. the following morning. How long did it last?

    (A) 7 hr. 15 min.   (B) 2 hr. 45 min.   (C) 3 hr. 45 min.
    (D) 4 hr. 45 min.   (E) 5 hr. 45 min.

4. An hour-and-a-half TV special was broken up into 5 equally-long segments. How long was each segment?

    (A) 6 minutes   (B) 10 minutes   (C) 12 minutes   (D) 18 minutes   (E) 20 minutes

5.   10 pounds 10 ounces
      + 7 pounds   8 ounces

6.   2 kilograms 500 grams
      − 1 kilogram   750 grams

7. Add 5 pounds 13 ounces and 4 pounds 7 ounces.   *Answer:*

8. What is the total weight (in kilograms) of fifty 250-gram envelopes? *Answer:*

9.   3 meters 20 centimeters
      + 2 meters 90 centimeters

10.   5 feet 6 inches
      − 2 feet 9 inches

11. A wire 3 feet 6 inches long is cut into 6 equal parts. How long is each part?
    Answer:

12. Find the area of a page that is 8 inches long and $4\frac{1}{2}$ inches wide.   *Answer:*

13. The area of a rectangular field is 1500 square yards. If the field is 60 feet long, how wide is it?

    (A) 9000 ft.^{2} (B) 90,000 ft.^{2} (C) 20 ft.   (D) 30 ft.   (E) 25 ft.

14. A rectangular cabinet is 6 feet long, 2 feet wide, and 4 feet high. Find the area of its top surface.

    (A) 12 ft.^{2} (B) 8 ft.^{2} (C) 12 ft.^{3} (D) 48 ft.^{2} (E) 48 ft.^{3}

15. Find the area of a square whose side length is 11 centimeters.   *Answer:*

16. The material for a tablecloth costs $2.50 per square foot. What is the cost of the material for a 10-foot by 6-foot tablecloth?   *Answer:*

17. Find the area of the floor of a room that is 20 feet long, 14 feet wide, and 9 feet high. *Answer:*

18. If carpeting costs $3.50 per square foot, how much would wall-to-wall carpeting cost for the floor in Exercise 17?   *Answer:*

19. Find the perimeter of a rectangle that is 8 feet by 5 feet.   *Answer:*

20. The length of a rectangle is 12 meters and the perimeter is 30 meters. Find the width. *Answer:*

21. Fencing for a rectangular field costs $3 per foot. If the field is 80 feet long and 60 feet wide, how much does it cost to enclose the field with fencing?   *Answer:*

22. The width of a rectangle is 5 centimeters and the perimeter is 40 centimeters. Find the area.   *Answer:*

23. The length of a rectangle is 9 inches and the area is 54 square inches. Find the perimeter.   *Answer:*

24. The base of a triangle is 6 inches and the height is 3 inches. Find the area.

    (A) 6 in.^{2} (B) 9 in.^{2} (C) 12 in.^{2} (D) 18 in.^{2} (E) 24 in.^{2}

25. A triangle has base 5 centimeters and area 40 square centimeters. Find its height.

    (A) 4 cm.   (B) 8 cm.   (C) 16 cm.   (D) 100 cm.   (E) 200 cm.

26. Find the perimeter of a triangle whose side lengths are 7.1 inches, 8 inches, and 8.9 inches.   *Answer:*

For Exercises 27–29, suppose that a circle has a radius of 5 centimeters.

27. Find its diameter.   *Answer:*

28. Find its circumference (to the nearest centimeter).   *Answer:*

29. Find its area (in terms of π).   *Answer:*

For Exercises 30–32, suppose that a circle has a diameter of 14 inches.

30. Find its radius.   *Answer:*

31. Find its circumference (to the nearest tenth of an inch).   *Answer:*

32. Find its area (in terms of π).   *Answer:*

For Exercises 33 and 34, suppose that the circumference of a circle is 12π inches.

33. Find its radius.   *Answer:*

34. Find its area (in terms of π).   *Answer:*

35. Find the total area of the region at the right, which consists of a rectangle and a semicircle. *Answer:*

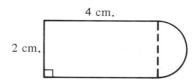

For Exercises 36–38, use the graph at the right, which shows the number of students who entered Patrick Henry Institute each year from 1981 to 1985.

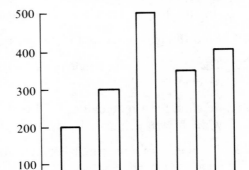

ENROLLMENT AT PATRICK HENRY INSTITUTE

36. In which year did the fewest students enter.

(A) 1981   (B) 1982
(C) 1983   (D) 1984
(E) 1985

37. The largest increase in any year was

(A) 100   (B) 200   (C) 300   (D) 400   (E) 500

38. The *total* number of students entering the institute from 1981 to 1985 was

(A) 1500   (B) 1700   (C) 1750   (D) 1800   (E) 2000

For Exercises 39–41, use the graph at the right which shows a family's annual budget.

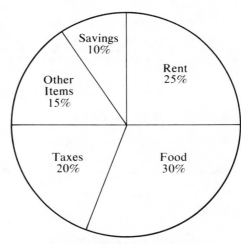

ANNUAL BUDGET: $25,000

39. How much money did the family allocate for food?

(A) $500   (B) $7000
(C) $7500   (D) $9000
(E) $10,000

40. $5000 was allocated for

(A) rent   (B) taxes
(C) savings   (D) savings and taxes
(E) other items

41. How much more did the family plan to spend on food than on rent?

(A) $1000   (B) $1250   (C) $1500   (D) $2000   (E) $2500

For Exercises 42–44, use the graph at the right, which shows the annual earnings of LMN Enterprises for the period 1981–1985.

ANNUAL EARNINGS

42. What was the increase in earnings from 1981 to 1982?

    (A) $1,000,000 (B) $1,500,000
    (C) $2,000,000 (D) $2,500,000
    (E) $3,000,000

43. What was the decrease in earnings from 1983 to 1984?

    (A) $3,000,000 (B) $4,000,000
    (C) $5,000,000 (D) $7,000,000
    (E) $10,000,000

44. The total earnings for the period 1981–1985 was

    (A) $23,000,000   (B) $30,000,000   (C) $37,000,000
    (D) $43,000,000   (E) $45,000,000

Review Exercises on Units I–IV

1. Express 105,020 in verbal form.   *Answer:*

2. Find *i.* the quotient and *ii.* the remainder: $385 \div 27$
   Answer: i.   *ii.*

3. Express 312 as the product of primes.   *Answer:*

4. Find *lcm* (8, 12, 50).   *Answer:*   5. $\dfrac{4}{9} \times \dfrac{3}{8} =$

6. $\dfrac{10}{21} \div \dfrac{15}{49} =$ 7. $2\dfrac{1}{2} + 5\dfrac{1}{4} =$ 8. $3\dfrac{3}{4} \div 1\dfrac{1}{2} =$

9. Express .003 as a fraction in lowest terms.   *Answer:*

10. Express $\dfrac{2}{9}$ as an infinite repeating decimal.   *Answer:*

11. $1.25 \times .008 =$ 12. $\dfrac{.0016}{.08} =$

13. Express $2\dfrac{1}{2}\%$ as a decimal.   *Answer:*

14. Express 1.09 as a percent.   *Answer:*

15. Express 24% as a fraction in lowest terms.   *Answer:*

16. A $40 sweater is reduced by 20%. What is the sale price?   *Answer:*

Practice Exam on Unit V

1. A movie begins at 8:20 P.M. and ends at 10:05 P.M. How long does it last?
   Answer:

2.   6 pounds 3 ounces
     – 2 pounds 10 ounces
   ‾‾‾‾‾‾‾‾‾‾‾‾‾‾‾‾‾‾‾‾

3.   4 feet 8 inches
     + 5 feet 9 inches
   ‾‾‾‾‾‾‾‾‾‾‾‾‾‾‾‾‾‾

4. The area of a rectangle is 40 square centimeters and the length is 20 centimeters. Find the perimeter.

   (A) 60 cm.   (B) 22 cm.   (C) 24 cm.   (D) 44 cm.   (E) 2 cm.

5. Carpeting costs $4 per square foot. How much does wall-to-wall carpeting cost for a room that is 15 feet long, 12 feet wide, and 10 feet high?

   (A) $720   (B) $480   (C) $600   (D) $7200   (E) $60

6. Find the area of a triangle with base 5 centimeters and height 8 centimeters.
   Answer:

7. The circumference of a circle is 9π inches. Find the area in terms of π.
   Answer:

8. Find the volume of a rectangular box that is 8 feet by 6 feet by 9 feet.
   Answer:

For Exercises 9–11, the graph at the right indicates the annual expenditures for a college.

9. What is the annual expenditure for salaries?

   (A) $2,500,000 (B) $3,500,000
   (C) $4,000,000 (D) $5,000,000
   (E) $10,000,000

10. For which item(s) does the college spend $3,000,000?

    (A) salaries   (B) scholarships
    (C) construction (D) maintenance
    (E) maintenance and construction

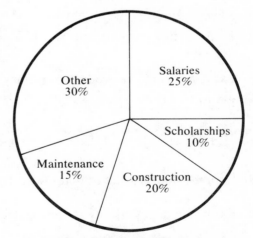

ANNUAL EXPENDITURES: $20,000,000

11. What is the combined annual expenditure for salaries and scholarships?

    (A) $5,000,000   (B) $6,000,000   (C) $7,000,000
    (D) $9,000,000   (E) $15,000,000

24. Arithmetic with Negative Integers

An understanding of negative numbers, and how to add, subtract, multiply, and divide them is essential in most subsequent work in algebra and scientific studies.

A. THE NUMBER LINE

The numbers with which we have been concerned thus far, other than 0, are called the **positive numbers**. Thus 4, 39, $\frac{1}{2}$, .8, and 1.03 are all positive. In particular, the *counting numbers,* 1, 2, 3, 4, . . . are also known as the **positive integers**. Certain concepts, however, are not expressed by positive numbers.

Negative numbers express such notions as being *below* 0 or *losing* money. Thus

$-5°$F.   describes the temperature (in Fahrenheit) on a cold day

$-\$10$ describes a loss of ten dollars

DEFINITION

> The numbers $-1, -2, -3, -4, \ldots$ are known as **negative integers**. Other **negative numbers** are $-\frac{1}{2}, -.3, -.75$, and so on. The **integers** consist of the positive integers (or counting numbers) together with 0 and the negative integers. Thus the integers are
>
> $$\ldots -4, -3, -2, -1, 0, 1, 2, 3, 4, \ldots$$

Negative numbers allow you to subtract a larger integer from a smaller one. For example,

$$3 - 5 = -2$$

Before developing the basic rules for arithmetic with negative numbers, we consider how to picture numbers along a horizontal line. This graphic picture will enhance your understanding of the basic rules of arithmetic.

Consider a horizontal line that extends without limit both to the left and to the right. This line will be called the **number line**. Choose any point on this line and call it the **origin**. This point corresponds to the number 0. Now choose another point to the *right* of 0. This second point corresponds to the number 1.

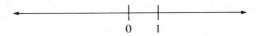

The length of the segment of the line between 0 and 1 determines one **unit of distance**. This is represented by the thickened portion of the line below. The number 2 corresponds to the point one (distance) unit to the right of 1. The number 3 corresponds to the point one unit further to the right. Continue this process to obtain the numbers 4, 5, 6, and so on, as in the figure below.

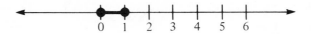

Positive numbers are pictured to the *right* of 0 along the number line. It seems reasonable that negative numbers should be pictured to the *left* of 0. Thus go one distance unit to the *left* of 0. This point corresponds to -1. Go one unit further to the left to locate -2. Continue to the left to locate -3, $-4, -5$, and so on, as in the figure below.

Observe that $3 < 5$ and 3 lies to the left of 5 on the x-axis. In general, for *any* two numbers a and b (whether positive, negative, or 0),

$$a < b \text{ when } a \text{ lies to the } left \text{ of } b$$

Thus $-5 < -3$ because -5 lies to the left of -3.

Midway between 0 and 1 on the number line is the point corresponding to the fraction $\frac{1}{2}$. (See the figure below.) The point corresponding to $\frac{7}{4}$, that is, to $1\frac{3}{4}$, lies three-fourths of the way from 1 to 2. Negative fractions lie to the left of 0. Thus $-\frac{1}{2}$ lies midway between -1 and 0.

Example 1 ▶ Consider the number line, pictured below. Label the points that correspond to the following numbers:

   i. 4   *ii.* 7   *iii.* -1 *iv.* -5 *v.* $\frac{3}{2}$

Solution. In part *v*, note that $\frac{3}{2} = 1\frac{1}{2}$. Thus $\frac{3}{2}$ lies midway between 1 and 2. See the figure below.

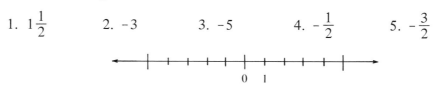

PRACTICE EXERCISES FOR TOPIC A

Consider the number line pictured below. Label the points that correspond to the following numbers.

   1. $1\frac{1}{2}$ 2. -3 3. -5 4. $-\frac{1}{2}$ 5. $-\frac{3}{2}$

Now try the exercises for Topic A on page 237.

B. ABSOLUTE VALUE

Two concepts related to arithmetic with negative numbers, the *opposite* and the *absolute value* of a number, can be represented on the number line.

DEFINITION

> It is useful to say that the numbers 5 and -5 are **opposites**. Thus -5 is the opposite of 5, and 5 is the opposite of -5. Also, let 0 be its own opposite.

We will use

$-a$

to indicate the opposite of any number *a* (whether positive, negative, or 0). Thus,

$$\underset{\underset{\text{opposite}}{\uparrow}}{-}(-5) = 5$$

(This says that the opposite of −5 is 5.)

On the number line a number and its opposite are the same distance from the origin, 0, but lie on *opposite* sides of 0. For example, the *positive* number 5 lies 5 units to the *right* of 0, whereas the *negative* number −5 lies 5 units to the *left* of 0.

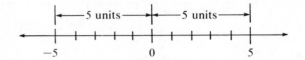

DEFINITION

> The **sign** of a positive number is + (*plus*).
> The **sign** of a negative number is − (*minus*).

Thus the sign of 5 is + ;   the sign of −5 is − .

Sometimes we want to ignore the sign of a negative number. For this purpose we introduce *absolute value*.

DEFINITION

> The **absolute value** of a *positive* number or of 0 is the number itself, whereas the absolute value of a negative number is its opposite, that is, the corresponding positive number.

Write

$|a|$ for the absolute value of the number *a*

Thus

$|5| = 5$ (The absolute value of the *positive* number 5 is 5 itself.)

$|0| = 0$ (The absolute value of 0 is 0 itself.)

$|-5| = 5$ (The absolute value of the *negative* number −5 is its opposite, 5.)

The absolute value of a number measures its distance from 0. Absolute value ignores whether the number lies to the right or left of 0. Thus, both 5 and -5 lie 5 units from 0, so that

$$|5| = |-5| = 5$$

Example 2 ▶   Find: *i.* $|13|$ *ii.* $|-12|$ *iii.* $|-.8|$

Solution.

i. 13 is positive. Thus $|13| = 13$

ii. -12 is negative. Thus $|-12| = 12$. (The absolute value of -12 is its opposite, 12.)

iii. $-.8$ is negative. Thus $|-.8| = .8$ ◀

PRACTICE EXERCISES FOR TOPIC B

Find each absolute value.

6. $|10| =$ 7. $|-15| =$ 8. $\left|-\dfrac{1}{2}\right| =$

Now try the exercises for Topic B on page 237.

C. ADDITION WITH NEGATIVE NUMBERS

If a football team *loses* 4 yards on one play and then *loses* 2 more yards, altogether it has *lost* 6 yards. Because negative numbers convey the notion of *loss*, we would want

$$(-4) + (-2) = -6$$

To add two negative numbers:

1. Add their absolute values.
2. Prefix their sum with a minus sign.

For example,

$$(-5) + (-3) = -(5 + 3) = -8$$

Example 3 ▶   Add. $(-7) + (-5)$

Solution.

1. Add their absolute values, 7 and 5.

$$7 + 5 = 12$$

2. Prefix their sum with a minus sign. Thus

$$(-7) + (-5) = -\underbrace{(7 + 5)}_{12} = -12 \qquad \blacktriangleleft$$

Addition of numbers can be illustrated on the number line.

When adding a *positive* number to a, move to the *right* of a

When adding a *negative* number to a, move to the *left* of a

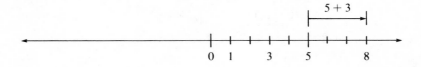

5 + 3 = 8   Move 3 units to the *right* of 5.

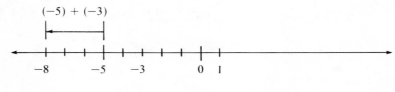

(-5) + (-3) = -8   Move 3 units to the *left* of -5.

These notions apply when adding numbers with different signs. If Steve *loses* $4 and then *gains* $4, his total gain is 0 (dollars). Thus

$$(-4) + 4 = 0$$

This can also be illustrated on the number line.

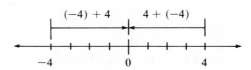

-4 + 4 = 0   Move 4 units to the *right* of -4.
4 + (-4) = 0   Move 4 units to the *left* of 4.

For any number a,

$$a + 0 = 0 + a = a$$
$$a + (-a) = (-a) + a = 0$$

Thus,

$$6 + 0 = 6$$
$$0 + (-8) = -8$$
$$5 + (-5) = 0$$
$$(-3) + 3 = 0$$

If the temperature is 4 degrees *below* 0°F, and it then *rises* 6°, the new temperature is 2° *above* 0°F.

$$(-4) + 6 = 2$$

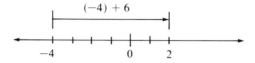

$(-4) + 6 = 2$ Move 6 units to the *right* of –4.

In practice, you add numbers with different signs as follows.

To add a positive number p and a negative number, other than –p:

1. Subtract the smaller absolute value from the larger absolute value.

2. Prefix the difference with the sign of the number with the *larger* absolute value.

Example 4 ▶   Add.   *i.* $8 + (-5)$ *ii.* $3 + (-9)$

Solution.

i. $\underset{8}{\underbrace{|8|}} > \underset{5}{\underbrace{|-5|}}$

Here the *positive* number, 8, has the larger absolute value. Prefix the difference, 8 – 5, with the sign of 8, which is +. (Usually, + is *not* written.)

$$8 + (-5) = 8 - 5 = 3$$

ii. $\underset{9}{\underbrace{|-9|}} > \underset{3}{\underbrace{|3|}}$

Here the negative number, –9, has the larger absolute value. Prefix the difference, 9 – 3, with the sign of –9, which is –.

$$3 + (-9) = -\underset{6}{\underbrace{(9 - 3)}} = -6$$ ◀

When adding several numbers (whether positive, negative, or 0), the following rules of addition are important. Let a, b, and c be any numbers. Then

$$a + b = b + a \qquad \text{(Commutative Law of Addition)}$$
$$(a + b) + c = a + (b + c) \qquad \text{(Associative Law of Addition)}$$

For example,

$$(-5) + (-4) = (-4) + (-5) = -9 \qquad \text{by the commutative law}$$
$$\underbrace{(5 + 2)}_{7} + 4 = 5 + \underbrace{(2 + 4)}_{6} = 11 \qquad \text{by the associative law}$$

The commutative and associative laws enable you to rearrange numbers when adding them. Thus

$$
\begin{aligned}
4 + (-8) + 3 &= 4 + [(-8) + 3] \\
&= 4 + [3 + (-8)] \qquad \text{by the commutative law} \\
&= [4 + 3] + (-8) \qquad \text{by the associative law} \\
&= 7 + (-8) \\
&= -(8 - 7) \\
&= -1
\end{aligned}
$$

To add three or more numbers, use the commutative and associative laws.

1. Rearrange, and add the positive numbers and the negative numbers separately.

2. Let p be the sum of the positive numbers, and let $-n$ be the sum of the negative numbers. Then the total sum is $p + (-n)$, or $p - n$.

Example 5 ▶   Add. $\begin{array}{r} 17 \\ 14 \\ -18 \\ 23 \\ -19 \end{array}$

Solution. Rearrange, and add the positive numbers and the negative numbers separately.

$$
\begin{array}{ccc}
17 & -18 & 54 \\
14 & -19 & -37 \\
\underline{23} & \underline{-37} & \overline{17} \\
\overline{54} & &
\end{array}
$$

The resulting sum, $54 + (-37)$, equals 17.   ◀

PRACTICE EXERCISES FOR TOPIC C

Add.

9. $(-5) + 2 =$ 10. $(-7) + (-3) =$ 11.
$$
\begin{array}{r}
12 \\
-\ 9 \\
13 \\
-15 \\
\underline{-20}
\end{array}
$$

Now try the exercises for Topic C on page 237.

D. SUBTRACTION WITH NEGATIVE NUMBERS

DEFINITION

> Let a and b be *any* two numbers (whether positive, negative, or 0). Define
> $$a - b = a + (-b)$$

Thus *subtracting b yields the same result as adding −b.* For example,

$$
\begin{aligned}
4 - 6 &= 4 + (-6) \\
&= -(6 - 4) \\
&= -2
\end{aligned}
$$

Furthermore, it follows from the above definition that

$$
\begin{aligned}
a - 0 &= a + (-0) = a + 0 = a \\
0 - a &= 0 + (-a) = -a
\end{aligned}
$$

Example 6 ▶   Subtract.   *i.* $5 - (-4)$ *ii.* $-5 - (-4)$

Solution. Recall that the opposite of -4 is 4.

$$-(-4) = 4$$

Thus to subtract −4, change −4 to its opposite, 4, and add.

i. 5 − (−4) = 5 + 4 = 9

ii. −5 − (−4) = −5 + 4

$$= -(5 - 4)$$

$$= -1$$ ◄

Observe that

5 − 3 = 2   if   3 + 2 = 5

Thus subtraction *undoes* addition.

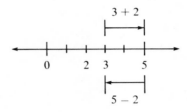

PRACTICES EXERCISES FOR TOPIC D

Subtract.

12. 4 − 12 =

13. (−4) − (−12) =

14. −9 − 6 =

15. 0 − (−10) =

Now try the exercises for Topic D on page 238.

E. MULTIPLICATION WITH NEGATIVE NUMBERS

When the factors of a product are denoted by letters, multiplication is often indicated by a *dot* rather than by a *cross.* Thus

$a \cdot b$ means $a \times b$

The commutative and associative laws also hold for multiplication, and enable you to rearrange the factors. Let a, b, and c be any numbers. Then

$a \cdot b = b \cdot a$ **(Commutative Law of Multiplication)**

$(a \cdot b) \cdot c = a \cdot (b \cdot c)$ **(Associative Law of Multiplication)**

Furthermore,

$a \cdot 1 = 1 \cdot a = a$

For example,

$$4 \times 5 = 5 \times 4 = 20 \qquad \text{by the commutative law}$$

$$\underbrace{(4 \times 5)}_{20} \times 2 = 4 \times \underbrace{(5 \times 2)}_{10} = 40 \qquad \text{by the associative law}$$

and, of course,

$$4 \times 1 = 1 \times 4 = 4$$

Multiplication by a positive integer amounts to repeated addition. If you *receive* $2 from each of 3 friends, you receive

$$2 \times 3 = 2 + 2 + 2 = 6 \text{ (dollars)}$$

However, if you *owe* $2 to each of 3 friends, your debt can be expressed by

$$(-2) \times 3 = (-2) + (-2) + (-2) = -6 \text{ (dollars)}$$

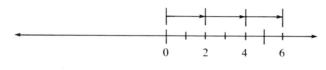

$$2 \times 3 = 6$$

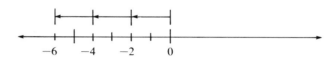

$$(-2) \times 3 = -6$$

Here are the rules for determining *the sign of a product:*

> If a and b are both positive, then
> $$a \cdot (-b) = (-a) \cdot b = -(a \cdot b)$$
> $$(-a) \cdot (-b) = a \cdot b$$
> $$a \cdot 0 = (-a) \cdot 0 = 0$$

(These rules actually apply to *any* numbers a and b, whether positive, negative, or 0.) Thus

$$2 \times (-3) = (-2) \times 3 = -(2 \times 3) = -6$$

$$(-2) \times (-3) = 6$$

$$2 \times 0 = (-2) \times 0 = 0$$

In other words, for *nonzero* factors:

---

1. when 1 factor is negative, the product is *negative*;
2. when 2 factors are negative, the product is *positive*.

More generally,

3. the product of an *odd* number (1, 3, 5, and so on) of negative factors is *negative*;
4. the product of an *even* number (2, 4, 6, and so on) of negative factors is *positive*.

Example 7 ▶   i. $\underbrace{(-2) \times (-2)}_{4} \times (-1) = -4$

Here there are 3 (an *odd* number of) negative factors. The product is *negative*.

   ii. $\underbrace{(-2) \times (-2)}_{4} \times \underbrace{(-1) \times (-1)}_{1} = 4$

Here there are 4 (an *even* number of) negative factors. The product is *positive*.   ◀

Often, the multiplication symbol is omitted before parentheses. For example, write

$$4(-3) \quad \text{instead of} \quad 4 \times (-3)$$

Here the parentheses indicate *multiplication* and *not* subtraction. Thus

$$4(-3) = -(4 \times 3) = -12$$

whereas, *without* parentheses about -3, subtraction is indicated.

$$4 - 3 = 1$$

Example 8 ▶   Multiply.   i. 7(-4)   ii. (-5)(-8)

Solution.

   i. $7(-4) = -(7 \times 4) = -28$

   There is one negative factor. The product is negative.

   ii. $(-5)(-8) = 5 \times 8 = 40$

   There are two negative factors. The product is positive.   ◀

PRACTICE EXERCISES FOR TOPIC E

Multiply.

16. $7(-2) =$ 17. $(-4)(-3) =$ 18. $(-12) \times 0 =$

19. $(-2)(-4)(-1) =$

Now try the exercises for Topic E on page 238.

F. DIVISION WITH NEGATIVE NUMBERS

Instead of writing $a \div b$, we often write $\frac{a}{b}$.

If a and b are both positive, then

$$\frac{-a}{b} = \frac{a}{-b} = -\frac{a}{b} \qquad (negative)$$

$$\frac{-a}{-b} = \frac{a}{b} \qquad (positive)$$

Thus, if the numerator and denominator have the *same sign*, the quotient is *positive*. If the numerator and denominator have *different signs*, the quotient is *negative*. Furthermore,

$$\frac{a}{a} = 1$$

$$\frac{a}{1} = a$$

$$\frac{a}{-1} = -a$$

$$\frac{0}{b} = \frac{0}{-b} = 0$$

and finally,

$$\frac{a}{0} \text{ and } \frac{-a}{0} \text{ are } undefined.$$

Observe that

$$\frac{6}{2} = 3 \quad \text{if} \quad 3 \times 2 = 6$$

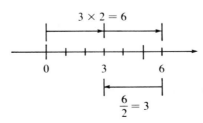

Thus division *undoes* multiplication, just as subtraction undoes addition. Note that 0 *divided by* any *nonzero* number yields 0. For example,

$$\frac{0}{3} = 0 \quad \text{because} \quad 0 \times 3 = 0$$

But division *by* 0 is undefined (*not* defined). For example,

$$\frac{3}{0} = a \quad \text{would mean} \quad a \times 0 \stackrel{?}{=} 3. \quad \text{But } a \times 0 = 0 \text{ (and } not \text{ 3)}$$

Example 9 ▶ Divide.   *i.* $\dfrac{-8}{4}$ *ii.* $\dfrac{12}{-3}$ *iii.* $\dfrac{-10}{-2}$

Solution.

i.   Numerator and denominator have different signs. The quotient is negative.

$$\frac{-8}{4} = -\frac{8}{4} = -2$$

ii.   Numerator and denominator have different signs. The quotient is negative.

$$\frac{12}{-3} = -\frac{12}{3} = -4$$

iii.   Numerator and denominator have the same sign. The quotient is positive.

$$\frac{-10}{-2} = \frac{10}{2} = 5$$

◀

Example 10 ▶ Divide, or indicate that division is undefined.

   i. $\dfrac{0}{4}$ *ii.* $\dfrac{0}{-4}$ *iii.* $\dfrac{-4}{0}$ *iv.* $\dfrac{0}{0}$

Solution. In *i.* and *ii.*, use the fact that 0 divided by any *nonzero* number yields 0.

   i. $\dfrac{0}{4} = 0$ *ii.* $\dfrac{0}{-4} = 0$

In *iii.* and *iv.*, use the fact that division by 0 is undefined.

iii. $\dfrac{-4}{0}$ is *undefined.*   *iv.* $\dfrac{0}{0}$ is *undefined.*

Note that in *iv.*, even though the numerator is 0, division by 0 is *not* defined.

◀

PRACTICE EXERCISES FOR TOPIC F

Divide, or indicate that division is undefined.

20. $\dfrac{10}{-5} =$ 21. $\dfrac{-18}{-6} =$ 22. $\dfrac{0}{-3} =$ 23. $\dfrac{-3}{0} =$

Now try the exercises for Topic F on page 238.

G. APPLICATIONS

Example 11 ▶ The temperature in Vancouver rises from $-9°$ Fahrenheit to $-2°$ Fahrenheit. How many degrees does it rise?

Solution. Subtract -9 from -2.

$$-2 - (-9) = -2 + 9 = 7$$

The temperature rises $7°$ Fahrenheit.   ◀

PRACTICE EXERCISES FOR TOPIC G

24. The temperature in Chicago is $-6°$F. If the temperature drops 4 degrees, what is the new reading?

Now try the exercises for Topic G on page 238.

EXERCISES

A. *Consider the number line pictured below. Label the points that correspond to the following numbers.*

1. 3   2. 6   3. -2 4. -4 5. -6 6. $\dfrac{1}{2}$ 7. $\dfrac{5}{2}$ 8. $-1\dfrac{1}{2}$

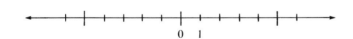

B. *Find each absolute value.*

9. $|5| =$ 10. $|-5| =$ 11. $|-15| =$ 12. $|0| =$

13. $|.73| =$ 14. $|-.73| =$ 15. $\left|\dfrac{1}{3}\right| =$ 16. $\left|-\dfrac{1}{3}\right| =$

C. *Add.*

17. $(-3) + (-2) =$ 18. $(-8) + (-6) =$ 19. $8 + (-6) =$

20. $(-8) + 6 =$ 21. $(-3) + 7 =$ 22. $9 + (-5) =$

23. $(-10) + (-10) =$ 24. $-10 + 10 =$ 25. $4 + (-8) =$

26. $8 + (-8) =$ **27.** $9 + (-12) =$ **28.** $(-15) + (-7) =$

29. $\begin{array}{r} 12 \\ -\ 9 \\ 15 \\ \underline{-17} \end{array}$ **30.** $\begin{array}{r} -34 \\ -18 \\ 19 \\ 26 \\ \underline{-\ 7} \end{array}$ **31.** $\begin{array}{r} 52 \\ -28 \\ -37 \\ 41 \\ \underline{-16} \end{array}$ **32.** $\begin{array}{r} 28 \\ -53 \\ 41 \\ -14 \\ \underline{12} \end{array}$

D. *Subtract.*

33. $4 - 5 =$ **34.** $8 - 10 =$ **35.** $6 - (-1) =$

36. $-6 - (-1) =$ **37.** $(-6) - 1 =$ **38.** $7 - (-2) =$

39. $-10 - (-3) =$ **40.** $9 - (-6) =$ **41.** $12 - (-11) =$

42. $(-12) - 11 =$ **43.** $-9 - (-9) =$ **44.** $0 - (-4) =$

E. *Multiply.*

45. $6(-2) =$ **46.** $(-10)5 =$ **47.** $(-3)(-4) =$

48. $8(-5) =$ **49.** $(-7)(-5) =$ **50.** $7(-5) =$

51. $(-6)(-6) =$ **52.** $12(-3) =$ **53.** $(-4) \times 0 =$

54. $0 \times 0 =$ **55.** $(-4)(-3)(-2) =$ **56.** $(-5)(-2)(-1)(-3) =$

F. *Divide, or indicate that division is undefined.*

57. $\dfrac{-6}{3} =$ **58.** $\dfrac{6}{-3} =$ **59.** $\dfrac{-6}{-3} =$ **60.** $\dfrac{15}{-5} =$

61. $\dfrac{-20}{4} =$ **62.** $\dfrac{27}{-3} =$ **63.** $\dfrac{32}{-8} =$ **64.** $\dfrac{-63}{7} =$

65. $\dfrac{6}{0} =$ **66.** $\dfrac{0}{6} =$ **67.** $\dfrac{-6}{0} =$ **68.** $\dfrac{0}{-6} =$

69. $\dfrac{0}{0} =$ **70.** $\dfrac{-12}{-12} =$

G. **71.** The temperature at 8 A.M. in Burlington is $-12°$ Fahrenheit. If the temperature drops 4 degrees the next hour, what is the reading at 9 A.M.?   *Answer:*

72. The temperature in Buffalo at 6 P.M. is $2°$ Celsius. By midnight the temperature has dropped 15 degrees. What is the reading at midnight?   *Answer:*

73. A bus travels 12 miles east and then 15 miles west. How far from its starting point is the bus? Is it east or west of the starting point?   *Answer:*

74. The temperature change over a 4-hour period is -20 degrees Celsius. Assuming a constant rate of change, how many degrees Celsius does it change each hour?   *Answer:*

75. Subtract 5 from -9.   *Answer:*   **76.** Subtract -5 from -9.   *Answer:*

25. Several Operations

Frequently, you will have to work with several different operations—addition, subtraction, multiplication, division, and raising to a power—in the same example. How can you tell which operation is to be done first in a given example?

A. SQUARES OF NEGATIVE NUMBERS

Recall that

$$a^2 = a \cdot a$$

Thus

$$6^2 = 6 \times 6 = 36$$

and

$$0^2 = 0 \times 0 = 0$$

If p is positive, then $-p$ is negative, and

$$(-p)^2 = (-p)(-p) = p \times p = p^2$$

The square of a negative number is positive because the product of two negative factors is positive. For example,

$$(-4)^2 = (-4)(-4) = 4 \times 4 = 16$$

Note that $(-4)^2$ means $(-4) \times (-4)$. Sometimes you will want to indicate

$$\underbrace{\text{the opposite of } \quad 4^2}_{- \qquad 4^2}$$

Because $4^2 = 16$, it follows that

$$-4^2 = -16$$

Recall that

$$(-4)^2 = (-4)(-4) = 16$$

Thus

$$\underbrace{-4^2}_{-16} \neq \underbrace{(-4)^2}_{16}$$

Example 1 ▶ Find: *i.* $(-7)^2$ *ii.* $(-8)^2$ *iii.* -8^2

Solution.

i. $(-7)^2 = (-7)(-7) = 7 \times 7 = 49$

ii. $(-8)^2 = (-8)(-8) = 8 \times 8 = 64$

iii. $-8^2 = -(8 \times 8) = -64$ ◀

PRACTICE EXERCISES FOR TOPIC A

1. $(-5)^2 =$ 2. $-5^2 =$ 3. $(-11)^2 =$

Now try the exercises for Topic A on page 244.

B. POWERS OF NEGATIVE NUMBERS

Observe that

$$(-2)^3 = \underbrace{(-2)(-2)(-2)}_{\underbrace{4 \ \times \ (-2)}_{-8}} = -8 \qquad (3 \text{ negative factors: } negative)$$

$$(-2)^4 = \underbrace{(-2)(-2)}_{4} \underbrace{(-2)(-2)}_{4} = 16 \qquad (4 \text{ negative factors: } positive)$$

$$(-2)^5 = \underbrace{\underbrace{(-2)(-2)(-2)}_{(-8)} \underbrace{(-2)(-2)}_{4}}_{-32} = -32 \qquad (5 \text{ negative factors: } negative)$$

In general, a negative number raised to an *odd* power is negative; a negative number raised to an *even* power is positive. Thus, if $-p$ is a negative number, then

$(-p)^n$ is negative if n is odd

$(-p)^n$ is positive if n is even

Example 2 ▶ Find: *i.* $(-10)^4$ *ii.* $(-10)^5$ *iii.* -10^6 *iv.* $-(-10)^7$

Solution.

i. 4 is *even*; thus $(-10)^4$ is *positive.*

$$(-10)^4 = \underbrace{10,000}_{4 \text{ zeros}}$$

ii. 5 is *odd*; thus $(-10)^5$ is *negative.*

$$(-10)^5 = -\underbrace{100,000}_{5 \text{ zeros}}$$

iii. -10^6 indicates the *opposite* of 10^6. Thus

$$-10^6 = -\underbrace{1,000,000}_{6 \text{ zeros}}$$

iv. $-(-10)^7$ indicates the opposite of $(-10)^7$. Because an *odd* power of a negative number is *negative*, $(-10)^7$ is negative. Thus $-(-10)^7$, its *opposite*, is *positive.*

$$-(-10)^7 = -(-\underbrace{10,000,000}_{7 \text{ zeros}}) = 10,000,000 \qquad ◀$$

PRACTICE EXERCISES FOR TOPIC B

4. $(-4)^3 =$ 5. $(-4)^4 =$ 6. $-2^6 =$ 7. $-(-2)^7 =$

Now try the exercises for Topic B on page 244.

C. ORDER OF OPERATIONS

By convention, the expression

$$4 \times 3^2 \qquad \text{means} \qquad 4 \text{ times the square of } 3$$

Thus

$$4 \times 3^2 = 4 \times 9 = 36$$

Here the exponent 2 refers only to the base 3. If you want to indicate the square of the product of 4 and 3, you must use parentheses. Thus

$$(4 \times 3)^2 = 12^2 = 144$$

Here the exponent applies to both factors 4 and 3.

The expression

5 + 4 × 2   means   add the product of 4 and 2 to 5

Thus

$$5 + 4 \times 2 = 5 + 8 = 13$$

If the sum of 5 and 4 is to be multiplied by 2, use parentheses. Thus in the expression

$$(5 + 4)2$$

first add within the parentheses:

$$\underbrace{(5 + 4)}_{9}2 = 9 \times 2 = 18$$

In general, the order in which addition, subtraction, multiplication, division, and raising to a power are applied in an example is often crucial. If parentheses are given, *first perform the operations within parentheses.* Otherwise:

1. First raise to a power.
2. Then multiply or divide from left to right.
3. Then add or subtract from left to right.

To remember the order of operations think of:

Rest please, my dear Aunt Sally

Raise to a power; multiply or divide; add or subtract

Example 3 ▶   4(6 − 3) =

(A) 24   (B) 21   (C) 12   (D) −12   (E) −21

Solution. Here parentheses are given. First perform the operation within parentheses.

$$4\underbrace{(6 - 3)}_{3} = 4(3) = 12$$

The correct choice is (C).   ◀

Example 4 ▶ $3 + 2 \times 5^2 =$

(A) 53   (B) 125   (C) 103   (D) 169   (E) 13

Solution. First raise to a power (*square*); then multiply; then add.

$$3 + 2 \times \underbrace{5^2}_{25} = 3 + \underbrace{2 \times 25}_{50}$$

$$= 3 + 50$$
$$= 53$$

The correct choice is (A).   ◀

Example 5 ▶ Find: $(-3)^2 + 4(-5)$

Solution. First raise to a power; then multiply. (In this particular example, these operations can be done in the same step.) Then add.

$$\underbrace{(-3)^2}_{9} + \underbrace{4(-5)}_{-20} = 9 + (-20) = -11$$ ◀

Example 6 ▶ Find: $2(-4)^2 + 3(-7)$

Solution. First raise to a power; then multiply; then add.

$$2\underbrace{(-4)^2}_{16} + 3(-7) = \underbrace{2(16)}_{32} + \underbrace{3(-7)}_{-21}$$

$$= 32 + (-21)$$
$$= 11$$ ◀

Example 7 ▶ Find: $\dfrac{8 - 2}{3} - 2^2$

Solution. Here the fraction line indicates that the entire expression $8 - 2$ is to be divided by 3. From this, subtract 2^2. Thus

$$\dfrac{\overbrace{(8 - 2)}^{6}}{3} - \overbrace{2^2}^{4} = \dfrac{6}{3} - 4$$

$$= 2 - 4$$
$$= -2$$ ◀

PRACTICE EXERCISES FOR TOPIC C

In Exercises 8 and 9, draw a circle around the correct letter.

8. $4(1 - 3) =$

(A) 1   (B) -1   (C) 8   (D) -8   (E) -2

9. $10(6 - 3) - 7(2 + 1) =$

    (A) 9   (B) 46   (C) 44   (D) 17   (E) 13

10. $5^2 - 2^3 =$ 11. $8 - 4^2 =$ 12. $(9 - 7)^2 =$

13. $\dfrac{9 - 1}{4} - 2^2 =$

Now try the exercises for Topic C on this page.

D. APPLICATIONS

Example 8▶   A woman buys six 2-dollar glasses and four 5-dollar frying pans. She pays for her purchases with a $50 bill. How much change should she receive?

    Solution.

$$50 - \big(6(2) + 4(5)\big) = 50 - (12 + 20)$$
$$= 50 - 32$$
$$= 18$$

Her purchases amount to $32, so that she receives $18 change.   ◀

PRACTICE EXERCISES FOR TOPIC D

14. A 250-pound man wishes to reduce his weight to 210 pounds in 8 weeks. How many pounds per week must he lose?

Now try the exercises for Topic D on page 245.

EXERCISES

A. *Evaluate.*

    1. $(-2)^2 =$ **2.** $(-3)^2 =$ **3.** $(-6)^2 =$ **4.** $(-9)^2 =$ **5.** $(-10)^2 =$

    6. $(-1)^2 =$ **7.** $-11^2 =$ **8.** $(-12)^2 =$ **9.** $-0^2 =$ **10.** $-(-3)^2 =$

B. *Evaluate.*

    11. $(-3)^3 =$ **12.** $(-3)^4 =$ **13.** $-3^4 =$ **14.** $-(-3)^3 =$ **15.** $(-2)^5 =$

    16. $(-2)^6 =$ **17.** $(-2)^7 =$ **18.** $(-2)^8 =$ **19.** $-10^7 =$ **20.** $-(-10^7) =$

C. *When several choices are given, draw a circle around the correct letter.*

    21. $5(3 + 1) =$

        (A) 16   (B) 20   (C) 9   (D) 54   (E) 25

22. $7(5 - 2) =$

(A) 35   (B) 33   (C) 73   (D) 21   (E) -21

23. $10(4 - 6) =$

(A) 34   (B) 46   (C) 20   (D) -20   (E) 8

24. $-4(9 - 4) =$

(A) -20   (B) -9   (C) 1   (D) -40   (E) -1

25. $-7(3 - 10) =$ **26.** $2 + 5(1 + 8) =$ **27.** $1 - 4(7 - 2) =$

28. $5 - 2(-10) =$

29. $2 + 4(3 + 7) =$

(A) 21   (B) 38   (C) 42   (D) 80   (E) 60

30. $5(3 + 2) + 2(3 - 1) =$

(A) 22   (B) 21   (C) 30   (D) 29   (E) 38

31. $2 \cdot 5^2 =$

(A) 50   (B) 100   (C) 49   (D) 20   (E) 64

32. $5(-3)^2 =$

(A) 4   (B) -4   (C) 45   (D) -45   (E) -30

33. $1 - 2(-10)^2 =$

(A) -201   (B) -199   (C) 201   (D) 401   (E) -399

34. $(-5)^2 + 2(-6) =$

(A) 13   (B) -37   (C) 21   (D) -29   (E) -13

35. $4^2 - 3(-2) =$ **36.** $10^2 - (-9)^2 =$ **37.** $2(-4)^2 + 3(-6) =$

38. $5^2 - 2(-1)^2 =$ **39.** $(4^2 - 1)5 =$ **40.** $(2 + 5)^2 - 1 =$

41. $2 + 5^2 - 1 =$ **42.** $2^2 + 5^2 - 1 =$ **43.** $(-4)5 - (-4)^2 =$

44. $(-3)(-2) + (-5)^2 =$ **45.** $(-2)(-5)^2 - 5(-2)^2 =$ **46.** $7(2 - 4)^2 - 3 =$

47. $(-4)(3 - 5)^2 - 2(1 - 2) =$ **48.** $(5 - 8)^2 - (3 + 2)^2 =$ **49.** $\dfrac{10 - 2}{2} + 6 =$

50. $\left(\dfrac{4 - 9}{5}\right)^2 - 1 =$ **51.** $\dfrac{15 - 5}{5} - \dfrac{4 - 12}{-2} =$ **52.** $\dfrac{(16 - 12)^2}{8} + (3 - 1)^3 =$

53. $\dfrac{(2 - 4)^3}{2} \times \left(\dfrac{9 - 3}{2}\right)^2 =$ **54.** $\dfrac{10 - 3^2}{2} + \dfrac{(8 - 10)^2}{8} =$

D. 55. A bookstore sells 10 books at $6 each and 5 books at $8 each. How much money does it receive?   *Answer:*

56. A theater sells 42 orchestra seats at $10, 58 mezzanine seats at $8, and 102 balcony seats at $6. How much money does it receive?   *Answer:*

57. Ann reads 40 pages per hour. If she has read 135 pages of a 375 page book, how long will it take her to finish it?   *Answer:*

58. A 300-pound man wishes to reduce to 220 pounds. If he is to lose 4 pounds per week, for how many weeks must he diet?   *Answer:*

59. A woman buys five pounds of nuts at $3 per pound, six pounds of cookies at $5 per pound, and three quarts of ice cream at $2 per quart. If she pays for these items with a $100 bill, how much change should she receive?   *Answer:*

60. A business makes a profit of $100,000. Of this, $78,000 is reinvested. If the rest of the money is split equally among four partners, how much is each partner's share? *Answer:*

61. A football team must gain 10 yards to obtain a first down. In three plays it gains 7 yards, loses 12 yards, and gains 11 yards. How many yards does it need for a first down?   *Answer:*

62. A typist charges $3 per page. If she types 5 pages per hour, how long will it take her to earn $120?   *Answer:*

26. Averaging

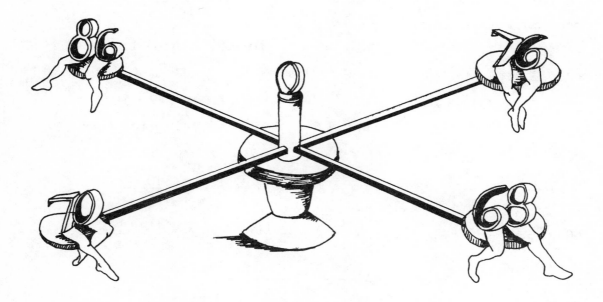

The *average* of several numbers is a notion that has wide application in business, the social sciences, and everyday life. For example, we commonly speak of average income, average height, and averages on exams.

A. CALCULATING AVERAGES

When we want to consider *several* numbers at a time—perhaps three numbers, or four numbers, or possibly ten numbers, we often use the letter n to indicate how many numbers are involved.

DEFINITION

> The **average** of n numbers is the sum of these numbers divided by n, that is,
>
> $$\frac{\text{the sum}}{n}$$

Example 1 ▶ Find the average of 66, 78, and 96.

Solution. Here $n = 3$. First add these 3 numbers.

$$
\begin{array}{r}
66 \\
78 \\
\underline{96} \\
240
\end{array}
$$

Now divide by 3.

$$\frac{240}{3} = 80$$

that is,

$$\frac{66 + 78 + 96}{3} = \frac{240}{3} = 80$$

The average of the 3 numbers is 80.   ◀

Example 2 ▶ Find the average of 40, 50, 52, and 56.

Solution. Here $n = 4$. First add these four numbers.

$$
\begin{array}{r}
40 \\
50 \\
52 \\
\underline{56} \\
198
\end{array}
$$

Now divide by 4.

$$\frac{198}{4} = \frac{198.0}{4} = 49.5$$

In other words,

$$\frac{40 + 50 + 52 + 56}{4} = \frac{198}{4} = 49.5$$

Thus the average of the 4 numbers is 49.5.   ◀

PRACTICE EXERCISES FOR TOPIC A

Find the average of the given numbers.

1. 10, 12, and 17.   2. 24, 26, 30, 40.

3. 46, 50, 62, 68.

Now try the exercises for Topic A on page 249.

B. APPLICATIONS

Example 3 ▶ A student had scores of 86, 68, 70, and 76 on her hour exams. Find the average of these grades.

Solution. There were 4 exams. Thus $n = 4$. Add the 4 grades.

$$
\begin{array}{r}
86 \\
68 \\
70 \\
\underline{76} \\
300
\end{array}
$$

Now divide by 4.

$$\frac{300}{4} = 75$$

The average of these four grades is 75.   ◀

Example 4 ▶ Suppose that in Example 3, the student receives a 0 on her fifth hour exam. Find the average of her grades of 86, 68, 70, 76, and 0.

Solution. There were 5 exams. (The 0 counts as one of the exam scores.) Thus $n = 5$. The sum is the same as before, namely, 300. But now divide by 5, instead of by 4.

$$\frac{300}{5} = 60$$

The average of these five grades is 60.   ◀

PRACTICE EXERCISES FOR TOPIC B

4. A college baseball team wins 12 games, 16 games, 14 games, and 20 games in four successive years. What is the average number of games the team has won?

Now try the exercises for Topic B on page 249.

C. MULTIPLE CHOICE

Example 5 ▶ The average of 43, 49, 57, and 71 is

(A) 53   (B) 54   (C) 55   (D) 44   (E) 220

Solution. Here $n = 4$.

$$\begin{array}{r} 43 \\ 49 \\ 57 \\ \underline{71} \\ 220 \end{array} \qquad \frac{220}{4} = 55$$

The average of these numbers is 55. Thus (C) is the correct choice.   ◀

Some averaging problems use algebraic methods. These problems will be discussed in Section 32.

PRACTICE EXERCISES FOR TOPIC C

Draw a circle around the correct letter.

5. The average of 15, 19, and 26 is

(A) 18   (B) 19   (C) 20   (D) 20.3   (E) 22

6. The average of 6, 7, 8, and 9 is

(A) 7   (B) 7.5   (C) 8   (D) 8.25   (E) 8.5

Now try the exercises for Topic C on page 250.

EXERCISES

A. *Find the average of the given numbers.*

1. 14, 16, 21 *Answer:*	**2.** 71, 79, 81 *Answer:*
3. 37, 62, 84 *Answer:*	**4.** 92, 105, 112 *Answer:*
5. 29, 36, 41, 54 *Answer:*	**6.** 37, 41, 44, 46 *Answer:*
7. 60, 68, 73, 87 *Answer:*	**8.** 20, 34, 48, 62 *Answer:*
9. 77, 93, 98, 100 *Answer:*	**10.** 59, 83, 89, 93 *Answer:*
11. 66, 72, 0, 80 *Answer:*	**12.** 53, 59, 69, 79 *Answer:*
13. 58, 82, 83, 91 *Answer:*	**14.** 94, 97, 99, 101 *Answer:*
15. 37, 42, 44, 51, 61 *Answer:*	**16.** 28, 40, 51, 52, 59 *Answer:*

B. **17.** Melissa had scores of 82, 84, and 92 on her Spanish exams. Find the average of these grades.   *Answer:*

18. Pablo had grades of 76, 81, 57, and 93 on his math quizzes. Find the average of these grades.   *Answer:*

19. In their first four games the Knicks scored 92 points, 94 points, 102 points, and 96 points. Find the average number of points they scored.   *Answer:*

20. The heights of four first-graders are 46 inches, 49 inches, 50 inches, and 51 inches. Find the average height of these children.   *Answer:*

21. A saleswoman makes 12 sales on Monday, 13 sales on Tuesday, 8 sales on Wednesday, and 15 sales on Thursday. Find the average number of sales she makes per day for these four days.   *Answer:*

22. On four successive nights a salesman spends $38, $44, $39, and $47 per night for his motel room. What is the average cost per night for his room?   *Answer:*

23. The average monthly rainfall for a town is as follows: January: 17 centimeters; February: 19 centimeters; March: 25 centimeters; April: 27 centimeters. Find the average monthly rainfall for the town during this period.   *Answer:*

24. In four practice tests, Ann answered 46 questions correctly, then 54 correctly, then 60 correctly, then 64 correctly. What was the average number of questions she answered correctly?   *Answer:*

C. *Draw a circle around the correct letter.*

25. The average of 17, 23, and 26 is

    (A) 22   (B) 23   (C) 24   (D) 20.6   (E) 68

26. The average of 59, 93, and 97 is

    (A) 249   (B) 24.9   (C) 90   (D) 89   (E) 83

27. The average of 20, 50, and 59 is

    (A) 49   (B) 129   (C) 43   (D) 34   (E) 34.25

28. The average of 59, 61, and 72 is

    (A) 60   (B) 65   (C) 61   (D) 64   (E) 192

29. The average of 40, 42, 45, and 49 is

    (A) 42   (B) 43   (C) 44   (D) 45   (E) 176

30. The average of 39, 42, 48, and 51 is

    (A) 42   (B) 45   (C) 48   (D) 50   (E) 180

31. The average of 57, 72, 79, and 80 is

    (A) 288   (B) 28.8   (C) 70   (D) 72   (E) 96

32. The average of 45, 65, 70, and 90 is

    (A) 70   (B) 80   (C) 67.5   (D) 78.5   (E) 77.5

33. A student has scores of 80, 82, 60, and 94 on her biology exams. What was her average score?

    (A) 80   (B) 79   (C) 81   (D) 82   (E) 4

34. A student scored 66, 70, 0, and 84 on his math tests. What was his average score?

(A) 70   (B) 220   (C) 73.3   (D) 55   (E) 65

35. On different trips the time it takes a train to travel between two stations is 48 minutes, 46 minutes, 47 minutes, and 54 minutes. What is the average number of minutes per trip?

(A) 38.75   (B) 45   (C) 48.75   (D) 47.7   (E) 50

36. The cost of producing four items is $52, $55, $56, and $61. The average cost per item is

(A) $55   (B) $55.50   (C) $56   (D) $60   (E) $224

37. In a small factory, twenty of the workers each earn $15,000 a year. The remaining five workers each earn $20,000 per year. The average annual wage of the workers in this factory is

(A) $15,000   (B) $16,000   (C) $17,000   (D) $17,500   (E) $18,000

Review Exercises for Unit VI

In Exercises 1–4, consider the number line pictured below. Label the points that correspond to the following numbers:

1. 8   **2.** −6   **3.** $\frac{1}{2}$ **4.** $-\frac{3}{2}$

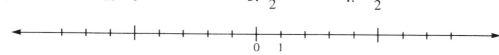

In Exercises 5–7, find each absolute value.

5. $|-10| =$ **6.** $\left|\frac{1}{4}\right| =$ **7.** $|0| =$

In Exercises 8–20, perform the indicated operations. Indicate when division is undefined.

8. $(-3) + (-7) =$ **9.** $4 - (-2) =$ **10.** $5(-4) =$

11. $(-2) + (-6) =$ **12.** $(-9) - (-10) =$ **13.** $\frac{-6}{-3} =$

14. $\frac{12}{-6} =$ **15.** $\frac{0}{-5} =$ **16.** $(-9)(-3) =$

17. $\frac{-8}{0} =$ **18.** $\frac{0}{0} =$ **19.** $\begin{array}{r} 12 \\ -10 \\ -9 \\ -6 \\ \hline \end{array}$ **20.** $\begin{array}{r} -3 \\ 20 \\ -9 \\ 10 \\ \hline \end{array}$

21. When the temperature drops 5 degrees from $3°$ below zero (Fahrenheit), the temperature reading is then _____

22. Subtract -8 from -9.   *Answer:*

23. $(-6)^2 =$ 24. $(-5)^3 =$ 25. $-(-1)^{10} =$ 26. $-2^6 =$

In Exercises 27–30, draw a circle around the correct letter.

27. $4(2 - 5) =$
    (A) 3   (B) -3 (C) 12   (D) -12 (E) 5

28. $9^2 - (4 - 2)^2 =$
    (A) 77   (B) 79   (C) 83   (D) 85   (E) 73

29. $\dfrac{18 - 12}{-3} - \left(\dfrac{4}{2}\right)^2 =$
    (A) -2 (B) -4 (C) -6 (D) 4   (E) 6

30. $4 - 2^2 + (-3)^3 =$
    (A) 27   (B) -27 (C) -23 (D) -31 (E) 0

31. $\left(\dfrac{-4 - 2}{6}\right)^2 =$ 32. $(12 - 3)^2 - 5 - (1 - 2)^2 =$ 33. $\left(\dfrac{9 - 3}{2}\right)^3 + 10^2 =$

34. $\dfrac{12}{2^2} - \left(\dfrac{4}{2}\right)^3 =$ 35. $(-2)(4 - 8) - (-3)(-2) =$ 36. $10 - 4(-2) + 6(-3)(-1) =$

37. $-2(5 - 3) - (-2)(-1) =$ 38. $\dfrac{4^2 - 1^2}{-3} \times (4 - 5)^3 =$

39. A woman pays for six three-dollar items at a supermarket with a twenty-dollar bill. How much change should she receive?   *Answer:*

40. A theater sells 200 tickets to a play at $6 per ticket. If costs for the play total $3000, how many more tickets must be sold for the company to break even?   *Answer:*

In Exercises 41–44, find the average of the given numbers.

41. 36, 42, 45   *Answer:*   42. 12, 20, 28, 34,   *Answer:*

43. 68, 69, 74, 75   *Answer:*   44. 0, 38, 40, 42, 60   *Answer:*

45. A student receives grades of 58, 60, 68, 72, and 92. Find the average of these grades.   *Answer:*

Review Exercises on Units I–V

1. Express 264 as the product of primes.   *Answer:*

2. $5^3 \times 2^4 =$

3. $\frac{1}{5} \times \frac{3}{7} =$

4. $\frac{3}{20} - \frac{1}{50} =$

5. $6\frac{1}{2} - 2\frac{5}{8} =$

6. $.027 \times 1.1 =$

7. $\frac{1.08}{8} =$

8. Express 1.12 as a fraction in lowest terms.   *Answer:*

9. Express $\frac{7}{50}$ as a percent.   *Answer:*

10. Express 10.5% as a decimal.   *Answer:*

11. A city of 520,000 has a 4% increase in population. What is its new population? *Answer:*

12. In a two-hour television program there are eight 3-minute commercials. How much programming time is there?   *Answer:*

13. A five-foot clothes line is divided into ten equal segments. How long is each segment? *Answer:*

14.   3 pounds 12 ounces
    + 2 pounds   8 ounces
    _____

15. An 8-foot by 3-foot mirror costs $.75 per square foot. What is the total cost? *Answer:*

16. The base of a triangle is 10 inches and the area is 90 square inches. What is the height? *Answer:*

17. Find the area of a circle whose circumference is 10π inches. Express your answer to the nearest tenth of a square inch.   *Answer:*

18. Find the surface area of a cube whose side length is 4 centimeters.   *Answer:*

Practice Exam on Unit VI

1. $(-3) + (-8) =$

2. $(-5) \times (-4) =$

3. $\frac{-18}{3} =$

4. $\frac{0}{-6} =$

5.   14
      - 9
      - 12
      - 21
        15
      _____

6. The temperature at 8 A.M. in Minneapolis is $2°F$. The next hour the temperature drops 5 degrees. What is the reading in Minneapolis at 9 A.M.?   *Answer:*

7. $2^2 - (-3)^3 =$

8. $\dfrac{8 - 2}{(-2)(-3)} =$

9. $\left(\dfrac{14 - 10}{2}\right)^2 + \dfrac{8 - 4}{-2} =$

10. $-2(5 - 3 + 1) - 3\left(4 + (-2)\right) =$

11. Find the average of 19, 29, 39, 42, and 49.   *Answer:*

12. Jim scores 18 points, 15 points, 18 points, and 29 points in his first four games as the center for his college basketball team. What is the average number of points he scores in these four games?

   (A) 18   (B) 19   (C) 20   (D) 21   (E) 22

13. A woman buys two $3-items and three $5-items at a pharmacy. If she pays for these items with two $20-bills, how much change should she receive?   *Answer:*

UNIT VII INTRODUCTION TO ALGEBRA

27. Algebraic Expressions

In this section you will be introduced to some of the fundamental algebraic terminology that will be used throughout the rest of the book. You will first learn how to express certain numerical concepts in terms of algebraic expressions.

A. VARIABLES

Sometimes, instead of adding, subtracting, multiplying, or dividing *individual* numbers, the same arithmetic process can apply to *any number* under discussion. It is important to have symbols that can stand for any one of several numbers. For example, if you want to add 5 to a number (whether that number is 1, 2, 10, or any other number), you can express this by

$x + 5$

If you want to multiply a number by 3, you can write

$3x$

In each case x stands for *any* number, and is called a *variable.*

255

DEFINITION

> **A variable** is a symbol that stands for any number being discussed.

Letters such as x, y, z, a, b, and c are frequently used as variables. Algebraic expressions such as

$$x + z$$
$$x - y$$
$$4ab$$
$$x^2 + 3x$$

are built up from variables.

Here is how we translate words to symbols.

Example 1 ▶ Suppose that x stands for some number. Express each of the following in terms of x.

   i. 4 more than the number
 ii. 1 less than the number
iii. 7 times the number
iv. the number divided by 5

Solution.

  i. First note that

       6 is 4 more than the number 2, and $6 = 2 + 4$

    Also,

       7 is 4 more than the number 3, and $7 = 3 + 4$

    In general, 4 more than the number means to add 4 to the number. Thus

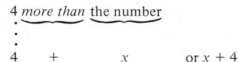

       4   + x or $x + 4$

 ii. Observe that

       3 is 1 less than the number 4, and $3 = 4 - 1$

    Also,

       4 is 1 less than the number 5, and $4 = 5 - 1$

    In general, 1 less than the number means to subtract 1 from the number. Thus

       1 *less than* the number   means   the number *minus* 1

                                x −   1

iii.   7 times the number

7   · x or $7x$

iv.   the number divided by 5

x ÷   5   or $\dfrac{x}{5}$ ◀

Example 2 ▶   Suppose that *a* stands for some number. Express each of the following in terms of *a*.

  i.   the number increased by 8
 ii.   the number decreased by 10
iii.   one-third of the number
iv.   2 more than one-third of the number
 v.   the square of the number

Solution.

  i.   *Increasing* a number by 8 means *adding* 8 to the number. Thus

the number increased by 8

a +   8

 ii.   *Decreasing* a number by 10 means *subtracting* 10 from the number. Thus

the number decreased by 10

a –   10

iii.   Here "of" indicates multiplication.

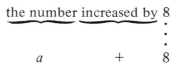

one-third of the number

$\dfrac{1}{3}$ · a or $\dfrac{1}{3}a$ or $\dfrac{a}{3}$

iv.   Use part *iii.*

2 more than one-third of the number

2 + $\dfrac{a}{3}$ or $\dfrac{a}{3} + 2$

 v.   the square of the number

a^2 ◀

PRACTICE EXERCISES FOR TOPIC A

In Exercises 1–5, suppose that x stands for some number. Express each of the following in terms of x.

1. 9 more than the number

2. the number decreased by 4

3. 6 divided by the number

4. 1 more than twice the number

5. 3 times the sum of the number and 2

Now try the exercises for Topic A on page 259.

B. APPLICATIONS

Example 3▶   Find the value, in cents, of n nickels and q quarters.

(A) $n + q$ (B) $n - q$ (C) $125nq$ (D) $5n + 25q$ (E) $5(n + 25q)$

Solution. One nickel is worth 5 cents. Thus

n nickels are worth $n \times 5$ cents,   or $5n$ cents

One quarter is worth 25 cents. Thus

q quarters are worth $q \times 25$ cents,   or $25q$ cents

Together, the value, in cents, is as indicated:

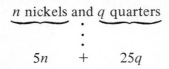

$$5n \quad + \quad 25q$$

The correct choice is (D).   ◀

If a car travels 50 miles per hour, then in 2 hours it travels 50×2, or 100, miles. Similarly, if a boy walks 3 miles per hour, then in 4 hours he walks 3×4, or 12, hours. In general, when an object travels at a *constant* rate, multiply the *rate* of travel by the *time* spent traveling to obtain the *distance* traveled.

$$\boxed{\text{Rate} \times \text{Time} = \text{Distance}}$$

Example 4▶   A car travels 40 miles per hour for y hours. It then increases its speed to 50 miles per hour for z more hours. Express the distance traveled in miles.

(A) $40 + y + 50 + z$ (B) $40y + 50 + z$ (C) $40 + y + 50z$
(D) $40(y + 50z)$ (E) $40y + 50z$

Solution. We can express miles *per* hour by the "fraction," $\frac{\text{miles}}{\text{hours}}$.

Rate $\times$ Time = Distance

$40 \frac{\text{miles}}{\cancel{\text{hours}}} \times y \ \cancel{\text{hours}} = 40y$ miles

$50 \frac{\text{miles}}{\cancel{\text{hours}}} \times z \ \cancel{\text{hours}} = 50z$ miles

The car first travels $40y$ miles (40 miles per hour for y hours) and then an *additional* $50z$ miles (50 miles per hour for z hours). The (*total*) distance it travels, in miles, is then given by choice (E),

$40y + 50z$ ◀

PRACTICE EXERCISES FOR TOPIC B

6. Find the value, in cents, of n nickels, d dimes, and q quarters.

7. Find the cost, in dollars, of a two-dollar items and b three-dollar items.

8. A plane travels x miles per hour for 5 hours. The distance it travels, in miles, is

(A) $5x$ miles (B) $(x + 5)$ miles (C) $\frac{x}{5}$ miles

(D) $\frac{5}{x}$ miles (E) $(5x + 5)$ miles

Now try the exercises for Topic B on page 260.

EXERCISES

A. *In Exercises 1–14, suppose that x stands for some number. Express each of the following in terms of x.*

1. the number plus 3   *Answer:*

2. the number minus 7   *Answer:*

3. 6 times the number   *Answer:*

4. the number divided by 10   *Answer:*

5. 2 more than the number   *Answer:*

6. 2 less than the number   *Answer:*

7. the number increased by 12   *Answer:*

8. the number decreased by 1   *Answer:*

9. the square of the number   *Answer:*

10. one-half of the number   *Answer:*

11. 1 more than 3 times the number   *Answer:*

12. 2 less than 5 times the number   *Answer:*

13. 6 less than one-half of the number   *Answer:*

14. 10 more than one-fourth of the number   *Answer:*

B. *When several choices are given, draw a circle around the correct letter.*

15. Find the value, in cents, of n nickels and d dimes.

   (A) $n + d$ (B) $5n + d$ (C) $n - d$ (D) $5n + 10d$ (E) $5(n + 10d)$

16. Find the value, in cents, of d dimes and p pennies.

   (A) $d + p$ (B) $d - p$ (C) $10d + p$ (D) $d + 10p$ (E) $10(d + p)$

17. Find the value, in cents, of q quarters and h half-dollars.

   (A) $q + \dfrac{h}{2}$ (B) $q + 2h$ (C) $25q + \dfrac{h}{2}$ (D) $25q + 2h$ (E) $25q + 50h$

18. Find the value, in cents, of n nickels, d dimes, and q quarters.

   (A) $n + d + q$ (B) $5n + d + q$ (C) $5n + 10d + 25q$

   (D) $5q + 10d + 25n$ (E) $5(n + d + q)$

19. Find the value, in dollars, of x five-dollar bills and y ten-dollar bills.
    Answer:

20. Find the value, in dollars, of m one-dollar bills and n five-dollar bills.
    Answer:

21. Find the value, in dollars, of x five-dollar bills, y twenty-dollar bills, and z fifty-dollar bills.   *Answer:*

22. Find the value, in *dollars*, of n nickels and h half-dollars.   *Answer:*

23. The cost, in cents, of x chocolate bars at 25 cents per chocolate bar is _____

24. The cost, in dollars, of r bottles of red wine at 5 dollars per bottle and w bottles of white wine at 6 dollars per bottle is _____

25. A train travels at 50 miles per hour for h hours. Express the distance it travels in miles.

   (A) $50h$ (B) $h + 50$ (C) $h - 50$ (D) $\dfrac{h}{50}$ (E) $\dfrac{50}{h}$

26. A train travels at m miles per hour for 4 hours. Express the distance it travels in miles.

   (A) $m + 4$ (B) m^4 (C) 4^m (D) $4m$ (E) $\dfrac{m}{4}$

27. A train travels m miles per hour for h hours. Express the distance it travels in miles.
    Answer:

28. A train travels at 60 miles per hour for h hours and then at 50 miles per hour for 2 hours. Express the distance it travels in miles.   *Answer:*

29. A train travels at 80 miles per hour for x hours and then at 60 miles per hour for y hours. Express the distance it travels in miles.   *Answer:*

30. A man travels in a train that is going at 50 miles per hour for x hours. When he leaves the train, he then walks 3 miles further. Express the total distance he has gone in miles.   *Answer:*

31. An envelope contains x 20-cent stamps and y 17-cent stamps. What is the total value, in cents, of these stamps?   *Answer:*

32. A television show has y 3-minute commercials and z 2-minute commercials. How many minutes of commercials are there?   *Answer:*

28. Substituting Numbers for Variables

Before considering how to combine algebraic expressions, let us consider how to evaluate an algebraic expression for given values of the variable or variables.

A. ONE VARIABLE

Recall that a variable is a symbol that stands for any number being discussed. Thus, in an algebraic expression such as $3x$, the variable x may stand for 2, or for 5, or for −3. The value of the expression $3x$ will be different in each case. For example,

when $x = 2$, then $3x = 3(2) = 6$

when $x = 5$, then $3x = 3(5) = 15$

when $x = -3$, then $3x = 3(-3) = -9$

In general, when you are given numbers that the variables of an algebraic expression represent, you can find the value of the expression by substituting the numbers for the variables.

Example 1 ▶   Find the value of $5t - 2$ when $t = 4$.

(A) 4   (B) 20   (C) 52   (D) 7   (E) 18

Solution. Substitute 4 for t in the expression $5t - 2$ to obtain

$5(4) - 2$

Now find this value.

$$\underbrace{5(4)}_{20} - 2 = 20 - 2 = 18$$

The correct choice is (E).   ◀

Example 2 ▶   Find the value of $10(5)^x$ when $x = 2$.

Solution. Substitute 2 for x in the expression $10(5)^x$ to obtain

$10(5)^2$

Raise to a power (here, square the 5) before multiplying.

$$10\underbrace{(5)^2}_{5 \times 5} = 10(5 \times 5)$$
$$= 10(25)$$
$$= 250$$ ◀

When a variable occurs more than once in an expression, *each time the variable occurs*, substitute the given number for that variable. Thus if $a = 3$, then the value of the expression

$a^2 + 5a$

is

$$\underbrace{3^2}_{9} + \underbrace{5(3)}_{15} = 9 + 15 = 24$$

Example 3 ▶ Find the value of $3y^2 - 2y + 1$ when $y = -2$.

(A) 17   (B) -15 (C) 9   (D) -7 (E) 31

Solution. Substitute -2 for *each occurrence* of y in the expression

$$3y^2 - 2y + 1$$

to obtain

$$3(-2)^2 - 2(-2) + 1$$

First raise to a power (here, *square* the -2), then multiply, then add and subtract.

$$3\underbrace{(-2)^2}_{4} - 2(-2) + 1 = \underbrace{3(4)}_{12} - \underbrace{2(-2)}_{-4} + 1$$
$$= 12 - (-4) + 1$$
$$= 12 + 4 + 1$$
$$= 17$$

In the third step, recall that *subtracting* -4 yields the same as *adding* 4. The correct choice is (A). ◀

PRACTICE EXERCISES FOR TOPIC A

1. Find the value of $10y - 3$ when $y = 5$.

   (A) 105   (B) 50   (C) 53   (D) 47   (E) 20

2. Find the value of $x^2 - 3x + 1$ when $x = 4$.

3. Find the value of $t^3 + 2t^2 + t - 2$ when $t = -1$.

4. Find the value of $2^a + 5$ when $a = 3$.

Now try the exercises for Topic A on page 266.

B. TWO VARIABLES

Sometimes the expressions that are to be evaluated contain two or more variables. Substitute each given number for the specified variable.

Example 4 ▶ Find the value of $x + 2y^2$ when $x = -1$ and $y = 3$.

(A) 2   (B) 17   (C) 35   (D) 25   (E) 37

Solution. Substitute -1 for x and 3 for y in the expression

$$x + 2y^2$$

Raise to a power, then multiply, then add.

$$-1 + 2\underbrace{(3)^2}_{9} = -1 + \underbrace{2(9)}_{18} = -1 + 18 = 17$$

The correct choice is (B). ◀

Example 5 ▶   What is the value of $4pq + q^2$ when $p = 2$ and $q = -3$?

> **Solution.** Substitute 2 for p and -3 for each occurrence of q in the expression
>
> $$4pq + q^2$$
>
> to obtain
>
> $$4(2)(-3) + \underbrace{(-3)^2}_{9} = \underbrace{4(2)}_{8}\underbrace{(-3)}_{} + 9$$
> $$\underbrace{\hphantom{4(2)(-3)}}_{-24}$$
> $$= -24 + 9$$
> $$= -15 \quad ◀$$

Example 6 ▶   If $t = 3xy^2$, find t when $x = 2$ and $y = -2$.

> **Solution.** In order to find the value of t, substitute 2 for x and -2 for y in the expression
>
> $$3xy^2$$
>
> Thus
>
> $$t = 3(2)\underbrace{(-2)^2}_{4}$$
> $$= \underbrace{3(2)}_{6}(4)$$
> $$= 24 \quad ◀$$

PRACTICE EXERCISES FOR TOPIC B

5. Find the value of $2ab - 3b$ when $a = 5$ and $b = 2$.

   (A) 20   (B) 14   (C) 26   (D) 4   (E) 24

6. What is the value of $x^2 - 3xy + 4$ when $x = 1$ and $y = -1$?

7. If $y = 3ab^2$, find y when $a = 4$ and $b = 2$.

8. Find the value of $x^2y - 1$ when $x = -2$ and $y = -3$.

Now try the exercises for Topic B on page 267.

C. FORMULAS

Formulas are algebraic equations that indicate the relationship between various quantities.

Example 7 ▶ A car travels at the rate of 55 miles per hour. The distance d (in miles) that it travels in t hours is given by the formula

$$d = 55t$$

Determine the distance it travels   *i.* in 2 hours;   *ii.* in 5 hours.

Solution.

i.   Substitute 2 for t in the formula $d = 55t$, and obtain

$$d = 55 \times 2 = 110$$

Thus in 2 hours the car travels 110 miles.

ii.   Substitute 5 for t:

$$d = 55 \times 5 = 275$$

In 5 hours the car travels 275 miles.   ◀

DEFINITION

> A **trapezoid** is a closed, 4-sided figure in which two opposite sides, the **bases**, are parallel.

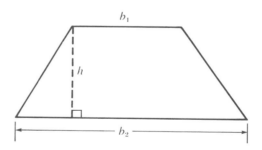

The area, A, of a trapezoid with bases b_1 and b_2 and height h, as in the accompanying figure, is given by the formula

$$A = \frac{b_1 + b_2}{2} \cdot h$$

Example 8 ▶ Find the area of a trapezoid whose bases are 5 inches and 9 inches and whose height is 6 inches.

Solution. Let $b_1 = 5$ inches, $b_2 = 9$ inches, and $h = 6$ inches. Then

$$A = \frac{b_1 + b_2}{2} \cdot h$$

$$= \frac{5 \text{ in.} + 9 \text{ in.}}{\overset{}{\underset{1}{2}}} \times \overset{3}{\cancel{6}} \text{ in.}$$

$$= \frac{14 \text{ in.}}{1} \times 3 \text{ in.}$$

$$= 42 \text{ in.}^2$$

The area of the trapezoid is 42 square inches.   ◄

PRACTICE EXERCISES FOR TOPIC C

9. Find the area of a trapezoid with bases 4 inches and 8 inches and height 5 inches.

10. The *time* a car has traveled (at a constant rate) is given by

$$t = \frac{d}{r}$$

where d is the distance it has gone and r is its rate. Find the time it takes to travel 200 miles at the rate of 50 miles per hour.

Now try the exercises for Topic C on page 268.

EXERCISES

A. *When several choices are given, draw a circle around the correct letter.*

1. Find the value of $4x$ when $x = 3$.
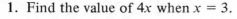

   (A) 43   (B) 12   (C) 7   (D) 64   (E) 1

2. Find the value of $y + 5$ when $y = 2$.

   (A) 2   (B) 7   (C) 25   (D) 32   (E) −3

3. Find the value of $a - 3$ when $a = 3$.

   (A) 3   (B) −3   (C) 0   (D) −6   (E) −9

4. Find the value of $b^2 + 1$ when $b = 5$.

   (A) 6   (B) 25   (C) 26   (D) 36   (E) 11

5. If $m = -3$, what is the value of $4m - 3$?

   (A) 1   (B) −1   (C) −6   (D) −15   (E) 36

6. If $n = -1$, find the value of $1 - n$.

   (A) 1   (B) −1   (C) 0   (D) 2   (E) −2

7. Find the value of 4^a when $a = 2$.   *Answer:*

8. Find the value of 10^b when $b = 2$.   *Answer:*

9. Find the value of -3^x when $x = 2$.   *Answer:*

10. Find the value of $(-5)^y$ when $y = 2$.   *Answer:*

11. Find the value of 10^m when $m = 4$.   *Answer:*

12. Find the value of $(-1)^n$ when $n = 6$.   *Answer:*

13. Find the value of $3(2)^x$ when $x = 2$.   *Answer:*

14. Find the value of $-5(3)^x$ when $x = 2$.   *Answer:*

15. Find the value of $1 - 3^x$ when $x = 2$.   *Answer:*

16. Find the value of $(1 - 3)^x$ when $x = 2$.   *Answer:*

17. Find the value of $x^2 + 6x$ when $x = 2$.
    (A) 10   (B) 12   (C) 14   (D) 16   (E) 20

18. Find the value of $y^2 - y$ when $y = -1$.
    (A) 1   (B) -1   (C) 2   (D) -2   (E) 0

19. If $a = 5$, what is the value of $2a^2 - 3a$?   *Answer:*

20. If $b = -4$, what is the value of $4 - b^2$?   *Answer:*

21. Evaluate $p^3 - 2p$ when $p = 10$.   *Answer:*

22. Evaluate $t^4 + 6t + 1$ when $t = -1$.   *Answer:*

B. *When several choices are given, draw a circle around the correct letter.*

23. Find the value of $p + q$ when $p = 5$ and $q = -7$.
    (A) 12   (B) -12   (C) 2   (D) -2   (E) -35

24. Find the value of $2a + 5b$ when $a = 3$ and $b = 4$.
    (A) 7   (B) 14   (C) 11   (D) 26   (E) 35

25. Find the value of $2c - 4d$ when $c = 6$ and $d = 4$.
    (A) -18   (B) 4   (C) -4   (D) -10   (E) -48

26. Find the value of $x + 3y$ when $x = 3$ and $y = -2$.
    (A) 3   (B) 6   (C) 1   (D) 0   (E) -3

27. Find the value of $5m - 2n$ when $m = -1$ and $n = -2$.   *Answer:*

28. Find the value of $10 - 2y - 3z$ when $y = 2$ and $z = -1$.   *Answer:*

29. Find the value of $a^2 + 2b$ when $a = 5$ and $b = 4$.   *Answer:*

30. Find the value of $x - y^2$ when $x = 4$ and $y = 3$.   *Answer:*

31. Find the value of $m + 3n^2$ when $m = 2$ and $n = 5$.   *Answer:*

32. Find the value of $2a - b^2$ when $a = 3$ and $b = -2$.   *Answer:*

33. Find the value of $5s + t^2$ when $s = -2$ and $t = -3$.   *Answer:*

34. Find the value of $x^2 - 2y^2$ when $x = 2$ and $y = -2$.   *Answer:*

35. Find the value of $a^2 + 2ab$ when $a = 2$ and $b = 5$.
    (A) 14   (B) 24   (C) 40   (D) 45   (E) 80

36. Find the value of $p^2 - 4pq$ when $p = -2$ and $q = 3$.
    (A) 24   (B) 28   (C) –15   (D) –20   (E) 0

37. What is the value of $x^2 + 10xy$ when $x = 10$ and $y = -1$?
    (A) 200   (B) 100   (C) 110   (D) 90   (E) 0

38. If $a = 4$ and $b = -2$, find the value of $a^2 + 4ab$.
    (A) 16   (B) 32   (C) 48   (D) –16   (E) 0

39. If $x = -2$ and $y = -3$, find the value of $x^2 - 3y^2$.   *Answer:*

40. If $m = 5$ and $n = 10$, find the value of $m^2 + 2mn + 1$.   *Answer:*

41. Find the value of $2ab + a^2 + b^2$ if $a = 2$ and $b = 3$.   *Answer:*

42. Find the value of $x^2 + 3xy$ when $x = -1$ and $y = 10$.   *Answer:*

43. Find the value of $x^2 + 2xy^2$ when $x = 3$ and $y = 2$.   *Answer:*

44. Find the value of $x^2 + 10xy$ when $x = -10$ and $y = -10$.   *Answer:*

45. If $c = 2ab^2$, find c when $a = 3$ and $b = 2$.   *Answer:*

46. If $z = 5xy^2$, find z when $x = 2$ and $y = -2$.   *Answer:*

47. If $a = 3m^2n$, find a when $m = -1$ and $n = -2$.   *Answer:*

48. If $t = 4x^2y^2$, find t when $x = 3$ and $y = -2$.   *Answer:*

C. 49. Find the area of a square if each side is of length *i.* 7 inches, *ii.* 12 inches.
    Answer: *i.*   *ii.*

50. Find the area of a triangle if *i.* $b = 10$ inches and $h = 17$ inches;
    ii. $b = 16$ inches and $h = 25$ inches.
    Answer: *i.*   *ii.*

51. An airplane travels at the rate of 400 miles per hour. The distance d (in miles) that
    it travels in t hours is given by the formula

    $$d = 400t$$

    Determine the distance it travels *i.* in 3 hours; *ii.* in 9 hours.
    Answer: *i.*   *ii.*

52. A piece of material cost $7 per square meter. The cost C (in dollars) for x square
    meters of this material is given by the formula

    $$C = 7x$$

    Determine the cost of *i.* 4 square meters; *ii.* 7 square meters of this material.
    Answer: *i.*   *ii.*

53. A ball is thrown upward from ground level with an initial velocity of 96 feet per second. Its height h (in feet) after t seconds is given by the formula

$$h = 96t - 16t^2$$

Determine its height *i.* after 1 second; *ii.* after 2 seconds; *iii.* after 6 seconds.
Answer: *i.*   *ii.*   *iii.*

54. The total value v (in dollars) of a number of coins is given by the formula

$$v = \frac{5n + 10d + 25q}{100}$$

where *n, d,* and *q* are, respectively, the number of nickels, dimes, and quarters. Find the total value of 10 nickels, 15 dimes, and 7 quarters.   *Answer:*

55. **A parallelogram** is a closed, 4-sided figure in which both pairs of opposite sides are parallel. The area, A, of a parallelogram with base b and height h is given by the formula

$$A = b \cdot h$$

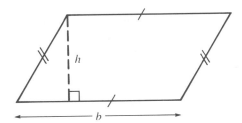

Find the area of a parallelogram with base 7 centimeters and height 6 centimeters. *Answer:*

56. Find the area of a trapezoid with bases 4 meters and 8 meters and with height 9 meters. (See Example 8.)

In Exercises 57–64, express your answers in terms of π.

57. **A sphere** (or ball) is a surface (in space) every point of which is at a fixed distance from a given point, known as the **center** of the sphere. The fixed distance is called the **radius** (length) of the sphere. The volume, V, of a sphere with radius r is given by the formula

$$V = \frac{4}{3}\pi r^3$$

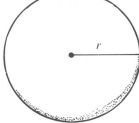

Find the volume of a sphere with radius 9 inches. Express your answer in terms of π. *Answer:*

58. The surface area, S, of a sphere with radius r (Exercise 57) is given by the formula

$$S = 4\pi r^2$$

Find the surface area of a sphere of radius 5 centimeters. Express your answer in terms of π. *Answer.*

59. A (right-circular) **cylinder,** as shown on the right, has bases that are circular regions each of which has radius r. The radius and the corresponding height form a right angle. The volume, V, of a cylinder with radius r and height h is given by the formula

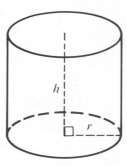

$$V = \underbrace{\pi r^2}_{\substack{\text{area of} \\ \text{circular} \\ \text{base}}} \overset{\searrow}{h}{\scriptstyle\text{height}}$$

Find the volume of a cylinder with radius 2 inches and height 10 inches. Express your answer in terms of π. *Answer:*

60. The *lateral* surface area (*excluding* top and bottom) of a cylinder (Exercise 59) is given by the formula

$$S = 2\pi rh$$

where S is the lateral surface, r is the radius of the base, and h is the height. How much paper is used to line the lateral surface of a cylindrical can in which the radius of the base is 7 centimeters and the height is 8 centimeters. Express your answer in terms of π. *Answer:*

61. The *total* surface area (*including* top and bottom) of a cylinder (Exercise 59) is given by the formula

$$T = \underbrace{2\pi rh}_{\substack{\text{lateral} \\ \text{surface}}} + \underbrace{2\pi r^2}_{\substack{\text{area of} \\ \text{top and} \\ \text{bottom} \\ \text{circles}}}$$

where T is the total surface area, r is the radius of the base, and h is the height. Find the total surface area of a tin can in which the radius of the base is 3 inches and the height is 7 inches. Express your answer in terms of π. *Answer:*

62. The figure at the right shows a (right-circular) cone in which the radius of the base is r, the height is h, and the slant height is s. The volume, V, of a cone is given by the formula

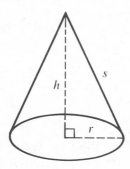

$$V = \frac{1}{3}\pi r^2 h$$

Find the volume of a cone in which the radius of the base is 4 centimeters and the height is 5 centimeters. *Answer:*

63. The lateral surface area of a cone (Exercise 62) is given by the formula

$$S = \pi rs$$

Find the lateral surface area of a cone in which the radius of the base is 6 inches and the slant height is 11 inches. *Answer:*

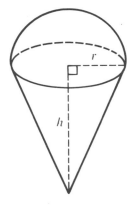

64. A cone is filled with ice cream and a hemisphere (half of a sphere) of ice cream is on top. How much ice cream is there altogether if the radius of the base of the cone is 2 inches and the height of the cone is 5 inches? (See Exercises 57 and 62.)
Answer:

29. Adding and Subtracting Polynomials

Here we consider how to add and subtract algebraic expressions known as *polynomials.* Before treating this topic, some preliminary notions must be introduced.

A. TERMS

DEFINITION

A **term** is a product of numbers and variables. A number, by itself, and a variable, by itself, are each considered to be terms.

For example, each of the following is a term.

$\quad$ 4 $\quad$ A number, by itself, is a term.

$\quad$ x $\quad$ A variable, by itself, is a term.

$\quad$ $4x$ $\quad$ This is the product of 4 and x.

$\quad$ $5a^2b$ $\quad$ $5a^2b = 5 \cdot a \cdot a \cdot b$

However, $x + 1$ and $a + b$ are *not* terms because the definition of a term does not permit adding a number and a variable, or adding two different variables.

DEFINITION

> The numerical factor (or the product of numerical factors) in a term is called the **numerical coefficient** or simply, the **coefficient**, of the term.

Example 1 ▶ What is the coefficient of each of the following terms?

$\quad$ *i.* $6x$ $\qquad\qquad$ *ii.* $-2x^3y$ $\qquad\qquad$ *iii.* 20

$\quad$ *iv.* $(5a) \cdot (2b)$ $\qquad\qquad$ *v.* a $\qquad\qquad$ *vi.* $-b$

Solution.

$\quad$ *i.* 6

$\quad$ *ii.* -2 $\quad$ Note that 3 is the *exponent* of x and is *not* the coefficient.

$\quad$ *iii.* 20 $\quad$ A (numerical) term is its own coefficient.

$\quad$ *iv.* 10 $\quad$ The product of the numerical factors 5 and 2 is 10.

$\quad$ *v.* 1 $\qquad$ $a = 1 \cdot a$

$\quad$ *vi.* -1 $\qquad$ $-b = (-1) \cdot b$ $\qquad\qquad\qquad\qquad\qquad\qquad\qquad$ ◀

It will be useful to say that the **sign of a term** is the sign of its coefficient. Thus the sign of $4x$ is $+$ and the sign of $-4x$ is $-$.

PRACTICE EXERCISES FOR TOPIC A

What is the coefficient of each term?

1. $12x^2$ $\qquad\qquad\qquad\qquad\qquad\qquad$ 2. $5ab^4$

3. -5

Now try the exercises for Topic A on page 278.

B. POLYNOMIALS

DEFINITION

> **A polynomial** is either a term or a sum of terms.

Each of the following is a polynomial.

$2x$ A term is a polynomial.

$x + 1$ A sum of terms is a polynomial.

$x^2 + 3x - 5$ Note that $x^2 + 3x - 5 = x^2 + 3x + (-5)$.

$ab + a + b$

Example 2 ▶ $x^2 + 2x - 3$ is the sum of the terms

⬚ , ⬚ , and ⬚

Solution. $x^2 + 2x - 3$ is the sum of the terms

$\boxed{x^2}$, $\boxed{2x}$, and $\boxed{-3}$ ◄

PRACTICE EXERCISES FOR TOPIC B

Fill in the terms.

4. $x^2 + 5x$ is the sum of the terms ⬚ and ⬚ .

5. $3a^4 - 5a + 1$ is the sum of the terms ⬚ , ⬚ , and ⬚ .

Now try the exercises for Topic B on page 278.

C. LIKE TERMS

Adding $\underbrace{2a + 3a}_{5a}$ is similar to adding $\underbrace{2 \text{ apples and } 3 \text{ apples}}_{5 \text{ apples}}$, whereas adding
$2a + 3b$ is similar to adding 2 apples and 3 bananas. You can combine terms when adding them only when they are alike in their variables and in the exponent of each variable.

DEFINITION

> Two or more nonzero terms are said to be **like terms** if
> 1. they contain exactly the same variables, and
> 2. each variable has the same exponent in each term.

In other words *like terms can differ only in their numerical coefficients or in the order of their variables.*

Here are some examples of *like terms*:

$10x$ and $2x$
y^2 and $-5y^2$
$3ab$ and $2ab$
x^2y and yx^2 Note that like terms can differ in the order of their variables.
5 and -4 Any two nonzero numbers are like terms because neither contains any variables.

Here are some examples of *unlike terms*:

$7p$ and $3q$ They differ in their variables.
$4x$ and $4x^2$ They differ in the exponents of x.
$5xy^2$ and $3xy$ They differ in the exponents of y.

Example 3 ▶   Which of the following pairs are like terms?

i. $5a^2$ and $-2a^2$ *ii.* $3b^2$ and $2b^3$ *iii.* $6st$ and $2su$

Solution.

i. like terms: Both terms have the same variable a, and the same exponent of a.

ii. unlike terms: They differ in the exponents of b.

iii. unlike terms: One contains the variable t and the other contains the variable u.   ◀

PRACTICE EXERCISES FOR TOPIC C

Which pairs are like terms?

6. $4xy$ and $-4yx$ 7. x^2y and xy^2

8. $10ab^3$ and $10ab^5$

Now try the exercises for Topic C on page 278.

D. ADDING AND SUBTRACTING LIKE TERMS

Observe that

$$3\underbrace{(4 + 2)}_{6} = 3(6) = 18$$

But we can also combine as follows:

$$\overbrace{3(4 + 2)} = \underbrace{3(4)}_{12} + \underbrace{3(2)}_{6}$$
$$= 12 + 6$$
$$= 18$$

This second method, though longer, illustrates the way *like terms* can be added and subtracted.

Let a and b be any numbers, and let x be a variable. Then according to the **Distributive Laws**,

$$ax + bx = (a + b)x$$

and

$$ax - bx = (a - b)x$$

Thus

$$5x + 3x = (5 + 3)x = 8x$$

and

$$5x - 3x = (5 - 3)x = 2x$$

The above rules also apply when x is replaced by a power of a variable, such as a^2, or by a product of variables, such as xy.

Example 4 ▶   Add or subtract:

i. $10a^2 + 4a^2$ *ii.* $6xy - 2xy$

Solution. In each case the terms to be combined are *like* terms.

i. $10a^2 + 4a^2 = (10 + 4)a^2 = 14a^2$

ii. $6xy - 2xy = (6 - 2)xy = 4xy$ ◀

Unlike terms cannot be added or subtracted in this way. For example, $2x + 3y$ cannot be further simplified.

PRACTICE EXERCISES FOR TOPIC D

Add or subtract. If the given expression cannot be simplified, just write down the expression.

9. $5a + 3a =$ 10. $12b^2 - 6b^2 =$

11. $10a - 7b =$

Now try the exercises for Topic D on page 278.

E. ADDING POLYNOMIALS

To add or subtract polynomials, *group like terms together* and add in columns.

Example 5 ▶   Add $4x + 5y$ and $4y - 3x$.

(A) $8x + 2y$ (B) $7x + 9y$ (C) $x + 9y$
(D) $16xy - 15xy$ (E) $x^2 + 9y^2$

Solution.   Rearrange the terms so that like terms are in the same column. Note that

$$4y - 3x = -3x + 4y$$

Thus

$$\begin{array}{r} 4x + 5y \\ -3x + 4y \\ \hline x + 9y \end{array}$$

The correct choice is (C).   ◀

Example 6 ▶   Find the sum of $2a^2 + 8a$ and $5 - 4a$.

(A) $7a^2 + 4a$ (B) $7a^2 - 4a$ (C) $7a^2 + 4a + 5$
(D) $2a^2 + 4a + 5$ (E) $2a^2 - 4a + 5$

Solution.   Line up the terms as indicated.

$$\begin{array}{r} 2a^2 + 8a \\ -4a + 5 \\ \hline 2a^2 + 4a + 5 \end{array}$$

The sum is $2a^2 + 4a + 5$, choice (D).   ◀

PRACTICE EXERCISES FOR TOPIC E

12. Add $2x - 3$ and $3x + 2$.

13. Find the sum of $4a + 3b - 1$ and $5a - b$.

Now try the exercises for Topic E on page 279.

F. SUBTRACTING POLYNOMIALS

For any polynomial P, let $-P$ denote the polynomial obtained by changing the sign of every term of P. For example, let

$$P = 5x - 3y + 2$$

Then

$$-P = -5x + 3y - 2$$

Now add P and $-P$.

$$\begin{array}{ll} P: & 5x - 3y + 2 \\ -P: & -5x + 3y - 2 \\ \hline & 0 \end{array}$$ Note that $0x + 0y + 0 = 0$.

In general, for any polynomial P,

$$P + (-P) = (-P) + P = 0$$

Recall that for numbers a and b,

$$a - b = a + (-b)$$

This is also the way we subtract polynomials. For any polynomials P and Q, define

$$P - Q = P + (-Q)$$

Thus, *to subtract the polynomial Q from P, add −Q.*

Example 7 ▶ Subtract $4x + 7$ from $x^2 - 3$.

(A) $5x^2 + 4$ (B) $3x^2 + 4$ (C) $x^2 + 4x + 4$
(D) $-x^2 + 4x + 10$ (E) $x^2 - 4x - 10$

Solution. Here $P = x^2 - 3$ and $Q = 4x + 7$. (You are asked to subtract Q from P.) Thus you want $P - Q$, or $P + (-Q)$.

$$-Q = -4x - 7$$

Line up like terms in the same column and add:

$$
\begin{array}{lr}
P: & x^2 \quad\quad - 3 \\
-Q: & \underline{\quad -4x - 7} \\
& x^2 - 4x - 10
\end{array}
$$

The correct choice is (E). ◀

Example 8 ▶ What is 5 less than $2a + 8$?

Solution.

5 less than

indicates to *subtract* 5. Thus

$$\underbrace{5 \text{ less than}}_{-5} 2a + 8$$

means

$$(2a + 8) - 5 \quad \text{or} \quad 2a + \underbrace{(8 - 5)}_{3}$$

or finally,

$$2a + 3$$ ◀

PRACTICE EXERCISES FOR TOPIC F

14. Subtract $5x + y$ from $10x + 3y$.

   (A) $15x + 4y$ (B) $5x + 2y$ (C) $5x + 4y$
   (D) $5x - 2y$ (E) $-5x - 2y$

15. Subtract $2a - 3b$ from $2a + 7b$.

16. What is 4 less than $3x + 9$?

Now try the exercises for Topic F on page 279.

EXERCISES

A. *What is the coefficient of each term?*

   1. $20x^2$ *Answer:*   **2.** $-3ab$ *Answer:*   **3.** x *Answer:*

   4. $-t^3$ *Answer:*   **5.** 16   *Answer:*   **6.** $(2x) \cdot (3y)$ *Answer:*

B. *Fill in the terms.*

   7. $x^2 + 4x$ is the sum of the terms ⬚ and ⬚.

   8. $a + 5$ is the sum of the terms ⬚ and ⬚.

   9. $3b^2 - 5$ is the sum of the terms ⬚ and ⬚.

   10. $2t^2 - 5t - 8$ is the sum of the terms ⬚, ⬚, and ⬚.

C. *Which pairs are like terms?*

   11. $20x$ and $20y$ *Answer:*   **12.** $3a$ and $7a$ *Answer:*

   13. $5b$ and $-5b$ *Answer:*   **14.** $2c^2$ and $-c^2$ *Answer:*

   15. xy and $2yx$ *Answer:*   **16.** xy^2 and $2xy$ *Answer:*

   17. $3a^2$ and $2a^3$ *Answer:*   **18.** 9 and -9 *Answer:*

   19. $a^2 b^2$ and $a \cdot b \cdot a \cdot b$ *Answer:*   **20.** $c^3 d$ and cd^3 *Answer:*

D. *Add or subtract. If the given expression cannot be simplified, just write down this expression.*

   21. $8a + 4a =$ **22.** $10b^2 + 10b^2 =$ **23.** $8x - 7x =$

   24. $6x - 5y =$ **25.** $3x - 3x =$ **26.** $4y^2 + (-4)y^2 =$

   27. $3xy + 9xy =$ **28.** $4xy^2 - 2x^2y =$

E. *In Exercises 29–32, draw a circle around the correct letter.*

29. Add $4x + 7y$ and $2x + 3y$.

    (A) $6x + 10y$ (B) $2x + 4y$ (C) $8x^2 + 21y^2$

    (D) $6x^2 + 10y^2$ (E) $16xy$

30. Add $5a + 3b$ and $2a - 6b$.

    (A) $7a + 3b$ (B) $10a^2 - 18b^2$ (C) $7a^2 - 3b^2$

    (D) $7a - 3b$ (E) $10a - 3b$

31. Add $4x + 9a$ and $3x - 9a$.

    (A) x (B) $7x + 18a$ (C) $7x - 18a$ (D) $7x$ (E) $7x^2$

32. Add $3x^3 + 5x$ and $6x^3 - x$.

    (A) $9x^6 + 4x^2$ (B) $9x^3 + 4x$ (C) $9x^3 - 4x$

    (D) $18x^3 - 5x$ (E) $18x^6 - 5x^2$

33. Add $4a + 1$ and $-3a - 2$.   *Answer:*

34. Add $3a + 5b$ and $10b - 2a$.   *Answer:*

35. Add $9x - 2y$ and $4y - x$.   *Answer:*

36. Find the sum of $4x^2 + 3$ and $6 - x^2$.   *Answer:*

37. Find the sum of $3x + 2y$ and $2x - 3y$.   *Answer:*

38. Find the sum of $9p - 3q$ and $6q - p$.   *Answer:*

39. Find the sum of $4x^2 + 3x - 1$ and $2x^2 - x + 4$.   *Answer:*

40. Find the sum of $6x + 3y - 2$ and $2x - xy + 5$.   *Answer:*

F. *When several choices are given, draw a circle around the correct letter.*

41. Subtract $4x + 2y$ from $9x + 5y$.

    (A) $13x + 7y$ (B) $36x + 10y$ (C) $36x^2 + 10y^2$

    (D) $-5x - 3y$ (E) $5x + 3y$

42. Subtract $3a + 2b$ from $8a + 2b$.

    (A) $5a + 2b$ (B) $5a + 4b$ (C) $5a - 4b$ (D) $5a$ (E) $11a + 4b$

43. Subtract $2p - 3q$ from $6p + 6q$.

    (A) $8p + 3q$ (B) $4p + 3q$ (C) $4p - 3q$ (D) $4p + 9q$ (E) $-4p - 9q$

44. Subtract $6x - 3$ from $9x - 2$.

    (A) $3x - 5$ (B) $3x + 5$ (C) $3x + 1$ (D) $3x - 1$ (E) $3x - 6$

45. Subtract $2x - 5$ from $3x^2 + 2$.   *Answer:*

46. Subtract $8a - 3b$ from $2a - 3$.   *Answer:*

47. Subtract $5a - 2b$ from $a + 3c$.   *Answer:*

48. Subtract $x - y$ from $x + y$.   *Answer:*

49. Subtract $4x + 1$ from $x^2 - 3x$.   *Answer:*

50. Subtract $5a - 1$ from $a^2 - 3$.   *Answer:*

51. What is 5 less than $10x + 3$?

(A) $5x + 3$ (B) $10x + 8$ (C) $10x - 5$ (D) $10x - 2$ (E) $10x - 8$

52. What is 3 less than $4a^2 - 6$?

(A) $a^2 - 6$ (B) $a^2 - 9$ (C) $a^2 - 3$ (D) $4a^2 - 3$ (E) $4a^2 - 9$

53. Let x represent a number. Express the sum of two more than the number and three less than twice the number.

(A) $x + 2$ (B) $3x - 2$ (C) $4x$ (D) $3x + 5$ (E) $3x - 1$

54. Suppose that Barbara has n nickels and Harriet has d dimes. How much do they have together (in cents)?

(A) $n + d$ (B) $5n + d$ (C) $n + 10d$ (D) $5n + 10d$ (E) $10d - 5n$

30. Solving Equations

Equations were introduced briefly in Section 14 in order to help solve certain types of percent problems. Here we will review the basic concepts involved in order to be able to solve an equation such as

$$5x - 2 = 18$$

or

$$7x + 3 = 9x - 3$$

A. ROOTS OF EQUATIONS

The statement

$$3x = 6$$

is known as an *equation*. It asserts that

Three times a number is equal to 6

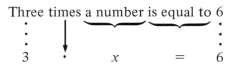

$$3 \quad \cdot \quad x \quad = \quad 6$$

> An **equation** is a statement of equality. In the equation
>
> $$3x = 6$$
>
> $3x$ is the **left side** of the equation and 6 is the **right side**. A number is a **root** or a **solution** of an equation (in one variable) if a true statement results when that number is substituted for the variable.

Thus 2 is a root of the equation

$$3x = 6$$

because when you substitute 2 for x you obtain the *true* statement

$$3(2) = 6$$

Also, 1 is *not* a root of the equation because when you substitute 1 for x, you obtain the *false* statement $3(1) = 6$.

The equations we will consider have exactly one root. To check whether a number is the root of an equation, substitute the number for *each occurrence* of the variable.

Example 1 ▶ Is 3 the root of the equation

$$4x - 2 = x + 7$$

Solution. Substitute 3 for each occurrence of x, to see whether the resulting statement is true.

$$4(3) - 2 = 3 + 7$$
$$12 - 2 = 10 \qquad true$$

Thus 3 is the root of the given equation.   ◀

Example 2 ▶ Is -2 the root of the equation

$$2t + 1 = 5t - 5$$

Solution. Substitute −2 for each occurrence of t to see whether the resulting statement is true.

$$2(-2) + 1 = 5(-2) - 5$$

$$-4 + 1 = -10 - 5$$

$$-3 = -15 \qquad \textit{false}$$

Thus −2 is *not* the root of this equation.   ◄

PRACTICE EXERCISES FOR TOPIC A

Answer "yes" or "no."

1. Is 5 a root of the equation

$$4x = 20$$

2. Is 3 a root of the equation

$$2y - 2 = y + 1$$

3. Is −2 a root of the equation

$$10t + 4 = 8t + 2$$

Now try the exercises for Topic A on page 286.

B. MULTIPLICATION PROPERTY

You can think of an equation, such as

$$2x = 6$$

as a balanced scale, as in the top figure to the right. You want to isolate x on one side of the equation. Because the 2 *multiplies* the x, and you want to isolate the unknown, you must divide the left side by 2. But now the scale is unbalanced. (See the middle figure.) If you change one side, you must do the same thing to the other side in order to preserve the balance. Thus divide both sides by 2. (See the bottom figure.) In symbols:

$$2x = 6$$

$$\frac{2x}{2} = \frac{6}{2}$$

$$x = 3$$

Thus 3 is the root of the equation $2x = 6$.

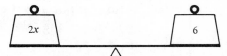

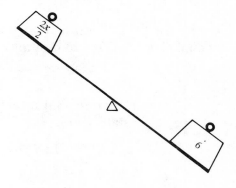

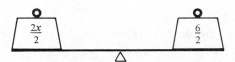

The **Multiplication Property** asserts:

> To simplify an equation, you can multiply or divide both sides by the same (nonzero) number without changing the root.

Example 3 ▶ If $\frac{x}{3} = 4$, find x.

Solution. What can you do to $\frac{x}{3}$ to obtain x?

$$\frac{x}{3} = 4 \qquad \text{Multiply both sides by 3.}$$

$$\frac{x}{3} \times 3 = 4 \times 3$$
$$x = 12$$

The root, x, is 12.

You can *check* that 12 is the root of the equation by substituting 12 for x in the given equation.

$$\frac{12}{3} = 4 \qquad true$$

The check shows that you have the correct root.   ◀

Example 4 ▶ If $4y = 1$, find y.

Solution. What can you do to $4y$ to obtain y?

$$4y = 1 \qquad \text{Divide both sides by 4.}$$

$$\frac{4y}{4} = \frac{1}{4}$$

$$y = \frac{1}{4}$$

The root, y, is the fraction $\frac{1}{4}$.   ◀

PRACTICE EXERCISES FOR TOPIC B

For each equation, find the root.

4. If $5x = 30$, find x.

5. Find y if $-4y = 28$.

6. Solve for z: $\frac{z}{3} = -6$

Now try the exercises for Topic B on page 287.

C. ADDITION PROPERTY

The **Addition Property** asserts:

> To simplify an equation, you can add the same number to, or subtract the same number from, both sides of an equation without changing the root.

Example 5 ▶   If $t - 3 = 4$, find t.

Solution. What can you do to $t - 3$ to obtain t?

$$t - 3 = 4 \qquad \text{Add 3 to both sides.}$$
$$t - 3 + 3 = 4 + 3$$
$$t = 7$$

The root, t, is 7.   ◀

Example 6 ▶   If $x + 5 = 3$, find x.

Solution. What can you do to $x + 5$ to obtain x?

$$x + 5 = 3 \qquad \text{Subtract 5 from both sides.}$$
$$x + 5 - 5 = 3 - 5$$
$$x = -2.$$

The root, x, is the negative integer -2.   ◀

PRACTICE EXERCISES FOR TOPIC C

For each equation, find the root.

7. If $x + 8 = 12$, find x.

8. Find y if $y - 3 = 4$.

9. Solve for t: $2 + t = -3$

Now try the exercises for Topic C on page 287.

D. USING BOTH PROPERTIES

You will often have to use both the Multiplication and the Addition Properties to solve some of the equations you will be given. To simplify these equations, *bring variables to one side, numerical terms to the other.*

Example 7 ▶ Find y if $2y + 7 = 11$.

Solution.

$$2y + 7 = 11 \qquad \text{Subtract 7 from both sides.}$$
$$2y + 7 - 7 = 11 - 7$$
$$2y = 4 \qquad \text{Divide both sides by 2.}$$
$$\frac{2y}{2} = \frac{4}{2}$$
$$y = 2$$

We can *check* that 2 is, indeed, the correct choice by substituting 2 for y in the given equation. Thus

$$2(2) + 7 = 11$$
$$4 + 7 = 11 \qquad true \qquad ◀$$

Example 8 ▶ If $5x - 1 = 2x + 8$, then $x =$

(A) 1   (B) 0   (C) –1   (D) 3   (E) $\dfrac{7}{3}$

Solution. Bring variables to one side, numerical terms to the other. Because the coefficient of x is larger on the left side, it is a good idea to bring the variables to the left side in this problem.

$$5x - 1 = 2x + 8 \qquad \text{Subtract } 2x \text{ from both sides.}$$
$$5x - 2x - 1 = 2x - 2x + 8$$
$$3x - 1 = 8 \qquad \text{Add 1 to both sides.}$$
$$3x - 1 + 1 = 8 + 1$$
$$3x = 9 \qquad \text{Divide both sides by 3.}$$
$$\frac{3x}{3} = \frac{9}{3}$$
$$x = 3$$

The correct choice is (D). ◀

Example 9 ▶ If $3y + 2 = 4y - 7$, find y.

Solution. Here the larger coefficient of y is on the right side. Bring variables to the right side.

$$3y + 2 = 4y - 7 \qquad \text{Subtract } 3y \text{ from both sides.}$$
$$3y - 3y + 2 = 4y - 3y - 7$$
$$2 = y - 7 \qquad \text{Add 7 to both sides.}$$
$$2 + 7 = y - 7 + 7$$
$$9 = y \qquad ◀$$

Example 10 ▶ Solve for t. $\frac{t}{6} + 1 = 3$

>   ***Solution.*** On the left side, 1 is added to $\frac{t}{6}$. Thus begin by subtracting 1 from both sides.

$$\frac{t}{6} + 1 = 3$$

$$\frac{t}{6} + 1 - 1 = 3 - 1$$

$$\frac{t}{6} = 2 \qquad \text{Multiply both sides by 6.}$$

$$\frac{t}{6} \times 6 = 2 \times 6$$

$$t = 12 \qquad\qquad\qquad\qquad ◀$$

PRACTICE EXERCISES FOR TOPIC D

When several choices are given, draw a circle around the correct letter.

10. If $3t + 1 = 7$, then $t =$

    (A) 2   (B) 4   (C) –3   (D) 10   (E) 11

11. Find y if $5y - 3 = 2$.

12. Solve for x: $3x + 12 = 4x$

    (A) 2   (B) 4   (C) –2   (D) –4   (E) 12

13. Find t if $6t - 2 = 2t + 10$.

Now try the exercises for Topic D on page 287.

EXERCISES

A. *Answer "yes" or "no."*

1. Is 3 the root of the equation
$$5t = 15 \qquad \textit{Answer:}$$

2. Is 6 the root of the equation
$$t + 3 = 9 \qquad \textit{Answer:}$$

3. Is 10 the root of the equation
$$2x + 20 = 0 \qquad \textit{Answer:}$$

4. Is –3 the root of the equation
$$\frac{y}{3} = -1 \qquad \textit{Answer:}$$

5. Is 4 the root of the equation
$$5x - 4 = 12 \qquad \textit{Answer:}$$

6. Is 3 the root of the equation
$$2y - 5 = 1 \qquad \textit{Answer:}$$

7. Is −4 the root of the equation

$$10 - 2t = 2 \qquad \textit{Answer:}$$

8. Is $\frac{1}{2}$ the root of the equation

$$6u = 3 \qquad \textit{Answer:}$$

9. Is 12 the root of the equation

$$3v - 9 = 2v + 5 \qquad \textit{Answer:}$$

10. Is 4 the root of the equation

$$8x - 12 = 5x + 4 \qquad \textit{Answer:}$$

11. Is −1 the root of the equation

$$1 - 2y = 3 - 3y \qquad \textit{Answer:}$$

12. Is 0 the root of the equation

$$5z + 4 = 10z - 4 \qquad \textit{Answer:}$$

B. *For each equation, find the root.*

13. If $2x = 20$, find x.   *Answer:*

14. If $4x = 28$, find x.   *Answer:*

15. If $7y = 35$, find y.   *Answer:*

16. If $10z = 45$, find z.   *Answer:*

17. Find t if $\frac{t}{2} = 6$.   *Answer:*

18. Find u if $\frac{u}{5} = 8$.   *Answer:*

19. Find x if $\frac{x}{9} = -2$.   *Answer:*

20. Find y if $\frac{y}{4} = 0$.   *Answer:*

21. Solve for t: $-4t = 16$
     Answer:

22. Solve for u: $-5u = -15$
     Answer:

23. Solve for x: $\frac{x}{-4} = 2$

     Answer:

24. Solve for y: $\frac{y}{-10} = -3$

     Answer:

C. *For each equation, find the root.*

25. If $x + 3 = 10$, find x.
     Answer:

26. If $y - 7 = 2$, find y.
     Answer:

27. If $z + 4 = -2$, find z.
     Answer:

28. If $u - 5 = -6$, find u.
     Answer:

29. Find x if $x + 8 = -3$.
     Answer:

30. Find y if $y - 7 = 0$.
     Answer:

31. Find t if $t + 6 = 0$.
     Answer:

32. Find u if $u - 8 = 8$.
     Answer:

33. Solve for x: $x + 12 = 11$
     Answer:

34. Solve for y: $9 + y = 6$
     Answer:

35. Solve for z: $-5 + z = 0$
     Answer:

36. Solve for t: $7 = t + 9$
     Answer:

D. *When several choices are given, draw a circle around the correct letter.*

37. If $2x + 1 = 7$, then $x =$

(A) 8   (B) 6   (C) 4   (D) 3   (E) $\frac{7}{2}$

38. If $5x - 2 = 13$, then $x =$

    (A) 15   (B) 11   (C) $\frac{11}{5}$ (D) $\frac{13}{5}$ (E) 3

39. If $6 - y = 0$, then $y =$
    (A) 6   (B) −6   (C) 0   (D) 1   (E) −1

40. If $8 - z = 4$, then $z =$
    (A) 8   (B) −8   (C) 4   (D) −4   (E) 0

41. Find t if $4t - 3 = 17$.
    Answer:

42. Find u if $5u - 25 = 0$.
    Answer:

43. Find v if $6 - 3v = 0$.
    Answer:

44. Find x if $10 - 2x = 3x$.
    Answer:

45. Solve for y: $7y - 3 = 4y$

    (A) 3   (B) $-\frac{3}{4}$ (C) 1   (D) −1   (E) −3

46. Solve for z: $10z + 2 = 8z + 6$
    (A) 2   (B) 8   (C) −2   (D) 1   (E) 0

47. Solve for x: $6x - 3 = 9x - 9$
    (A) −12   (B) −4   (C) 4   (D) 2   (E) −2

48. Solve for y: $10 - 3y = 4 + 3y$
    (A) 0   (B) 10   (C) 6   (D) 1   (E) −1

49. If $\frac{t}{2} - 1 = 0$, find t.
    Answer:

50. If $\frac{x}{3} - 1 = 5$, find x.
    Answer:

51. If $\frac{y}{5} + 2 = -3$, find y.
    Answer:

52. If $\frac{z}{6} + 8 = 2$, find z.
    Answer:

53. Find x if $9x - 4 = 7x + 1$.
    Answer:

54. Find y if $8 - 3y = 6y - 1$.
    Answer:

55. Find t if $2t + 3 = 4t - 13$.
    Answer:

56. Find u if $9 - 5u = 8 + u$.
    Answer:

57. Find x if $6 + x = 3 - 4x$.
    Answer:

58. Find y if $9y + 3 = 2 - y$.
    Answer:

31. Ratios and Proportions

Many equations involve fractions. One such type of equation, known as a *proportion*, states that the *ratios* of various quantities are equal.

A. RATIOS

A **ratio** is simply a quotient. A ratio of 4 to 1 is expressed as the fraction

$$\frac{4}{1}$$

A ratio of 4 to 1 indicates that there is four times as much of one quantity as of another. A ratio of 5 to 2 is expressed by

$$\frac{5}{2}$$

and indicates that there are five units of one quantity for every two units of a second quantity. Finally, a ratio of 4 to 5 is expressed by

$$\frac{4}{5}$$

and indicates that there are four units of one quantity for every five units of a second quantity.

Example 1 ▶ Express the ratio of 2 to 7 as a fraction.

Solution.

$$\frac{2}{7}$$

◀

PRACTICE EXERCISES FOR TOPIC A

Express each of the following ratios as fractions.

1. 2 to 1   2. 5 to 4   3. 3 to 4

Now try the exercises for Topic A on page 294.

B. SOLVING PROPORTIONS

An equation, such as

$$\frac{2}{4} = \frac{1}{2}$$

that expresses the equivalence of fractions is an example of a *proportion*.

DEFINITION

> **A proportion** is an equation of the form
>
> $$\frac{a}{b} = \frac{c}{d}$$

A proportion says that the ratio of a to b is equal to the ratio of c to d. In Example 2, the equation is a proportion.

Example 2 ▶ If $\frac{x}{4} = \frac{3}{5}$, find x.

Solution. Multiply both sides by 4.

$$\frac{x}{4} = \frac{3}{5}$$

$$\frac{x}{4} \times 4 = \frac{3}{5} \times 4 \qquad \text{The right side equals } \frac{3}{5} \times \frac{4}{1}.$$

$$x = \frac{12}{5}$$

◀

When the variable is in the denominator, as in Example 3, a technique known as *cross-multiplying* is useful.

Example 3 ▶ If $\frac{4}{x} = \frac{2}{3}$, find x.

(A) $\frac{8}{3}$ (B) $\frac{3}{8}$ (C) $\frac{4}{3}$ (D) 12   (E) 6

Solution. Cross-multiply, as indicated by the arrows.

$$\frac{4}{x} = \frac{2}{3}$$ Cross-multiply. $\frac{4}{x} \diagdown\diagup \frac{2}{3}$

$$4 \times 3 = 2x$$

$$12 = 2x$$ Divide both sides by 2.

$$\frac{12}{2} = \frac{2x}{2}$$

$$6 = x$$

The correct choice is (E).   ◄

Applications involving proportions will be discussed in Section 32.

PRACTICE EXERCISES FOR TOPIC B

In Exercise 4, draw a circle around the correct letter.

4. If $\frac{x}{15} = \frac{2}{3}$, then $x =$

(A) 5   (B) 6   (C) 9   (D) 10   (E) 25

5. Find y if $\frac{y}{20} = \frac{-1}{4}$.   6. If $\frac{3}{u} = \frac{5}{15}$, then $u =$

Now try the exercises for Topic B on page 294.

C. EQUATIONS IN THE FORM OF A PROPORTION

In Examples 4 and 5, the equations, though more complicated, have the form of a proportion. Cross-multiplying is also useful in these examples.

Example 4 ▶ If $\frac{t}{4} = \frac{t-3}{3}$, find t.

Solution.

$$\frac{t}{4} = \frac{t-3}{3}$$ Cross-multiply. $\frac{t}{4} \diagdown\diagup \frac{t-3}{3}$

$$3t = 4(t-3)$$ Use the distributive laws on the right side.
Thus $4(t-3) = 4t - 4 \cdot 3 = 4t - 12$.

$$3t = 4t - 12$$ Bring variables to the right side by
subtracting $3t$ from both sides.

$$3t - 3t = 4t - 3t - 12$$

$$0 = t - 12$$ Add 12 to both sides.

$$0 + 12 = t - 12 + 12$$

$$12 = t$$ ◄

Example 5 ▶   Find x if $\dfrac{x+1}{5} = \dfrac{3x-4}{8}$.

Solution.

$$\dfrac{x+1}{5} = \dfrac{3x-4}{8}$$ Cross-multiply. $\dfrac{x+1}{5} \diagup\!\!\!\!\diagdown \dfrac{3x-4}{8}$

$$8(x+1) = 5(3x-4)$$ Use the distributive laws.

$$8x + 8 = 15x - 20$$ Bring variables to the right side by subtracting $8x$ from both sides.

$$8x - 8x + 8 = 15x - 8x - 20$$

$$8 = 7x - 20$$ Add 20 to both sides.

$$8 + 20 = 7x - 20 + 20$$

$$28 = 7x$$ Divide both sides by 7.

$$\dfrac{28}{7} = \dfrac{7x}{7}$$

$$4 = x$$ ◀

PRACTICE EXERCISES FOR TOPIC C

In Exercise 7, draw a circle around the correct letter.

7. If $\dfrac{2x-1}{10} = \dfrac{1}{2}$, then $x =$

   (A) 1   (B) 2   (C) 3   (D) 4   (E) 5

8. Find y if $\dfrac{y}{3} = \dfrac{y+8}{5}$.

9. If $\dfrac{2-t}{7} = \dfrac{t+4}{14}$, then $t =$

Now try the exercises for Topic C on page 295.

D. MULTIPLYING BY THE *lcd*

In Examples 4 and 5, when we cross-multiplied, we actually multiplied both fractions by the *lcd* (Section 7). Some equations are not proportions, but nevertheless involve fractions. Often, multiplying both sides by the *lcd* of the fractions is the best method of attack.

Example 6 ▶   If $\dfrac{x}{2} - 1 = \dfrac{x}{4}$, then $x =$

   (A) 2   (B) 4   (C) 6   (D) 8   (E) 12

Solution. The *lcd* of $\frac{x}{2}$ and $\frac{x}{4}$ is 4. Multiply both sides by 4.

$$\frac{x}{2} - 1 = \frac{x}{4}$$

$$4 \times \left(\frac{x}{2} - 1\right) = 4 \times \frac{x}{4} \qquad \text{Use the distributive laws on the left side}$$

and note that $4 \times \frac{x}{2} = 2x$.

$$2x - 4 = x$$

$$2x - x - 4 = x - x$$

$$x - 4 = 0$$

$$x - 4 + 4 = 0 + 4$$

$$x = 4$$

The correct choice is (B). ◀

Example 7▶ Find y if $\frac{y}{5} + 3 = \frac{y}{2}$.

Solution. The *lcd* of $\frac{y}{5}$ and $\frac{y}{2}$ is 10. Multiply both sides by 10.

$$\frac{y}{5} + 3 = \frac{y}{2}$$

$$10 \times \left(\frac{y}{5} + 3\right) = 10 \times \frac{y}{2} \qquad \text{Use the distributive laws and note}$$

that $10 \times \frac{y}{5} = 2y$.

$$2y + 30 = 5y \qquad \text{Subtract } 2y \text{ from both sides.}$$

$$2y - 2y + 30 = 5y - 2y$$

$$30 = 3y$$

$$\frac{30}{3} = \frac{3y}{3}$$

$$10 = y \qquad\qquad ◀$$

PRACTICE EXERCISES FOR TOPIC D

In Exercise 10, draw a circle around the correct letter.

10. If $\frac{x}{3} + 1 = \frac{x}{6} + 2$, then $x =$

(A) 1   (B) 3   (C) 6   (D) 9   (E) 12

11. Find y if $\dfrac{y}{5} - 1 = \dfrac{y}{2} - 7$

12. If $\dfrac{t+4}{2} = \dfrac{t}{8} + 8$, then $t =$

Now try the exercises for Topic D on page 296.

EXERCISES

A. *In Exercises 1–6, express each of the following ratios as fractions.*

1. 5 to 1   *Answer:*   2. 1 to 5   *Answer:*   3. 7 to 2   *Answer:*

4. 4 to 3   *Answer:*   5. 3 to 5   *Answer:*   6. 10 to 7   *Answer:*

B. *In Exercises 7–12, draw a circle around the correct letter.*

7. If $\dfrac{x}{4} = \dfrac{5}{2}$, find x.

   (A) 4   (B) 8   (C) 10   (D) $\dfrac{5}{8}$ (E) $\dfrac{4}{5}$

8. If $\dfrac{x}{3} = \dfrac{4}{6}$, find x.

   (A) 2   (B) 13   (C) 18   (D) $\dfrac{2}{9}$ (E) 12

9. If $\dfrac{y}{7} = \dfrac{28}{4}$, find y.

   (A) 7   (B) −7   (C) 14   (D) 35   (E) 49

10. If $\dfrac{z}{10} = \dfrac{3}{2}$, find z.

   (A) 5   (B) 10   (C) 15   (D) 30   (E) $\dfrac{3}{20}$

11. If $\dfrac{t}{8} = \dfrac{3}{4}$, then $t =$

   (A) 2   (B) 4   (C) 6   (D) 12   (E) $\dfrac{3}{32}$

12. If $\dfrac{u}{9} = \dfrac{-1}{3}$, then $u =$

   (A) −1   (B) −3   (C) 3   (D) −9   (E) 27

13. If $\dfrac{x}{25} = \dfrac{2}{5}$, then $x =$ 14. If $\dfrac{x}{12} = \dfrac{-5}{6}$, then $x =$

15. Find y if $\dfrac{y}{3} = \dfrac{1}{5}$.   *Answer:*

16. Find x if $\dfrac{x}{4} = \dfrac{2}{3}$.   *Answer:*

17. Find t if $\dfrac{t}{9} = \dfrac{1}{6}$.   *Answer:*

18. Find x if $\dfrac{x}{12} = \dfrac{5}{8}$.   *Answer:*

19. Find x if $\dfrac{3}{x} = \dfrac{1}{3}$.   *Answer:*

20. Find y if $\dfrac{8}{y} = \dfrac{2}{3}$.   *Answer:*

21. If $\dfrac{9}{t} = \dfrac{3}{2}$, find t.   *Answer:*

22. If $\dfrac{-10}{t} = \dfrac{5}{2}$, find t.   *Answer:*

23. If $\dfrac{5}{2x} = \dfrac{1}{4}$, find x.   *Answer:*

24. If $\dfrac{4}{3y} = \dfrac{1}{8}$, find y.   *Answer:*

C. *When several choices are given, draw a circle around the correct letter.*

25. If $\dfrac{x}{2} = \dfrac{x+3}{3}$, find x.

(A) 1   (B) 2   (C) 3   (D) 6   (E) 12

26. If $\dfrac{y}{4} = \dfrac{y+4}{12}$, find y.

(A) 1   (B) 2   (C) 4   (D) 8   (E) 24

27. If $\dfrac{t}{5} = \dfrac{t-6}{4}$, find t.   *Answer:*

28. If $\dfrac{u}{4} = \dfrac{u-2}{2}$, find u.   *Answer:*

29. Find x if $\dfrac{x+1}{2} = \dfrac{x+4}{3}$.

(A) 2   (B) 3   (C) 4   (D) 5   (E) 15

30. Find y if $\dfrac{y+5}{4} = \dfrac{y+3}{3}$.

(A) 0   (B) 3   (C) −3   (D) 8   (E) −1

31. If $\dfrac{x+1}{8} = \dfrac{x-2}{4}$, find x.
Answer:

32. If $\dfrac{x-3}{7} = \dfrac{x+4}{14}$, find x.
Answer:

33. If $\dfrac{3x}{8} = \dfrac{x+1}{4}$, find x.
Answer:

34. Find x if $\dfrac{5x}{4} = \dfrac{x+3}{2}$.
Answer:

35. Find y if $\dfrac{y+3}{7} = \dfrac{2y-3}{5}$.
Answer:

36. Find x if $\dfrac{2x+1}{3} = \dfrac{3x+2}{6}$.
Answer:

37. If $\dfrac{u+1}{2} = \dfrac{2u+2}{5}$, then $u =$

38. If $\dfrac{v-3}{3} = \dfrac{v+6}{12}$, then $v =$

39. If $\dfrac{2x + 1}{8} = \dfrac{4x - 1}{4}$, then $x =$ **40.** If $\dfrac{x + 4}{8} = \dfrac{3x + 5}{10}$, then $x =$

D. *In Exercises 41–44, draw a circle around the correct letter.*

41. If $\dfrac{x}{2} + 1 = \dfrac{x}{3} + 2$, then $x =$

(A) 6   (B) 12   (C) 24   (D) −2   (E) −12

42. If $\dfrac{x}{3} - 2 = \dfrac{x}{9}$, then $x =$

(A) 3   (B) −3   (C) 0   (D) 9   (E) 18

43. If $\dfrac{x}{5} - 2 = \dfrac{x}{10}$, then $x =$

(A) 5   (B) −5   (C) 10   (D) 15   (E) 20

44. If $\dfrac{y}{9} + 1 = \dfrac{y}{6}$, then $y =$

(A) 3   (B) 9   (C) 18   (D) 36   (E) 54

45. Find x if $\dfrac{x}{5} + 3 = \dfrac{x}{2}$.
Answer:

46. Find y if $\dfrac{y}{12} + 1 = \dfrac{y}{10}$.
Answer:

47. Find t if $\dfrac{t}{8} = \dfrac{t}{12} - 1$.
Answer:

48. Find u if $\dfrac{3u}{4} - 1 = \dfrac{2u}{3}$.
Answer

49. Find x if $\dfrac{x}{2} - 6 = \dfrac{x}{3}$.
Answer:

50. Find y if $\dfrac{y}{5} + 1 = \dfrac{y + 1}{3}$.
Answer:

51. If $\dfrac{x}{3} - 2 = \dfrac{x}{5}$, then $x =$ **52.** If $\dfrac{x}{6} - 1 = \dfrac{x}{4}$, then $x =$

53. If $\dfrac{x}{3} + 1 = \dfrac{x}{6}$, then $x =$ **54.** If $\dfrac{x}{7} + 2 = \dfrac{x + 4}{2}$, then $x =$

32. Applied Problems

Some percent problems in Section 18 were stated verbally. Now we consider other types of verbal problems. You must formulate the problem in terms of an equation, and then solve the equation in order to solve the given problem.

A. APPLICATIONS OF EQUATIONS

Example 1 ▶ A drama group charges $8 a ticket to see Eliot's *A Cocktail Party.* If its expenses are $1200, how many tickets must it sell to make a profit of $800?

(A) 100   (B) 125   (C) 150   (D) 200   (E) 250

Solution. Let x be the number of tickets the group must sell. Then $8x$ is the income (in dollars) from the sale of these tickets. Use the formula

Income − Expenses = Profit

to set up the equation.

$$8x - 1200 = 800 \qquad \text{Add 1200 to both sides.}$$
$$8x = 2000$$
$$x = \frac{2000}{8}$$
$$x = 250$$

Thus 250 tickets must be sold [choice (E)].   ◀

Example 2 ▶ A rope 16 feet long is cut into two pieces. If one piece is seven times as long as the other, find the length of the *shorter* piece.

Solution. Let x be the length (in feet) of the shorter piece. The longer piece is seven times as long. Thus its length is expressed by $7x$.

Length of shorter piece $+$ length of longer piece $= 16$ feet

$$x + 7x = 16$$
$$8x = 16$$
$$x = 2$$

The length of the shorter piece is 2 feet.   ◀

PRACTICE EXERCISES FOR TOPIC A

In Exercise 1, draw a circle around the correct letter.

1. A drama club charges 5 dollars for tickets to its performance of *Medea*. If expenses are $550, how many tickets must be sold to make a profit of $250?
   (A) 50   (B) 110   (C) 150   (D) 160   (E) 250

2. A 52-inch board is cut into two pieces in such a way that one piece is three times as long as the other piece. Find the length of the *longer* piece.

Now try the exercises for Topic A on page 300.

B. PROPORTIONS

Recall that a *proportion* is an equation of the form

$$\frac{a}{b} = \frac{c}{d}$$

Proportions were consider in Section 31. Many applied problems are solved by setting up proportions.

Example 3 ▶ Three out of every 5 students who take freshman psychology pass the final. If 120 students pass the final, how many take the course?
   (A) 24   (B) 72   (C) 200   (D) 360   (E) 600

Solution. Let x be the number of students who take freshman psychology. We are told that 3 *out of* 5 students pass the course. In other words, $\frac{3}{5}$ of the students pass the course. Thus we can set up the proportion

$$\frac{\text{number who pass the final}}{\text{number who take freshman psych}} = \frac{3}{5}$$

Because 120 students pass the final and x students take freshman psych, this proportion can be written as

$$\frac{120}{x} = \frac{3}{5} \qquad \text{Cross-multiply.}$$

$$120 \times 5 = 3x$$

$$600 = 3x$$

$$200 = x$$

The correct choice is (C).   ◀

Example 4 ▶ 25% of the 80 students at a school of music are women. The school decides to admit just enough additional women so that 50% of the students will then be women. How many additional women must be admitted?

Solution. Here the first sentence tells us how many women are presently at the music school. Recall that "of" in a sentence like this indicates multiplication.

25% of the 80 students are women

$$.25 \ \times \ \ 80 \ \ \ = \ \text{(number of) women}$$

$$
\begin{array}{r}
.25 \ \longleftarrow \ \text{2 decimal digits} \\
\times\, 80 \ \longleftarrow \ + 0 \text{ decimal digits} \\
\hline
20.00 \ \longleftarrow \ \text{2 decimal digits}
\end{array}
$$

Thus presently, there are 20 women.

Now let x be the number of *additional* women that must be admitted. (No additional men are to be admitted.) Thus the *totals* for women and for *all* students are as follows:

	Number at Present	+	Additional Number
Women	20	+	x
All students	80	+	x

Because 50% of the students will then be women and because $50\% = \frac{1}{2}$, set up the proportion

$$\frac{\text{total number of women}}{\text{total number of students}} = \frac{1}{2}$$

$$\frac{20 + x}{80 + x} = \frac{1}{2} \qquad \text{Cross-multiply.}$$

$$2(20 + x) = 1(80 + x) \qquad \text{Use the distributive laws.}$$

$$40 + 2x = 80 + x$$

$$40 + 2x - x = 80 + x - x$$

$$40 + x = 80$$

$$x = 40 \qquad\qquad\qquad\qquad\qquad \blacktriangleleft$$

PRACTICE EXERCISES FOR TOPIC B

3. A survey finds that 7 out of every 10 women in a town plan to vote for the Republican candidate for mayor. If 126 of the women surveyed plan to vote for the Republican candidate, how many women are surveyed?

4. Twenty-five percent of the 400 workers in a factory are from minority groups. How many additional minority-group workers must be hired in order that 50 percent of the workers be from minority groups?

5. A student gets grades of 84, 92, and 88 on her first three chemistry exams. What grade must she obtain on her fourth exam in order to have a 90 average?

Now try the exercises for Topic B on page 301.

EXERCISES

A. *In Exercises 1–4, draw a circle around the correct letter.*

1. At a football game each ticket costs $5. If expenses for the game total $2500, how many tickets must be sold in order to make a profit of $2000?

   (A) 100   (B) 400   (C) 500   (D) 900   (E) 1000

2. A publisher charges bookstores $9 per dictionary. If expenses amount to $4500, how many dictionaries must be sold in order for the publisher to earn $8100 in profits?

   (A) 500   (B) 900   (C) 1400   (D) 2000   (E) 9000

3. An opera company charges $12 per ticket to its performance of *Aida*. Its expenses total $7200, and the company *loses* $2700 on this venture. How many tickets does it sell?

   (A) 225   (B) 375   (C) 600   (D) 825   (E) 975

4. A 32-foot board is sawed into two pieces. If one piece is 5 feet longer than the other, find the length of the *longer* piece.

   (A) 27 feet   (B) 21 feet   (C) 19 feet

   (D) 18 feet 6 inches   (E) 19 feet 6 inches

5. A 24-meter cable is sawed into two pieces. The longer piece is twice as long as the shorter piece. Find the length of the *shorter* piece.   *Answer:*

6. It costs a manufacturer $10 to produce each video game. In addition, there is a general overhead cost of $20,000. He decides to produce 5000 games. If he receives $20 per game from a wholesaler, how many games must he sell to break even? *Answer:*

7. In Exercise 6, how many games must the manufacturer sell to make a profit of $10,000?   *Answer:*

8. A novelty manufacturer finds that it costs her $2 to produce an item, and that there is a general overhead cost of $500. If she is willing to spend $5000 on this venture, how many items can she produce?   *Answer:*

9. Bill has grades of 84, 89, and 96 on his first 3 math exams. What grade must he obtain on his fourth exam in order to have a 91 average for the 4 exams? *Answer:*

10. A saleswoman makes sales totalling $840, $880, $920, and $1055 on four successive days. What must her sales total be on the fifth day to average $1000 per day for this period?   *Answer:*

B. *When several choices are given, draw a circle around the correct letter.*

11. The width and length of an 8 X 10 photograph are enlarged proportionally. If the width of the enlargement is 12 inches, what is the length?

   (A) 9 inches   (B) 12 inches   (C) 15 inches   (D) 16 inches   (E) 20 inches

12. A saleswoman receives a $36 commission on a $200 sale. At this rate, how much does she receive on a $450 sale?

   (A) $64   (B) $72   (C) $80   (D) $81   (E) $90

13. A typist makes 4 errors on 14 pages. At this rate, how many errors will he make on 49 pages?

   (A) 10   (B) 12   (C) 14   (D) 16   (E) 18

14. One month an agency finds that 2 out of every 5 cars sold are sports cars. If the agency sells 70 sports cars that month, how many cars does it sell?

   (A) 35   (B) 90   (C) 105   (D) 175   (E) 350

15. A hostess makes 20 cups of coffee for 14 guests. At this rate, how many cups should she make for 35 guests?   *Answer:*

16. A basketball player makes 3 out of 5 free throws. At this rate, how many free throws must she attempt in order to make 15 of them?   *Answer:*

17. Three out of every 10 students at a college take French. If there are 660 students at the college, how many take French?   *Answer:*

18. Two out of every 9 medical students drop out before graduation. In order to graduate 350 doctors, how many medical students should be admitted?

    (A) 375   (B) 400   (C) 425   (D) 450   (E) 500

19. Thirty percent of the 500 students at a college are from California. In order to have half the students from California, how many additional California students must be admitted?

    (A) 150   (B) 200   (C) 500   (D) 650   (E) 700

20. Ten percent of the 600 employees in a factory are women. In order to have 50% women employees in the factory, how many additional women must be hired?   *Answer:*

21. Forty percent of the 1200 students at a college are from minority groups. In order to have 50% of the students from minority groups, how many additional minority-group students must be admitted?   *Answer:*

33. Square Roots and Right Triangles

The Pythagorean Theorem for right triangles, which will be our final topic, requires a working ability with square roots.

A. SQUARE ROOTS

You know that

$$2^2 = 2 \times 2 = 4$$

Suppose you are asked,

What *positive* number times itself is 4?

In other words, you are asked to find a *positive* number such that

(the number) × (the number) = 4

Clearly, 2 satisfies this condition. 2 is called the "positive square root of 4." Note that the *negative* number –2 also works because

$$(-2) \times (-2) = 4$$

–2 is called the "negative square root of 4."

DEFINITION

> **Square Root.** Let *a* be positive. Then *b* is called the **positive square root of *a*** if *b* is positive and if
>
> $$b^2 = a$$
>
> In this case, –*b* is called the **negative square root of *a*.** Also the **square root of 0** is 0.

Write

$$\sqrt{a} = b \qquad Read: \text{ The positive square root of } a \text{ equals } b,$$

if *b* is the *positive* square root of *a*. Thus

$$\sqrt{4} = 2 \qquad \text{The } positive \text{ square root of 4 is 2.}$$

Also, write

$$\sqrt{0} = 0$$

Example 1 ▶ Find: *i.* $\sqrt{9}$ *ii.* $\sqrt{25}$ *iii.* $\sqrt{144}$

Solution.

i. $\sqrt{9} = 3$ because 3 is positive and $3^2 = 9$.

ii. $\sqrt{25} = 5$ because 5 is positive and $5^2 = 25$.

iii. $\sqrt{144} = 12$ because 12 is positive and $12^2 = 144$.   ◀

To indicate that –2 is the *negative square root* of 4, write

$$-\sqrt{4} = -2$$

Similarly,

$$-\sqrt{9} = -3$$

Next, observe that

$$1^2 = 1 \qquad \text{and} \qquad 2^2 = 4$$

Is it possible to find a positive number a whose *square* equals 2? In other words, can we find a positive number a such that

$$a^2 = 2$$

and therefore

$$a = \sqrt{2}$$

Because

$$1^2 = 1, \quad (\sqrt{2})^2 = 2, \quad \text{and} \quad 2^2 = 4$$

as well as

$$1 < 2 \quad \text{and} \quad 2 < 4$$

it follows that

$$1 < \sqrt{2} \quad \text{and} \quad \sqrt{2} < 2$$

Furthermore,

$$1.4^2 = 1.96 \quad \text{and} \quad 1.5^2 = 2.25$$

Surely,

$$1.4^2 < 2 \quad \text{and} \quad 2 < 1.5^2$$

Thus

$$1.4 < \sqrt{2} \quad \text{and} \quad \sqrt{2} < 1.5$$

Next, observe that

```
    1.41            1.42
  X 1.41          X 1.42
  ------          ------
    1 41            2 84
   56 4            56 8
  1 41            1 42
  -------         -------
  1.98 81         2.01 64
```

Thus

$$(1.41)^2 = 1.9881 \quad \text{and} \quad (1.42)^2 = 2.0164$$

It follows that

$$(1.41)^2 < 2 \quad \text{and} \quad 2 < (1.42)^2$$

and therefore that

$$1.41 < \sqrt{2} \quad \text{and} \quad \sqrt{2} < 1.42$$

Is there such a number $\sqrt{2}$? Consider the three squares in the following figure.

1.9881 square inches	2 square inches	2.0164 square inches
1.41 inches		1.42 inches

With very precise instruments, we might be able to draw the first and third squares whose side lengths are 1.41 inches and 1.42 inches, respectively. Whether or not we can actually draw the middle square, doesn't it seem reasonable that there should be a square whose area is 2 squares inches? The length of a side is then $\sqrt{2}$ inches.

It can be shown that $\sqrt{2}$, like the number π, is an *irrational* number and cannot be written as an "ordinary" or an infinite repeating decimal. But for most purposes, it can be approximated closely enough by using

$$\sqrt{2} \approx 1.414 \quad \textit{Read:} \quad \sqrt{2} \text{ is approximately equal to } 1.414.$$

Similarly,

$$\sqrt{3} \approx 1.732$$

PRACTICE EXERCISES FOR TOPIC A

Find each square root.

1. $\sqrt{64} =$ 2. $\sqrt{100} =$

Now try the exercises for Topic A on page 311.

B. PRODUCTS AND QUOTIENTS OF ROOTS

Observe that

$$\sqrt{9 \times 4} = \sqrt{36} = 6$$

Furthermore,

$$\sqrt{9} \times \sqrt{4} = 3 \times 2 = 6$$

It follows that

$$\sqrt{9 \times 4} = \sqrt{9} \times \sqrt{4}$$

This suggests that (with suitable restrictions) the *square root of a product equals the product of the square roots.* More precisely, let a and b be positive. Then

$$\sqrt{ab} = \sqrt{a}\,\sqrt{b}$$

Example 2 ▶ $\sqrt{2500} =$

(A) 5   (B) 50   (C) 500   (D) 62,500   (E) 6,250,000

Solution. Let $a = 25$ and $b = 100$. Use the above rule.

$$\sqrt{2500} = \sqrt{25 \times 100}$$
$$= \sqrt{25} \times \sqrt{100}$$
$$= 5 \times 10$$
$$= 50$$

The correct choice is (B).   ◀

Sometimes we use the preceding rule in the reverse order. Thus, for positive numbers a and b,

$$\sqrt{a}\ \sqrt{b} = \sqrt{ab}$$

The product of the roots is the root of the product.

Example 3 ▶ $\sqrt{18} \times \sqrt{2} =$

Solution. Let $a = 18$ and $b = 2$. Use

$$\sqrt{a}\ \sqrt{b} = \sqrt{ab}$$
$$\sqrt{18} \times \sqrt{2} = \sqrt{\underbrace{18 \times 2}_{36}}$$
$$= \sqrt{36}$$
$$= 6$$ ◀

Next, observe that

$$\sqrt{\frac{100}{25}} = \sqrt{4} = 2$$

Also,

$$\frac{\sqrt{100}}{\sqrt{25}} = \frac{10}{5} = 2$$

It follows that

$$\sqrt{\frac{100}{25}} = \frac{\sqrt{100}}{\sqrt{25}}$$

The square root of a quotient equals the quotient of the square roots. More precisely, if a and b are both positive, then

$$\sqrt{\frac{a}{b}} = \frac{\sqrt{a}}{\sqrt{b}}$$

Example 4 ▶ $\sqrt{\dfrac{4}{9}} =$

(A) $\dfrac{16}{81}$ (B) $\dfrac{4}{81}$ (C) $\dfrac{4}{9}$ (D) $\dfrac{2}{9}$ (E) $\dfrac{2}{3}$

Solution. Let $a = 4$ and $b = 9$ in the preceding rule.

$$\sqrt{\frac{4}{9}} = \frac{\sqrt{4}}{\sqrt{9}} = \frac{2}{3}$$

The correct choice is (E).   ◀

PRACTICE EXERCISES FOR TOPIC B

In Exercise 3, draw a circle around the correct letter.

3. $\sqrt{1600} =$

  (A) 4   (B) 40   (C) 100   (D) 400   (E) 4000

4. $\sqrt{27} \times \sqrt{3} =$ 5. $\sqrt{\dfrac{9}{4}} =$

Now try the exercises for Topic B on page 311.

C. SQUARE ROOTS AND OTHER OPERATIONS

Example 5 ▶ Find $\sqrt{9 + 16}$.

(A) 3   (B) 4   (C) 5   (D) 7   (E) 25

Solution. You are asked to find the (positive) square root of the sum, $9 + 16$. Thus first add 9 and 16; then find the square root of the sum.

$$\sqrt{\underbrace{9 + 16}_{25}} = \sqrt{25} = 5$$

The correct choice is (C).   ◀

Observe that

$$\sqrt{9} = 3 \quad \text{and} \quad \sqrt{16} = 4$$

so that

$$\sqrt{9} + \sqrt{16} = 3 + 4 = 7$$

But, in Example 5 we saw that

$$\sqrt{9 + 16} = \sqrt{25} = 5$$

Therefore,

$$\sqrt{9 + 16} \neq \sqrt{9} + \sqrt{16}$$

(Read: $\sqrt{9 + 16}$ does *not* equal $\sqrt{9} + \sqrt{16}$.) In general, for positive numbers a and b,

$$\sqrt{a + b} \neq \sqrt{a} + \sqrt{b}$$

and if $a > b$,

$$\sqrt{a - b} \neq \sqrt{a} - \sqrt{b}$$

To sum up, for positive numbers a and b,

$$\sqrt{ab} = \sqrt{a}\,\sqrt{b} \quad \text{and} \quad \sqrt{\frac{a}{b}} = \frac{\sqrt{a}}{\sqrt{b}}$$

However,

$$\sqrt{a + b} \neq \sqrt{a} + \sqrt{b} \quad \text{and} \quad \sqrt{a - b} \neq \sqrt{a} - \sqrt{b}$$

Example 6 will illustrate a process that has an important application to geometry [*Topic* (D)].

Example 6 ▶ Find $\sqrt{10^2 - 6^2}$.

(A) 2   (B) 4   (C) 8   (D) 64   (E) 4096

Solution. First square; then subtract; then find the square root of the difference.

$$\sqrt{\underset{100}{10^2} - \underset{36}{6^2}} = \sqrt{\underset{64}{100 - 36}}$$

$$= \sqrt{64}$$

$$= 8$$

The correct choice is (C).   ◀

PRACTICE EXERCISES FOR TOPIC C

In Exercise 6, draw a circle around the correct letter.

6. $\sqrt{25 - 9} =$

   (A) 2   (B) 3   (C) 4   (D) 5   (E) 16

7. $\sqrt{6^2 + 8^2} =$ 8. $\sqrt{5^2 - 0^2} =$

Now try the exercises for Topic C on page 312.

D. PYTHAGOREAN THEOREM

A **right triangle** is a triangle in which one of the angles is 90°. This angle is called a **right angle**. The side opposite the right angle is called the **hypotenuse**. The ancient Greeks discovered that the sides of lengths a, b, and c of a right triangle are related by the formula

$$c^2 = a^2 + b^2$$

Here c is the length of the hypotenuse. This relationship is known as the **Pythagorean Theorem**. (*See the figure below.*) Obtain the square root of each side of the above equation. Because c is positive,

$$c = \sqrt{a^2 + b^2}$$

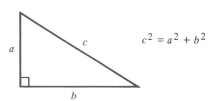

Example 7 ▶ In the right triangle shown here, find c.

   (A) 5   (B) 7   (C) 9
   (D) 25   (E) 49

Solution.

$$c = \sqrt{a^2 + b^2}$$

Substitute 3 for a and 4 for b.

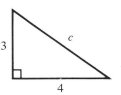

$$
\begin{aligned}
c &= \sqrt{3^2 + 4^2} \\
&= \sqrt{9 + 16} \\
&= \sqrt{25} \\
&= 5
\end{aligned}
$$

The correct choice is (A). ◀

Example 8 ▶ In the right triangle shown here, find *b*.

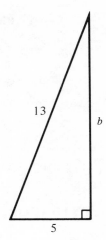

Solution. Here the right triangle is "turned around." We are given that $a = 5$ and that c, the length of the hypotenuse, is equal to 13.

$$a^2 + b^2 = c^2$$ Subtract a^2 from each side.

$$b^2 = c^2 - a^2$$ Find the (positive) square root of each side.

$$b = \sqrt{c^2 - a^2}$$

Substitute 13 for c and 5 for a.

$$b = \sqrt{13^2 - 5^2}$$
$$= \sqrt{169 - 25}$$
$$= \sqrt{144}$$
$$= 12$$

Example 9 ▶ In the right triangle shown here, find *c*.

(A) 2   (B) 4   (C) 8

(D) 6   (E) $\sqrt{34}$

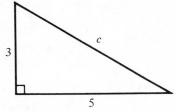

Solution.

$$c = \sqrt{a^2 + b^2}$$

Substitute 3 for a and 5 for b.

$$c = \sqrt{3^2 + 5^2}$$
$$= \sqrt{9 + 25}$$
$$= \sqrt{34}$$

$\sqrt{34}$ is *less than* 6 because $\sqrt{36} = 6$. Thus the correct choice is (E).   ◀

PRACTICE EXERCISES FOR TOPIC D

Find the length of the indicated side of the right triangle.

9.

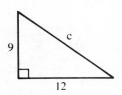

10.

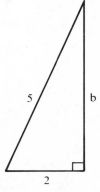

Now try the exercises for Topic D on page 313.

EXERCISES

A. *Find each square root.*

1. $\sqrt{16} =$ 2. $\sqrt{36} =$ 3. $\sqrt{49} =$ 4. $\sqrt{81} =$

5. $\sqrt{121} =$ 6. $\sqrt{169} =$ 7. $\sqrt{0} =$ 8. $\sqrt{1} =$

B. *When several choices are given, draw a circle around the correct letter.*

9. $\sqrt{400} =$
   (A) 15   (B) 20   (C) 25   (D) 100   (E) 200

10. $\sqrt{900} =$
    (A) 3   (B) 30   (C) 300   (D) 450   (E) 90

11. $\sqrt{8100} =$
    (A) 9   (B) 81   (C) 90   (D) 900   (E) 4050

12. $\sqrt{10,000} =$
    (A) 10   (B) 100   (C) 1000   (D) 5000   (E) 100,000

13. $\sqrt{640,000} =$ 14. $\sqrt{1,000,000} =$

15. $\sqrt{3} \times \sqrt{12} =$
    (A) 15   (B) $\sqrt{15}$ (C) 36   (D) 6   (E) 216

16. $\sqrt{8} \times \sqrt{2} =$
    (A) 16   (B) 256   (C) 10   (D) $\sqrt{10}$ (E) 4

17. $\sqrt{5} \times \sqrt{5} =$ 18. $\sqrt{20} \times \sqrt{5} =$

19. $\sqrt{32} \times \sqrt{2} =$ 20. $\sqrt{48} \times \sqrt{3} =$

21. $\sqrt{\dfrac{1}{16}} =$

    (A) $\dfrac{1}{2}$ (B) $\dfrac{1}{4}$ (C) $\dfrac{1}{8}$ (D) 4   (E) $\dfrac{1}{256}$

22. $\sqrt{\dfrac{9}{25}} =$

    (A) $\dfrac{3}{25}$ (B) $\dfrac{9}{5}$ (C) $\dfrac{3}{5}$ (D) $\dfrac{81}{5}$ (E) $\dfrac{81}{25}$

23. $\sqrt{\dfrac{4}{81}} =$ 24. $\sqrt{\dfrac{49}{100}} =$ 25. $\sqrt{\dfrac{25}{64}} =$

26. $\sqrt{\dfrac{49}{4}} =$ 27. $\sqrt{\dfrac{100}{81}} =$ 28. $\sqrt{\dfrac{121}{144}} =$

C. *When several choices are given, draw a circle around the correct letter.*

29. $\sqrt{4 + 5} =$

    (A) $\sqrt{6}$ (B) 6   (C) 9   (D) 3   (E) $\sqrt{45}$

30. $\sqrt{7 + 18} =$

    (A) 25   (B) 5   (C) 125   (D) $\sqrt{125}$ (E) $\sqrt{7} + \sqrt{18}$

31. $\sqrt{16 - 7} =$

    (A) 9   (B) 3   (C) $\sqrt{16} - \sqrt{7}$ (D) $\sqrt{\dfrac{16}{7}}$ (E) 81

32. $\sqrt{100 - 64} =$

    (A) 2   (B) 4   (C) 6   (D) 8   (E) 36

33. $\sqrt{4 + 9} =$ 34. $\sqrt{49 - 36} =$

35. $\sqrt{3^2 + 4^2} =$

    (A) 25   (B) 125   (C) 5   (D) 7   (E) 49

36. $\sqrt{8^2 + 6^2} =$

    (A) 10   (B) 100   (C) 14   (D) 48   (E) $\sqrt{48}$

37. $\sqrt{10^2 - 8^2} =$

    (A) 2   (B) 4   (C) 16   (D) 6   (E) 36

38. $\sqrt{13^2 - 12^2} =$

    (A) 11   (B) 121   (C) 5   (D) 25   (E) 125

39. $\sqrt{4^2 + 5^2} =$ 40. $\sqrt{4^2 - 3^2} =$

D. *Find the length of the indicated side of the right triangle. Draw a circle around the correct letter.*

41.

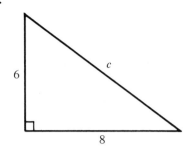

    (A) 10   (B) 12   (C) 14

    (D) 2   (E) 196

42.

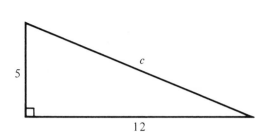

    (A) 13   (B) 15   (C) 17

    (D) 20   (E) 25

43.

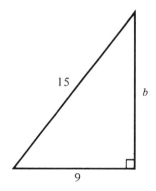

    (A) 6   (B) 9   (C) 10

    (D) 12   (E) 24

44.

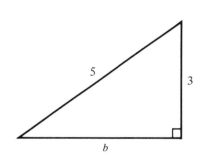

    (A) 2   (B) 3   (C) 4

    (D) $\sqrt{3}$ (E) 8

45.

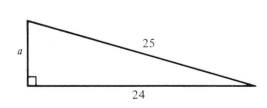

    (A) 1   (B) 2   (C) 5

    (D) 7   (E) 12

46.

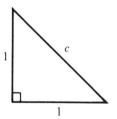

    (A) 1   (B) 2   (C) 3

    (D) $\sqrt{2}$ (E) $\sqrt{3}$

47.

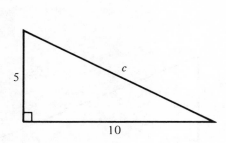

(A) 5   (B) 10   (C) 12

(D) 15   (E) $\sqrt{125}$

48.

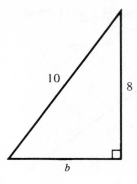

(A) 2   (B) 4   (C) 6

(D) $\sqrt{6}$ (E) 36

49.

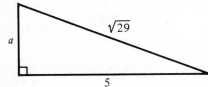

(A) 2   (B) 4   (C) 8

(D) 24   (E) $\sqrt{24}$

50.

(A) 3   (B) 4   (C) 5

(D) $\sqrt{5}$ (E) 25

Review Exercises for Unit VII

When several choices are given, draw a circle around the correct letter.

In Exercises 1 and 2, suppose that x stands for some number. Express each of the following in terms of x.

1. 10 less than the number. *Answer:*

2. 7 more than three times the number. *Answer:*

3. Find the value, in cents, of 7 dimes and 3 nickels. *Answer:*

4. A car travels for 2 hours at x miles per hour and then for 3 hours at y miles per hour. How many miles does it travel?

   (A) $2x$ (B) $3y$ (C) $2x + 3y$ (D) $2x - 3y$ (E) $6xy$

5. Find the value of $x^3 + 4x - 3$ when $x = 2$. *Answer:*

6. Find the value of $3^a - 1$ when $a = 2$.   *Answer:*

7. If $y = 3ab^2 - a + 1$, find y when $a = 3$ and $b = -1$.

   (A) -11 (B) -13 (C) 6 (D) 7 (E) 8

8. The height (in feet) of a helicopter is given by $t^2 + 2t$ (seconds). Find its height after 4 seconds.   *Answer:*

9. What is the coefficient of $12a^2b^3$?   *Answer:*

10. What is the coefficient of $(2x)(5y^3)$?   *Answer:*

11. Fill in the terms: $3x^2 + 4x - 1$ is the sum of the terms $\boxed{}$, $\boxed{}$, and $\boxed{}$.

In Exercises 12 and 13, answer "yes" or "no."

12. Are $4x^2y^3$ and $3x^3y^2$ like terms?   *Answer:*

13. Are $-2ab^4$ and ab^4 like terms?   *Answer:*

In Exercises 14–17, add or subtract. If the given expression cannot be simplified, just write down the expression.

14. $5a + 4a =$

15. $7b^2 + 2b =$

16. $10x + 3y - 2x + 5y =$

17. $4a - 2b + 1 - 3a =$

18. Find the sum of $10x + 5y$ and $3x - 2y$.   *Answer:*

19. Subtract $4xy + 1$ from $10xy + 3$.   *Answer:*

20. What is 3 less than $5a - b - 1$?   *Answer:*

In Exercises 21 and 22, answer "yes" or "no."

21. Is 3 a root of $4x - 8 = x + 2$?   *Answer:*

22. Is 7 a root of $10 - t = 2t - 11$?   *Answer:*

In Exercises 23–30, find each root.

23. $5x = 20$ *Answer:*

24. $x + 5 = 20$ *Answer:*

25. $x - 5 = 20$ *Answer:*

26. $\dfrac{x}{5} = 20$ *Answer:*

27. $4y + 3 = 8y - 1$ *Answer:*

28. $\dfrac{t}{4} = \dfrac{3}{2}$ *Answer:*

29. $\dfrac{t-1}{10} = \dfrac{3}{5}$ *Answer:*

30. $\dfrac{4}{x} = \dfrac{6}{15}$ *Answer:*

31. Find t if $4t - 3 = 3t + 7$.

   (A) 1 (B) 10 (C) 20 (D) 33 (E) 37

32. If $\dfrac{4x-3}{4} = \dfrac{2x-3}{4}$, then $x =$

(A) 0   (B) 1   (C) 3   (D) 4   (E) −3

33. If $\dfrac{x}{4} + 1 = \dfrac{x}{2} - 1$, then $x =$

(A) 1   (B) 2   (C) 4   (D) 8   (E) 16

34. It costs $4400 to publish a pamphlet. If the pamphlet sells for $2, how many copies must be sold to make a $2000 profit? *Answer:*

35. If 3 pounds of grain feed 50 rats, how many pounds of grain are needed to feed 125 rats? *Answer:*

36. Anita has grades of 76, 80, and 85 on her first three math quizzes. What grade does she need on her fourth quiz to obtain an 85 average? *Answer:*

37. Find $\sqrt{121}$.   *Answer:*   **38.** Find $\sqrt{9} + \sqrt{4}$. *Answer:*

39. Find $\sqrt{6^2 + 8^2}$. *Answer:*   **40.** Find $\sqrt{6^2} + \sqrt{8^2}$. *Answer:*

41. Find the length of the indicated side of the right triangle.

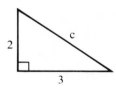

(A) 4   (B) 5   (C) $\sqrt{5}$ (D) 13   (E) $\sqrt{13}$

Review Exercises on Units I–VI

1. Express 54,000 as the product of primes. *Answer:*

2. Find *lcm* (12, 16, 20). *Answer:*

3. What is the largest power of 10 that divides 1,050,000. *Answer:*

4. $\dfrac{3}{14} \times \dfrac{7}{9} =$ **5.** $\dfrac{13}{20} - \dfrac{4}{5} =$ **6.** $1.08 \times .012 =$ **7.** $\dfrac{.0096}{400} =$

8. Express .012 as a fraction in lowest terms. *Answer:*

9. Express 250% as a mixed number. *Answer:*

10. A student answers 76% of the 50 questions on a test correctly. How many questions does she answer correctly? *Answer:*

11. Find the length of a rectangle with area 20 square centimeters and width 2 centimeters. *Answer:*

12. The circumference of a circle is 5π inches. Express its area in terms of π. *Answer:*

13. Find the volume of air in a room that is 20 feet by 14 feet by 9 feet. *Answer:*

14. A two-and-one-half hour concert begins at 8:15 in the evening. What time does it finish? *Answer:*

15. $(-5) - (-9) =$ 16. $\left(\dfrac{12 - 3 + 6}{5}\right)^2 =$ 17. $4(8 - 3) + 2 - 7(6 - 10) =$

18. Find the average of 12, 20, 35, 40, and 50. *Answer:*

Practice Exam on Unit VII

When several choices are given, draw a circle around the correct letter.

1. Suppose that x stands for a number. Express

    5 less than twice the number

in terms of x.

(A) $2x - 5$ (B) $5 - 2x$ (C) $5 + 2x$ (D) $(5 - x) - 2$ (E) $5 - \dfrac{x}{2}$

2. A woman buys x \$2-items and y \$5-items. How many dollars does she spend on these purchases?

(A) $x + y$ (B) $7x + y$ (C) $7xy$ (D) $10x + y$ (E) $2x + 5y$

3. Find the value of $4x^3 - 2x + 3$ when $x = -2$. *Answer:*

4. Find the value of $5xy - 3x^2$ when $x = -1$ and $y = -3$. *Answer:*

5. What is the coefficient of $(4x^2)(3y^2)$? *Answer:*

6. $5a - 2b + 4a + 2b =$

7. Find the sum of $4x + 3y - 1$ and $2x - 4y + 3$. *Answer:*

8. What is 6 more than $x + y - 1$? *Answer:*

In Exercises 9–11, find each root.

9. $\dfrac{x}{4} = -7$ *Answer:*   10. $4t - 8 = 2t + 12$ *Answer:*

11. $\dfrac{t - 1}{5} = \dfrac{t + 1}{6}$ *Answer:*

12. If four pounds of cheese suffice for 30 guests, at this rate how many pounds of cheese will suffice for 45 guests? *Answer:*

13. Find $\sqrt{10^2 - 6^2}$. *Answer:*

14. Find the length of the indicated side of the right triangle.

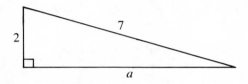

(A) 5   (B) $\sqrt{5}$ (C) 45   (D) $\sqrt{45}$ (E) $\sqrt{47}$

PRACTICE FINAL EXAMS

You are now given three sample final exams for practice. Complete solutions to these practice exam questions are given, beginning on page 384.

Practice Final Exam A

Choose the answer you think is correct. Then, in the answer column at the right, fill in the box underneath the corresponding letter.

1. Four hundred twenty thousand five hundred is written

    (A) 420,005   (B) 425,000   (C) 425,500

    (D) 420,500   (E) 420,000.05

2. 328 − 79 =

    (A) 249   (B) 259   (C) 407   (D) 149   (E) 159

3. 6512 ÷ 16 =

    (A) 470   (B) 407   (C) 47   (D) 40   (E) 40.7

	A	B	C	D	E
1	▯	▯	▯	▯	▯
2	▯	▯	▯	▯	▯
3	▯	▯	▯	▯	▯

4. $4^3 =$

(A) 43   (B) 444   (C) 16   (D) 64   (E) 81

5.   Express 162 as the product of primes.

(A) 2×81 (B) 2×9^2 (C) 2×3^4

(D) $(2 \times 3)^4$ (E) $2^4 \times 3$

6.   Which of these fractions is the smallest?

(A) $\frac{2}{3}$ (B) $\frac{3}{5}$ (C) $\frac{3}{4}$ (D) $\frac{5}{8}$ (E) $\frac{5}{6}$

7. $\frac{2}{9} + \frac{3}{4} =$

(A) $\frac{5}{13}$ (B) $\frac{5}{36}$ (C) $\frac{35}{36}$ (D) $\frac{36}{35}$ (E) $\frac{11}{36}$

8. $4\frac{1}{2} - 2\frac{1}{3} =$

(A) $2\frac{1}{3}$ (B) $2\frac{1}{6}$ (C) $1\frac{1}{6}$ (D) $2\frac{1}{5}$ (E) $1\frac{1}{5}$

9. $8\frac{1}{4} \div 2\frac{3}{4} =$

(A) 3   (B) $3\frac{1}{2}$ (C) 4   (D) $\frac{4}{33}$ (E) $\frac{363}{16}$

10. $\left(\frac{2}{3}\right)^2 =$

(A) $\frac{2}{9}$ (B) $\frac{4}{9}$ (C) $\frac{4}{3}$ (D) $\frac{8}{3}$ (E) $\frac{2}{81}$

11. $15.093 + 2.979 + 8 =$

(A) 25.072   (B) 26.072   (C) 26.062

(D) 25.962   (E) 26.909

12.   Which of these numbers is the smallest?

(A) .0404   (B) .0440   (C) .0309   (D) .0490   (E) .0390

A B C D E
4 ⊓ ⊓ ⊓ ⊓ ⊓

A B C D E
5 ⊓ ⊓ ⊓ ⊓ ⊓

A B C D E
6 ⊓ ⊓ ⊓ ⊓ ⊓

A B C D E
7 ⊓ ⊓ ⊓ ⊓ ⊓

A B C D E
8 ⊓ ⊓ ⊓ ⊓ ⊓

A B C D E
9 ⊓ ⊓ ⊓ ⊓ ⊓

A B C D E
10 ⊓ ⊓ ⊓ ⊓ ⊓

A B C D E
11 ⊓ ⊓ ⊓ ⊓ ⊓

A B C D E
12 ⊓ ⊓ ⊓ ⊓ ⊓

13. $87.8 - 7.96 =$

(A) 79.94   (B) 79.84   (C) 8.2   (D) .82   (E) 8.22

14. Change $\frac{12}{13}$ to a decimal rounded to the nearest hundredth.

(A) .92   (B) .923   (C) .93   (D) .9   (E) .90

15. Change $\frac{2}{9}$ to an infinite repeating decimal.

(A) .2   (B) .222   (C) .222 222 . . .

(D) .29 29 29 . . .   (E) .222 999 . . .

16. $(.04)^2 =$

(A) 1.6   (B) .016   (C) .16   (D) .0016   (E) .000 16

17. A store sells 16 copies of a textbook priced at $7.95. What are the total sales for this book?

(A) $102   (B) $128   (C) $127.20

(D) $12.70   (E) $128.20

18. What is 30% of 150?

(A) 15   (B) 25   (C) 40   (D) 45   (E) 50

19. If 60 is 80% of a certain number, find that number.

(A) 48   (B) 480   (C) 72   (D) 75   (E) 90

20. If a $45 dress is on sale reduced by 10%, its sale price is

(A) $4.50   (B) $49.50   (C) $40   (D) $40.50   (E) $41.50

21. A club sells 952 raffles at $3 each. The printing of the raffle books costs $97. If there are $52 in other expenses, the profit is

(A) $2856   (B) $2707   (C) $2717   (D) $3005   (E) $2759

	A	B	C	D	E
13	[]	[]	[]	[]	[]
14	[]	[]	[]	[]	[]
15	[]	[]	[]	[]	[]
16	[]	[]	[]	[]	[]
17	[]	[]	[]	[]	[]
18	[]	[]	[]	[]	[]
19	[]	[]	[]	[]	[]
20	[]	[]	[]	[]	[]
21	[]	[]	[]	[]	[]

22.   27 pounds 12 ounces
      −18 pounds 14 ounces

(A) 9 pounds 2 ounces   (B) 8 pounds 2 ounces

(C) 8 pounds 8 ounces   (D) 8 pounds 14 ounces

(E) 9 pounds 14 ounces

23.   Find the area of a rectangular table top that is 40 inches long and 22 inches wide.

(A) 800 in.²   (B) 880 in.²   (C) 62 in.

(D) 124 in.   (E) 808 in.²

24.   The base of a triangle is 6 inches and the height is 4 inches. The area of the triangle is

(A) 6 in.²   (B) 9 in.²   (C) 12 in.²   (D) 18 in.²   (E) 24 in.²

25.   Use the graph at the right to find the total number of copies of Health Digest sold from September through December, 1984.

1984 SALES
(Thousands)

(A) 11,000   (B) 12,000

(C) 12,500   (D) 13,000

(E) 14,000

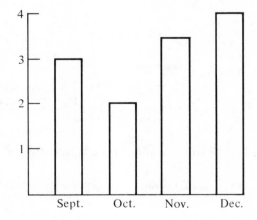

26. $2(-5)^2 + 4(-7) =$

(A) −78   (B) 22   (C) 72   (D) 6   (E) −128

27. $5(2 - 4) + (3 - 1)^2 =$

(A) −6   (B) 6   (C) 14   (D) −2   (E) 0

	A	B	C	D	E
22	[]	[]	[]	[]	[]
23	[]	[]	[]	[]	[]
24	[]	[]	[]	[]	[]
25	[]	[]	[]	[]	[]
26	[]	[]	[]	[]	[]
27	[]	[]	[]	[]	[]

28. $\dfrac{(10 - 1)^2}{3} =$

    (A) 3   (B) 9   (C) 27   (D) 33   (E) 81

29. The average of 78, 81, 82, and 91 is

    (A) 80   (B) 82.5   (C) 83   (D) 84   (E) 85

30. Find the sum of $4x + 9y$ and $6x - 9y$.

    (A) $10x$ (B) $18y$ (C) $10x - 18y$

    (D) $10x + 9y$ (E) $2x - 18y$

31. Add $3x^2 - 2x + 4$ and $x^2 - 5$.

    (A) $4x^2 - 2x - 1$ (B) $4x^2 - 2x + 1$ (C) $4x^2 + 2x - 1$

    (D) $4x^2 - 1$ (E) $2x^2 - 2x + 9$

32. Find the value, in dollars, of t ten-dollar bills and f five-dollar bills.

    (A) $f + t$ (B) $50\,ft$ (C) $5f + 10t$

    (D) $5(f + 10t)$ (E) $f + 10t$

33. Find x if $4x + 3 = 5x - 1$.

    (A) 1   (B) 2   (C) 3   (D) 4   (E) 5

34. If $x = 4$ and $y = -3$, find the value of $x^2 - 5xy$.

    (A) 76   (B) -54   (C) -44   (D) -51   (E) 69

35. If $\dfrac{x}{5} = \dfrac{-3}{4}$, then $x =$

    (A) 15   (B) -15   (C) $-\dfrac{15}{4}$ (D) $-\dfrac{3}{20}$ (E) $-\dfrac{20}{3}$

36. If $c = 8ab^2$, find c when $a = -1$ and $b = 2$.

    (A) -16   (B) 16   (C) -32   (D) 32   (E) 256

37. A batter gets 2 hits for every 7 times at bat. How many times must he come to bat in order to have 150 hits?

    (A) 350   (B) 420   (C) 500   (D) 525   (E) 550

	A	B	C	D	E
28	[ ]	[ ]	[ ]	[ ]	[ ]
29	[ ]	[ ]	[ ]	[ ]	[ ]
30	[ ]	[ ]	[ ]	[ ]	[ ]
31	[ ]	[ ]	[ ]	[ ]	[ ]
32	[ ]	[ ]	[ ]	[ ]	[ ]
33	[ ]	[ ]	[ ]	[ ]	[ ]
34	[ ]	[ ]	[ ]	[ ]	[ ]
35	[ ]	[ ]	[ ]	[ ]	[ ]
36	[ ]	[ ]	[ ]	[ ]	[ ]
37	[ ]	[ ]	[ ]	[ ]	[ ]

38. In the right triangle shown here, find x.

(A) 5   (B) 17   (C) 289

(D) $\sqrt{17}$ (E) $\sqrt{15}$

39. $\sqrt{4^2} - \sqrt{1^2} =$

(A) $\sqrt{2}$ (B) 2   (C) $\sqrt{3}$ (D) 3   (E) $\sqrt{15}$

40. It costs a toy manufacturer $5 to produce a hockey game. In addition, there is a general overhead cost of $15,000. He decides to produce 6000 games. If he receives $15 per game from a wholesaler, how many games must he sell to break even.

(A) 1000   (B) 3000   (C) 4000   (D) 4500   (E) 5000

Practice Final Exam B

Choose the answer you think is correct. Then, in the answer column at the right, fill in the box underneath the corresponding letter.

1. Sixty thousand four hundred five is written

(A) 60,045   (B) 64,105   (C) 60,405   (D) 64,405   (E) 60,045

2. $6061 \div 19 =$

(A) 309   (B) 319   (C) 39   (D) 320   (E) 31.9

3. $3^3 \times 10^2 =$

(A) 900   (B) 270   (C) 2700   (D) 27,000   (E) 9000

4. Express 640 as the product of primes.

(A) $2^7 \times 5$ (B) $2^6 \times 5$ (C) $2^6 \times 10$

(D) $2^5 \times 5$ (E) $2^5 \times 10$

5. Which of the following is a prime?

   (A) 29   (B) 39   (C) 49   (D) 69   (E) 99

6. Find the least common multiple of 20 and 25.

   (A) 5   (B) 10   (C) 40   (D) 100   (E) 200

7. $\frac{1}{8} + \frac{3}{5} =$

   (A) $\frac{4}{13}$ (B) $\frac{27}{40}$ (C) $\frac{4}{40}$ (D) $\frac{1}{10}$ (E) $\frac{29}{40}$

8. $2\frac{3}{4} + 1\frac{1}{4} =$

   (A) $3\frac{3}{4}$ (B) $3\frac{4}{5}$ (C) 3   (D) 4   (E) 5

9. $\left(\frac{1}{4}\right)^3 =$

   (A) $\frac{1}{64}$ (B) $\frac{1}{16}$ (C) $\frac{1}{251}$ (D) $\frac{1}{12}$ (E) $\frac{3}{16}$

10. $3.09 - 1.98 =$

    (A) 2.07   (B) 2.11   (C) 1.11   (D) 1.91   (E) .11

11. $(.2)^3 =$

    (A) .002   (B) .04   (C) .08   (D) .008   (E) .0008

12. It costs a food vendor 65 cents to make up each sandwich. He makes up 80 sandwiches and sells 70 for $1.50 each. What is his profit?

    (A) $59.50   (B) $68   (C) $52

    (D) $105   (E) $53

13. What is 9% of 8?

    (A) 72   (B) 7.2   (C) .72   (D) .072   (E) $\frac{8}{9}$

14. A sweater that normally sells for $32 is reduced in price by 25%. Its new price is

    (A) $8   (B) $24   (C) $7   (D) $25   (E) $25.60

15. If 8 ounces of cheese cost $2.80, how much will a 12-ounce slice of this cheese cost?

(A) $4.20   (B) $3.90   (C) $4.50   (D) $5.60   (E) $3.60

16. If pencils sell for 15 cents each, how many can you buy for $6?

(A) 90   (B) 4   (C) 40   (D) 400   (E) 900

17. $4\frac{1}{2} \div 2\frac{1}{4} =$

(A) $2\frac{1}{4}$ (B) 2   (C) $\frac{1}{2}$ (D) $\frac{1}{4}$ (E) $1\frac{3}{4}$

18. Which of the following fractions is the smallest?

(A) $\frac{1}{7}$ (B) $\frac{1}{10}$ (C) $\frac{2}{11}$ (D) $\frac{2}{15}$ (E) $\frac{3}{20}$

19. Which of the following numbers is the smallest?

(A) 2.04   (B) 2.39   (C) 2.009   (D) 2.102   (E) 2.010

20. $5.08 + 3.038 + 12 =$

(A) 20.46   (B) 23.16   (C) 21.18

(D) 20.118   (E) 20.046

21. A jacket sells for $85 plus a 6% sales tax. What is the total price?

(A) $90   (B) $90.10   (C) $91   (D) $5.10   (E) $51

22. A laborer earns $44 for 8 hours of work. At this rate how much will he earn for 28 hours of work?

(A) $144   (B) $154   (C) $164   (D) $140   (E) $180

23. Oaktag sells for $1.50 per square yard. What is the cost of a piece that is 6 yards by 5 yards?

(A) $3   (B) $40   (C) $45   (D) $20   (E) $60

15 A B C D E
16 A B C D E
17 A B C D E
18 A B C D E
19 A B C D E
20 A B C D E
21 A B C D E
22 A B C D E
23 A B C D E

24.   10 hours   6 minutes
     – 4 hours 51 minutes

24 A B C D E
   [] [] [] [] []

25 A B C D E
   [] [] [] [] []

(A) 6 hours 55 minutes   (B) 5 hours 55 minutes

26 A B C D E
   [] [] [] [] []

(C) 6 hours 15 minutes   (D) 5 hours 15 minutes

27 A B C D E
   [] [] [] [] []

(E) 5 hours 45 minutes

28 A B C D E
   [] [] [] [] []

25.   A 1-pound bag of candy is divided into 8 equal portions. Each portion
      weighs

29 A B C D E
   [] [] [] [] []

30 A B C D E
   [] [] [] [] []

(A) 1 oz.   (B) 1.5 oz.   (C) 2 oz.   (D) 4 oz.   (E) 8 oz.

26.   If the area of a triangle is 12 square inches and the base is 6 inches,
      then the height is

(A) 2 inches   (B) 4 inches   (C) 6 inches

(D) 1 foot   (E) 6 feet

27.   The graph at the right shows the number of
      automobiles sold by a dealer during the first
      four months of the year. The *total* number
      sold during this four-month period is
      closest to

(A) 30   (B) 40   (C) 45

(D) 48   (E) 55

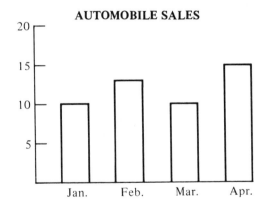

AUTOMOBILE SALES

28. $5(2^2) - (2 \times 5)^2 =$

(A) –80   (B) –40   (C) 0   (D) 50   (E) 100

29.   On four math exams, Sheila's scores are 72, 85, 80, and 87. What is
      her exam average?

(A) 78   (B) 79   (C) 80   (D) 81   (E) 82

30. $\dfrac{-12}{-3} =$

(A) 4   (B) –4   (C) –2   (D) 3   (E) undefined

31. $(-2)(-5) - (-3)^2 =$

    (A) 19   (B) 1   (C) -19   (D) 49   (E) 169

32. If the sales tax rate is 4% and if the sales tax on a table is $6, what is the price of the table (before the sales tax)?

    (A) $24   (B) $240   (C) $100   (D) $150   (E) $250

33. Subtract $4a - 5$ from $5a + 4$.

    (A) $9a - 1$ (B) $a - 1$ (C) $a - 9$

    (D) $a + 9$ (E) $9a + 9$

34. Suppose that gloves cost $9 a pair and scarves cost $7 apiece. The cost of x pairs of gloves and y scarves is

    (A) $x + y$ (B) $9x + y$ (C) $9(x + y)$

    (D) $9x + 7y$ (E) $9(x + 7y)$

35. What is the value of $5ab - b^2$ when $a = 4$ and $b = -1$?

    (A) 24   (B) -21   (C) -19   (D) 19   (E) 21

36. If $5x - 2 = 22$, then $x =$

    (A) 4   (B) 5   (C) $\dfrac{24}{5}$ (D) 120   (E) 100

37. If $\dfrac{x + 3}{2} = \dfrac{3x - 3}{3}$, then $x =$

    (A) 1   (B) 3   (C) 5   (D) -5   (E) 7

38. The width and length of a 5 × 7 photograph are enlarged proportionally. If the width of the enlargement is $7\frac{1}{2}$ inches, what is the length?

    (A) 9 inches   (B) 10 inches   (C) $10\frac{1}{2}$ inches

    (D) 11 inches   (E) 12 inches

39. Simplify $\sqrt{10^2 - 6^2}$.

    (A) 16   (B) 4   (C) 64   (D) 8   (E) $\sqrt{8}$

	A	B	C	D	E
31	[]	[]	[]	[]	[]
32	[]	[]	[]	[]	[]
33	[]	[]	[]	[]	[]
34	[]	[]	[]	[]	[]
35	[]	[]	[]	[]	[]
36	[]	[]	[]	[]	[]
37	[]	[]	[]	[]	[]
38	[]	[]	[]	[]	[]
39	[]	[]	[]	[]	[]

40. Expenses for a dance amount to $1260, and tickets to the dance sell for $8 each. If a profit of $1156 is made, how many tickets are sold?

(A) 200   (B) 144   (C) 300   (D) 302   (E) 312

40 A B C D E

Practice Final Exam C

Choose the answer you think is correct. Then, in the answer column at the right, fill in the box underneath the corresponding letter.

1. Eighty thousand and four-tenths is written

(A) 80,004   (B) 80,410   (C) 80,000.4

(D) 80,000.04   (E) 80,004.1

2. $602 - 98 =$

(A) 604   (B) 504   (C) 598   (D) 584   (E) 700

3. $5^3 =$

(A) 25   (B) 75   (C) 125   (D) 175   (E) 250

4. $2^3 \times 3^2 =$

(A) 24   (B) 32   (C) 36   (D) 54   (E) 72

5. Express 144 as the product of primes.

(A) $2^4 \times 3^2$ (B) $2^5 \times 3$ (C) $2^5 \times 3^2$

(D) $2^2 \times 3^3$ (E) $2^2 \times 3^4$

6. Find the least common multiple of 12 and 32.

(A) 48   (B) 64   (C) 96   (D) 144   (E) 384

7. $\frac{3}{8} - \frac{2}{5} =$

(A) $\frac{1}{3}$ (B) $\frac{1}{8}$ (C) $\frac{1}{40}$ (D) $-\frac{1}{40}$ (E) 0

1 A B C D E
2 A B C D E
3 A B C D E
4 A B C D E
5 A B C D E
6 A B C D E
7 A B C D E

8. $\dfrac{3}{5} \times \dfrac{10}{9} =$

(A) $\dfrac{13}{45}$ (B) $\dfrac{2}{3}$ (C) $\dfrac{13}{14}$ (D) $\dfrac{27}{50}$ (E) $\dfrac{30}{14}$

9. $\dfrac{3}{10} \div \dfrac{9}{20} =$

(A) $\dfrac{2}{3}$ (B) $\dfrac{3}{2}$ (C) $\dfrac{27}{200}$ (D) $\dfrac{2}{5}$ (E) 3

10. $3\dfrac{3}{4} + 1\dfrac{1}{4} =$

(A) 4   (B) 5   (C) $4\dfrac{1}{2}$ (D) $4\dfrac{3}{4}$ (E) $5\dfrac{1}{4}$

11. Which of these fractions is the smallest?

(A) $\dfrac{1}{2}$ (B) $\dfrac{4}{9}$ (C) $\dfrac{5}{11}$ (D) $\dfrac{5}{12}$ (E) $\dfrac{9}{20}$

12. Which of these numbers is the smallest?

(A) .0303   (B) .0299   (C) .0312   (D) .0289   (E) .0298

13. $6.003 + 5.809 + 11 =$

(A) 23.109   (B) 22.812   (C) 23.812

(D) 22.839   (E) 22.849

14. $(.08)^2 =$

(A) .008   (B) .0008   (C) .64   (D) .064   (E) .0064

15. A gallon of gasoline costs $1.25. What is the cost of 16 gallons of gasoline?

(A) $16   (B) $18   (C) $19   (D) $20   (E) $20.50

16. Find the cost of 5 pounds of potatoes at 22 cents per pound and 3 pounds of onions at 19 cents per pound.

(A) $1.10   (B) $.57   (C) $1.67   (D) $2.28   (E) $3.28

	A	B	C	D	E
8					
9					
10					
11					
12					
13					
14					
15					
16					

17. Change $\frac{4}{13}$ to a decimal rounded to the nearest hundredth.

    (A) .04   (B) .30   (C) .307   (D) .31   (E) .37

17 A B C D E
18 A B C D E
19 A B C D E
20 A B C D E
21 A B C D E
22 A B C D E
23 A B C D E
24 A B C D E

18. If 48 is 75% of a certain number, find that number.

    (A) 36   (B) 3600   (C) 12   (D) 16   (E) 64

19. An alloy contains 30% nickel. How much nickel is there in 120 tons of the alloy?

    (A) 30 tons   (B) 36 tons   (C) 360 tons

    (D) 40 tons   (E) 400 tons

20. A radio, which was selling for $60, was reduced by 25%. The new price is

    (A) $40   (B) $35   (C) $45   (D) $48   (E) $50

21.   10 pounds 11 ounces
      - 6 pounds 12 ounces

    (A) 4 pounds 1 ounce   (B) 3 pounds 1 ounce

    (C) 3 pounds 9 ounces   (D) 3 pounds 15 ounces

    (E) 4 pounds 15 ounces

22. How much does it cost to carpet a room that is 16 feet by 12 feet, if carpeting costs $4 per square foot?

    (A) $192   (B) $48   (C) $64   (D) $384   (E) $768

23. $\frac{4^2 - 2^3}{2} =$

    (A) 16   (B) 8   (C) 6   (D) 4   (E) 2

24. $(6 - 2)^2 - \left(\frac{4}{2}\right)^2 =$

    (A) 12   (B) 10   (C) 14   (D) 30   (E) 28

25. $(-2-1)^2 - 4(-3) =$

(A) -8 (B) 16 (C) -3 (D) 21 (E) -21

26.   The average of 10, 11, 12, and 13 is

(A) 11 (B) 11.5 (C) 12 (D) 12.5 (E) 46

27.   A box measures 4 inches by 3 inches by 2 inches. Its volume is

(A) 24 in.^{3} (B) 12 in.^{3} (C) 26 in.^{3}

(D) 18 in.^{3} (E) 144 in.^{3}

28.   A man is 6 feet 2 inches tall. His wife is 4 inches shorter. Her height is

(A) 5 feet 6 inches   (B) 5 feet 8 inches

(C) 5 feet 10 inches   (D) 6 feet

(E) 6 feet 6 inches

29.   The graph at the right indicates the number of students enrolled at Southwestern Tech from 1982 to 1985. The increase in enrollment from 1983 to 1984 was closest to

(A) 500 (B) 1000 (C) 3000

(D) 3500 (E) 5000

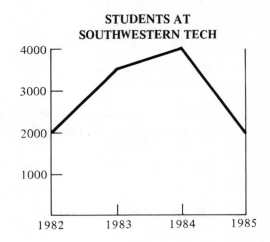

STUDENTS AT SOUTHWESTERN TECH

30.   If x represents a number, express 5 less than 4 times the number.

(A) $5 - 4x$ (B) $5 + 4x$ (C) $4x - 5$

(D) $4(x - 5)$ (E) $\dfrac{x - 5}{4}$

31.   Subtract $2x^2 - 3x + 1$ from $x^2 - 6$.

       (A) $3x^2 - 3x - 5$ (B) $x^2 - 3x - 5$ (C) $x^2 - 3x + 7$

       (D) $-x^2 + 3x - 7$ (E) $x^2 - 7$

32.   Find x if $4x - 7 = 2x + 9$.

       (A) $\dfrac{9}{2}$ (B) 8 (C) -8 (D) $\dfrac{8}{3}$ (E) $-\dfrac{8}{3}$

33.   Find t if $\dfrac{t + 3}{4} = \dfrac{t + 1}{3}$.

       (A) 2 (B) -3 (C) -1 (D) 0 (E) 5

34.   Find the value of $3p^2 - 2pq$ when $p = -2$ and $q = 3$.

       (A) 48 (B) 24 (C) -36 (D) -24 (E) 60

35.   Find the value of $\dfrac{3}{2^x}$ when $x = 3$.

       (A) 1 (B) $\dfrac{1}{2}$ (C) $\dfrac{3}{4}$ (D) $\dfrac{3}{8}$ (E) $\dfrac{27}{8}$

36.   Find the value, in cents, of n nickels, q quarters, and 3 pennies.

       (A) $n + q + 3$ (B) $3nq$ (C) $3(n + q)$

       (D) $5n + 25q + 3$ (E) $5(n + q + 3)$

37.   A psychologist needs 9 pounds of grain to feed 50 rats. If the rat
       population increases to 75, how much grain is needed to feed them?

       (A) 6 pounds   (B) 12 pounds   (C) 13.5 pounds

       (D) 15 pounds   (E) 24 pounds

	A	B	C	D	E
31	[]	[]	[]	[]	[]
32	[]	[]	[]	[]	[]
33	[]	[]	[]	[]	[]
34	[]	[]	[]	[]	[]
35	[]	[]	[]	[]	[]
36	[]	[]	[]	[]	[]
37	[]	[]	[]	[]	[]

38.   If $\frac{x}{3} - 2 = \frac{x - 2}{5}$, then $x =$

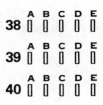

(A) 3   (B) 5   (C) 10   (D) 12   (E) 30

39.   In the right triangle shown here, find x.

(A) 3   (B) $\sqrt{3}$ (C) 9

(D) $\sqrt{17}$ (E) $\sqrt{15}$

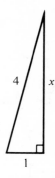

40.   Out of 60 guidance counsellors at a college, 25% are women. Without firing anyone, the college wishes to bring the percentage of women up to 50 percent. How many women must it hire?

(A) 15   (B) 25   (C) 30   (D) 45   (E) 60

Diagnostic Test, page 1

1. *i.* 53,808   *ii.* 58,000,002
iii. ninety thousand thirty-six

2. *i.* 292   *ii.* 122,385   *iii.* 606
iv. quotient = 21, remainder = 23

3. *i.* 125   *ii.* 144

4. *i.* $2^2 \times 3^2 \times 11$

5. *i.* no   *ii.* 10^4 or 10,000   *iii.* $2^3 \times 3^5 \times 5^2$

6. *i.* 6   *ii.* 120

7. *i.* $\dfrac{19}{20}$ *ii.* $\dfrac{17}{24}$

8. *i.* $\dfrac{1}{12}$ *ii.* $\dfrac{1}{6}$

9. *i.* $3\dfrac{5}{6}$ *ii.* $8\dfrac{1}{8}$ *iii.* 2

10. *i.* (D) $\dfrac{2}{7}$

11. *i.* .047   *ii.* .875   *iii.* .18 18 18...

12. *i.* (E) .4001

13. *i.* 24.985   *ii.* 9.81

14. *i.* .264   *ii.* \$388.05   *iii.* 6000   *iv.* 8

15. *i.* \$21.82   *ii.* \$2440   *iii.* \$100

16. *i.* 4,200,000   *ii.* 17.1   *iii.* 1000

17. *i.* .36   *ii.* $\dfrac{3}{250}$ *iii.* $\dfrac{14}{25}$

18. *i.* 24   *ii.* 75   *iii.* 60%   *iv.* 100 tons

19. *i.* \$1.26   *ii.* \$18.75   *iii.* \$187.20

20. *i.* 3 hours 33 minutes   *ii.* 4 pounds
15 ounces   *iii.* 8 feet 4 inches   *iv.* 16

21. *i.* \$648   *ii.* 26 feet   *iii.* 50 inches

22. *i.* 16 cm.²   *ii.* 78.5 in.²   *iii.* 62 ft.²

23. *i.* (D) 12,000   *ii.* (A) food   *iii.* (D) \$250,000

24. *i.* 13   *ii.* 30   *iii.* −3

25. *i.* −5   *ii.* −17   *iii.* 20

26. *i.* 80   *ii.* 60

27. *i.* $3x + 2$ *ii.* $5x + 10y$

28. *i.* 52   *ii.* 34

29. *i.* $11x + y$ *ii.* $8x^2 - 6x + 8$

30. *i.* 2

31. *i.* −10   *ii.* 6   *iii.* 20

32. *i.* 3100   *ii.* 25   *iii.* 240

33. *i.* $\sqrt{39}$ *ii.* $\sqrt{13}$

Section 1, page 13

Complete Solutions to the Practice Exercises for Section 1 begin on page 353.

1. 245

3. 780

5. 7602

7. 6500

9. 63,056

11. 400,212

13. 747,008

15. 198,240

17. 2,001,096

19. 10,163,000

21. five hundred twenty-eight

23. four thousand eight hundred

25. ninety thousand one hundred

27. eighty-five thousand eight hundred fifty-nine

29. three hundred eighty thousand nineteen

31. three hundred eighty-three thousand one hundred five

33. nine million two thousand five hundred eighty-five

35. forty-two million four hundred

37. (A) 29

39. (B) 60,000,000

41. (C) 66,480

43. (B) 177,000

45. (C) 22,600,000

Section 2, page 24

Complete Solutions to the Practice Exercises for Section 2 begin on page 353.

1. 1180

3. 25

5. 1360

7. 1662

9. 223

11. 667

13. 331

15. 109

17. 5

19. 991
Check:
998
+ 991
1989

21. 206
Check:
388
+ 206
594

23. 22,446

25. 73,485

27. 463,740

29. 4,994,703

31. 23,616,044

33. 60

35. 31

37. 26

39. 16

41. 310

43. 57

45. 106
Check:
106
× 33
318
318
3498

47. 45
Check:
45
× 63
135
270
2835

49. i. 6   ii. 11

51. i. 22   ii. 8

53. i. 15   ii. 12   iii.
15
× 39
135
45
585
+ 12
597

55. i. 114   ii. 10   iii.
114
× 78
912
798
8892
+ 10
8902

57. 280

59. 62,000

61. 120,000

63. $244

65. $27

67. 275

69. 15

Section 3, page 31

Complete Solutions to the Practice Exercises for Section 3 begin on page 354.

1. 4

3. 36

5. 81

7. 121

9. 900

11. 256

13. 216

15. 1

17. 729

19. 1331

21. 32

23. 128

25. 256

27. 160,000

29. 1

31. i. 5   ii. 6   iii. the 6th power

33. 28

35. 36

37. 275

39. 3969

41. 84

43. 360

45. 616

Section 4, page 37

Complete Solutions to the Practice Exercises for Section 4 begin on page 355.

1. 1, 2, 3, 6   **3.** 1, 2, 4, 8   **5.** 1, 19   **7.** 1, 2, 11, 22

9. 1, 2, 3, 5, 6, 10, 15, 30   **11.** *i.* (A) 5, (B) 10, (C) 25, (D) 50;   *ii.* (A) 5

13. *i.* (D) 40;   *ii.* (A) 4, (B) 5, (C) 10   **15.** *i.* (C) 9, (E) 36;   *ii.* (A) 3, (C) 9

17. 6, 12, 18, and so on   **19.** 18, 36, 54, and so on   **21.** 1 and 2   **23.** prime

25. composite   **27.** composite   **29.** prime   **31.** prime

33. prime   **35.** composite   **37.** 2×3 **39.** 2×7

41. $2^2 \times 5$ **43.** $2^2 \times 3^2$ **45.** $2^4 \times 3$ **47.** $2^2 \times 3 \times 5$

49. 3^4 **51.** $2^2 \times 5^2$ **53.** 13^2 **55.** $2^5 \times 3^2$

Section 5, page 44

Complete Solutions to the Practice Exercises for Section 5 begin on page 355.

1. even   **3.** even   **5.** divisible by 2   **7.** not divisible by 2

9. not divisible by 5   **11.** divisible by 5   **13.** divisible by 5   **15.** divisible by 10

17. divisible by 10   **19.** divisible by 100   **21.** divisible by 100   **23.** divisible by 1000

25. divisible by 1000   **27.** no   **29.** 10   **31.** 10^6 (or 1,000,000)

33. divisible by 3   **35.** divisible by 3   **37.** divisible by 3   **39.** divisible by 9

41. divisible by 9   **43.** not divisible by 9   **45.** 2^8 **47.** 5^5

49. $2^2 \times 3^2 \times 5^3$ **51.** $3^3 \times 5$ **53.** $2^3 \times 3^3$ **55.** $3^4 \times 7$

Section 6, page 51

Complete Solutions to the Practice Exercises for Section 6 begin on page 356.

1. 6   **3.** 2   **5.** 3   **7.** 5

9. 3   **11.** 1   **13.** $12 = 2^2 \times 3$ **15.** 7

17. $32 = 2^5$ **19.** $27 = 3^3$ **21.** 5   **23.** 6

25. 12   **27.** 6   **29.** 8   **31.** 18

33. 66   **35.** 60   **37.** 75   **39.** $30 = 2 \times 3 \times 5$

41. $48 = 2^4 \times 3$ **43.** $60 = 2^2 \times 3 \times 5$ **45.** $135 = 3^3 \times 5$ **47.** $336 = 2^4 \times 3 \times 7$

49. $900 = 2^2 \times 3^2 \times 5^2$ **51.** $396 = 2^2 \times 3^2 \times 11$

53. $1470 = 2 \times 3 \times 5 \times 7^2$ **55.** $1760 = 2^5 \times 5 \times 11$

57. 30   **59.** 12   **61.** 144   **63.** 300

65. 576   **67.** 4455

Review Exercises for Unit I, page 52

1. 5840
2. 9,600,000
3. 27,052

4. six thousand two hundred eight
5. one hundred fifty thousand ninety

6. (D) 3500
7. 1441
8. 16,070
9. 4998

10. 238,856
11. 409,000
12. 84
13. *i.* 19   *ii.* 15

14. $14
15. 1500
16. 64
17. 125

18. 243
19. 10,000,000
20. *i.* 4   *ii.* 6   *iii.* the 6th power

21. 144
22. 175,000
23. 1, 2, 3, 4, 6, 8, 12, 24

24. *i.* (C) 8, (E) 40;   *ii.* (A) 1, (B) 4, (C) 8
25. 14, 28, 42, and so on

26. 5
27. prime
28. prime
29. composite

30. $2^2 \times 7$
31. 2×3^3
32. $2^5 \times 3$
33. $2^4 \times 3 \times 5$

34. odd
35. yes
36. yes
37. 100, or 10^2

38. no
39. yes
40. $2^7 \times 5 \times 7$
41. $3^3 \times 7 \times 37$

42. 5
43. 12
44. 7
45. 27

46. 10
47. 21
48. 45
49. 280

50. 728
51. 720
52. 60
53. 1440

Practice Exam on Unit I, page 54
Complete Solutions begin on page 356.

Section 7, page 64
Complete Solutions to the Practice Exercises for Section 7 begin on page 357.

1. $\frac{1}{2}$
3. $\frac{5}{9}$
5. $\frac{2}{3}$
7. $\frac{3}{4}$
9. $\frac{2}{10}$

11. $\frac{16}{36}$
13. *i.* $\frac{10}{16}$ *ii.* $\frac{15}{24}$
15. *i.* $\frac{25}{60}$ *ii.* $\frac{30}{72}$
17. $\frac{5}{7}$
19. 1

21. $\frac{1}{3}$
23. $\frac{7}{10}$
25. $\frac{1}{2}$
27. *i.* 4   *ii.* $\frac{1}{2} = \frac{2}{4}$, $\frac{1}{4} = \frac{1}{4}$

29. *i.* 12   *ii.* $\frac{3}{4} = \frac{9}{12}$, $\frac{5}{6} = \frac{10}{12}$
31. *i.* 36   *ii.* $\frac{2}{9} = \frac{8}{36}$, $\frac{1}{12} = \frac{3}{36}$

33. *i.* 48   *ii.* $\frac{7}{12} = \frac{28}{48}$, $\frac{3}{16} = \frac{9}{48}$
35. *i.* 120   *ii.* $\frac{1}{30} = \frac{4}{120}$, $\frac{7}{40} = \frac{21}{120}$

37. $\frac{5}{6}$
39. $\frac{4}{15}$
41. $\frac{25}{63}$
43. $\frac{37}{40}$
45. $\frac{29}{60}$

47. $\frac{7}{24}$
49. $\frac{3}{50}$
51. $\frac{25}{36}$
53. $\frac{11}{12}$
55. $\frac{1}{20}$

57. $\dfrac{5}{9}$ **59.** $\dfrac{25}{24}$ **61.** $\dfrac{29}{16}$ **63.** $\dfrac{18}{25}$ **65.** (A) $\dfrac{13}{25}$

67. (C) $\dfrac{3}{8}$ **69.** (E) $\dfrac{7}{10}$

Section 8, page 73
Complete Solutions to the Practice Exercises for Section 8 begin on page 358.

1. $\dfrac{1}{6}$ **3.** $\dfrac{1}{5}$ **5.** $\dfrac{2}{7}$ **7.** $\dfrac{1}{6}$ **9.** $\dfrac{2}{3}$

11. $\dfrac{5}{14}$ **13.** $\dfrac{1}{36}$ **15.** $\dfrac{1}{30}$ **17.** $\dfrac{1}{30}$ **19.** $\dfrac{1}{24}$

21. 1 **23.** $\dfrac{1}{120}$ **25.** $\dfrac{1}{4}$ **27.** $\dfrac{9}{100}$ **29.** $\dfrac{25}{144}$

31. $\dfrac{8}{27}$ **33.** $\dfrac{2}{3}$ **35.** $\dfrac{1}{2}$ **37.** $\dfrac{5}{12}$ **39.** $\dfrac{3}{4}$

41. $\dfrac{10}{3}$ **43.** $\dfrac{5}{2}$ **45.** $\dfrac{1}{54}$ **47.** $\dfrac{1}{60}$ **49.** (B) $\dfrac{5}{28}$

51. (A) $\dfrac{3}{7}$ **53.** (B) $\dfrac{9}{10}$ **55.** (B) 10 **57.** (A) $\dfrac{5}{32}$ **59.** (B) $\dfrac{27}{2}$

61. (D) $\dfrac{1}{8}$ **63.** (E) $\dfrac{1}{6}$ cup **65.** (D) $\dfrac{1}{4}$ **67.** (B) 14

Section 9, page 83
Complete Solutions to the Practice Exercises for Section 9 begin on page 359.

1. $\dfrac{3}{2}$ **3.** $\dfrac{21}{5}$ **5.** $\dfrac{27}{5}$ **7.** $\dfrac{23}{3}$ **9.** $1\dfrac{3}{5}$

11. $7\dfrac{3}{4}$ **13.** $14\dfrac{5}{6}$ **15.** $5\dfrac{1}{2}$ **17.** 6 **19.** $7\dfrac{3}{4}$

21. $3\dfrac{9}{10}$ **23.** $9\dfrac{1}{8}$ **25.** $1\dfrac{3}{4}$ **27.** $2\dfrac{1}{2}$ **29.** $\dfrac{3}{8}$

31. $11\dfrac{7}{8}$ **33.** 10 **35.** $7\dfrac{7}{8}$ **37.** $4\dfrac{7}{32}$ **39.** $15\dfrac{5}{8}$

41. 4 **43.** $3\dfrac{1}{6}$ **45.** $3\dfrac{5}{13}$ **47.** 2 **49.** $\dfrac{91}{8}$

51. $\dfrac{20}{17}$ **53.** (B) $7\dfrac{1}{4}$ **55.** (D) $\dfrac{81}{8}$ **57.** (B) $\dfrac{9}{4}$ **59.** (E) $\dfrac{4}{9}$

61. (B) $3\dfrac{3}{4}$ **63.** (D) 27 **65.** (D) $2\dfrac{5}{8}$ yards

Section 10, page 91

Complete Solutions to the Practice Exercises for Section 10 begin on page 361.

1. (A) $\frac{2}{5}$ 3. (C) $\frac{1}{10}$ 5. (D) $\frac{2}{15}$ 7. $\frac{1}{2}$ 9. $\frac{1}{3}$

11. $\frac{5}{6}$ 13. $\frac{9}{11}$ 15. $\frac{7}{13}$ 17. (C) $\frac{1}{5}$ 19. (A) $\frac{1}{10}$

21. (E) $\frac{2}{11}$ 23. (C) $\frac{2}{5}$ 25. (D) $\frac{1}{10}$ 27. (B) $\frac{9}{100}$ 29. (E) $\frac{8}{7}$

31. (A) $\frac{7}{10}$ 33. (A) $\frac{8}{11}$ 35. (E) $\frac{4}{15}$ 37. (A) $\frac{1}{2}$ 39. (E) $\frac{7}{10}$

41. (D) $\frac{13}{16}$ 43. (A) $\frac{5}{3}$ 45. $\frac{1}{2}$ pound of feathers

47. The side that measures $\frac{7}{10}$ of a meter   49. The 8-ounce bar

Review Exercises for Unit II, page 95

1. $\frac{3}{5}$ 2. $\frac{3}{5}$ 3. *i.* $\frac{9}{24}$ *ii.* $\frac{21}{56}$ 4. $\frac{1}{2}$ 5. $\frac{1}{3}$

6. *i.* 18   *ii.* $\frac{4}{9} = \frac{8}{18}, \frac{5}{6} = \frac{15}{18}$.   7. *i.* 60   *ii.* $\frac{7}{12} = \frac{35}{60}, \frac{3}{20} = \frac{9}{60}$

8. $\frac{17}{20}$ 9. $\frac{27}{40}$ 10. $\frac{44}{75}$ 11. $\frac{41}{24}$ or $1\frac{17}{24}$ 12. $\frac{4}{5}$

13. (C) $\frac{1}{6}$ 14. $\frac{6}{35}$ 15. $\frac{2}{15}$ 16. $\frac{5}{64}$ 17. $\frac{9}{20}$

18. 3   19. $\frac{1}{4}$ 20. $\frac{1}{6}$ 21. 25   22. $\frac{27}{98}$

23. (A) $\frac{3}{8}$ 24. (B) 8   25. (A) $\frac{8}{15}$ 26. $\frac{13}{5}$ 27. $\frac{41}{4}$

28. $4\frac{5}{6}$ 29. 8   30. $7\frac{2}{3}$ 31. $4\frac{3}{8}$ 32. $1\frac{1}{6}$

33. $16\frac{1}{5}$ 34. $11\frac{11}{24}$ 35. $\frac{13}{40}$ 36. $1\frac{21}{23}$ 37. $\frac{323}{32}$

38. $\frac{105}{26}$ 39. (D) $8\frac{5}{12}$ 40. (C) $3\frac{1}{4}$ pounds   41. (A) $\frac{3}{16}$ 42. $\frac{3}{5}$

43. $\frac{2}{3}$ 44. (D) $\frac{1}{10}$ 45. (B) $\frac{2}{9}$ 46. (C) $\frac{3}{50}$ 47. (A) $\frac{3}{4}$

48. (D) $\frac{5}{9}$ 49. $\frac{3}{10}$ of a pound of chicken   50. the 14-ounce box for 90 cents

Review Exercises on Unit I, page 97

1. 72,045   2. one million four thousand eight hundred two   3. 107

4. 279,630   **5.** *i.* 28 *ii.* 13   **6.** 540   **7.** 288   **8.** $2^2 \times 3 \times 11$

9. 10^3 **10.** 150   **11.** 3   **12.** $2^5 \times 3^2 \times 5$

Practice Exam on Unit II, page 98
Complete Solutions begin on page 361.

Section 11, page 106
Complete Solutions to the Practice Exercises for Section 11 begin on page 363.

1. $\dfrac{\boxed{7}}{10}$ **3.** $\dfrac{\boxed{8}}{10} + \dfrac{\boxed{3}}{100}$ **5.** $\dfrac{\boxed{6}}{10} + \dfrac{\boxed{6}}{100} + \dfrac{\boxed{1}}{1000}$ **7.** 5

9. 1   **11.** *i.* 8 *ii.* 0 *iii.* 5   **13.** *i.* 1 *ii.* 9 *iii.* 5   **15.** .29

17. .6   **19.** .43   **21.** .015   **23.** nine tenths

25. four hundred one thousandths   **27.** seven thousandths   **29.** 4.3

31. 17.03   **33.** 100.01   **35.** 6.3   **37.** 100.005

39. .8   **41.** 2.5   **43.** .35   **45.** .28

47. .005   **49.** $\dfrac{3}{5}$ **51.** $\dfrac{6}{25}$ **53.** $\dfrac{1}{50}$

55. $\dfrac{5}{8}$ **57.** $\dfrac{6}{5}$ **59.** .666 666...   **61.** .63 63 63...

63. .58 333 333...   **65.** 1.1 666 666...   **67.** .0 333 333...

Section 12, page 113
Complete Solutions to the Practice Exercises for Section 12 begin on page 363.

1. (E) .1   **3.** (D) .11   **5.** (D) .38   **7.** (A) .09   **9.** (C) .39

11. (A) .451   **13.** (B) .575   **15.** (D) .100   **17.** (E) .136   **19.** (B) .703

21. (E) .601   **23.** (E) .7   **25.** (E) .1501   **27.** (D) .07   **29.** (E) 10.009

31. (B) 3.389   **33.** (A) .004   **35.** (C) .0039   **37.** (D) .0202   **39.** (B) $.39

41. (E) $9.98   **43.** (C) .87   **45.** (E) 2.11   **47.** (A) 6.1   **49.** (E) 6.0101

51. (B) $.91   **53.** (E) $12.01

Section 13, page 118
Complete Solutions to the Practice Exercises for Section 13 begin on page 363.

1. .9   **3.** 20.5   **5.** 14.73   **7.** 26.61   **9.** 39.828

11. 115.44   **13.** 133.211   **15.** 121.989   **17.** 64.75   **19.** .6

21. 1.4   **23.** 10.9   **25.** 6.04   **27.** 41.73   **29.** .183

31. 24.51   **33.** 44.77   **35.** 11.77   **37.** 33.33   **39.** 27.111

41. 117.483   **43.** 5.91   **45.** 16.14   **47.** 15.764   **49.** $21.36

51. $1641.97   **53.** $202.30   **55.** $4.25   **57.** $3.15   **59.** 58.4 miles

Section 14, page 125

Complete Solutions to the Practice Exercises for Section 14 begin on page 364.

1. 1.92   **3.** 5040   **5.** .52   **7.** 40.92   **9.** .15

11. .112   **13.** .08   **15.** .2668   **17.** 2.4   **19.** 114.8

21. 1081.45   **23.** .09   **25.** .0004   **27.** .001   **29.** .000 027

31. $7.80   **33.** $61.50   **35.** $3.40   **37.** 1.5 inches   **39.** $420.65

41. .054   **43.** .000 372   **45.** .000 053   **47.** 5.508   **49.** 50

51. 800   **53.** 40,000   **55.** 20   **57.** 3   **59.** 40

61. 18   **63.** 24   **65.** 15   **67.** 72   **69.** (A) $14.40

71. (B) $29.70   **73.** (A) .8 inch   **75.** $3.20

Section 15, page 130

Complete Solutions to the Practice Exercises for Section 15 begin on page 365.

1. (E) $44.16   **3.** (A) $596   **5.** (D) $620   **7.** $950   **9.** $4.00

11. (C) $3.60   **13.** (D) $1.90   **15.** (C) $1520   **17.** $1315   **19.** $588

21. $158.75

Section 16, page 138

Complete Solutions to the Practice Exercises for Section 16 begin on page 365.

1. 8000   **3.** 8000   **5.** 7000   **7.** 20,000,000

9. 20,000,000   **11.** *i.* 536,200 *ii.* 536,000 *iii.* 500,000   **13.** 400,000

15. 3,000,000   **17.** $79,000   **19.** $90,000   **21.** .1

23. .4   **25.** 100.0   **27.** .94   **29.** 4.90

31. .672   **33.** .600   **35.** *i.* .4 *ii.* .43 *iii.* .432

37. *i.* .4 *ii.* .44 *iii.* .437   **39.** 9.9 seconds   **41.** 4

43. 27   **45.** 1   **47.** $8   **49.** 240,000

51. 31,000,000   **53.** *i.* 8000 *ii.* 8443 *iii.* 8000 *iv.* yes   **55.** 5,100,000

57. 80   **59.** 10,001,000   **61.** 4   **63.** 21 miles per gallon

65. 1.33   **67.** 10.13

Review Exercises for Unit III, page 141

1. $\dfrac{\boxed{4}}{10} + \dfrac{\boxed{2}}{100} + \dfrac{\boxed{9}}{1000}$ 2. *i.* 0 *ii.* 4 *iii.* 1   3. .045   4. forty-six thousandths

5. 12.03   6. .12   7. $\dfrac{3}{8}$ 8. .2 333 333 . . .

9. (E) .401   10. (A) .03   11. (D) 1.001   12. (C) $48.99

13. (A) .662   14. (C) 4.110   15. 72.28   16. 1601.01

17. 39.1813   18. $1109.82   19. $44.30   20. 20.7 miles

21. 1.84   22. 7172.1   23. .1   24. .065

25. .000 064   26. $47.70   27. .505   28. .000 38

29. 30   30. .07   31. (A) $32.50   32. (B) 15

33. (E) $8.25   34. (A) $118.60   35. $12.65   36. $107,400

37. 17,000   38. *i.* 4,850,000 *ii.* 4,900,000 *iii.* 5,000,000   39. .79

40. *i.* .5 *ii.* .51 *iii.* .509   41. 184,000,000   42. 440

43. 9

Review Exercises on Units I and II, page 143

1. twenty-seven thousand eighty-five   2. 489,600   3. 72

4. *i.* 17 *iii.* 37 *iv.* 47   5. $2^5 \times 5^2 \times 7$ 6. 60   7. 4

8. (C) $\dfrac{1}{5}$ 9. $\dfrac{3}{4}$ 10. $\dfrac{7}{10}$ 11. $\dfrac{4}{11}$ 12. $\dfrac{5}{6}$

13. $3\dfrac{3}{4}$ or $\dfrac{15}{4}$ 14. $\dfrac{153}{16}$ or $9\dfrac{9}{16}$ 15. $3\dfrac{3}{4}$ pounds

Practice Exam on Unit III, page 144

Complete Solutions begin on page 365.

Section 17, page 150

Complete Solutions to the Practice Exercises for Section 17 begin on page 366.

1. $\dfrac{1}{2}$ 3. $\dfrac{3}{4}$ 5. $\dfrac{11}{50}$ 7. $\dfrac{3}{2}$ 9. $\dfrac{12}{25}$

11. (C) $\dfrac{3}{20}$ 13. $\dfrac{4}{5}$ 15. .53   17. .4   19. .02

21. 2.25   23. .1905   25. .1025   27. (E) .025   29. .78

31. 43%   33. 145%   35. 2%   37. .2%   39. 1000%

41. 255%   43. 53%   45. 80%   47. 450%   49. $8\dfrac{1}{3}$%

51. $33\frac{1}{3}\%$ **53.** $116\frac{2}{3}\%$ **55.** 150%   **57.** 410%   **59.** 1250%

61. 105%   **63.** $133\frac{1}{3}\%$

Section 18, page 157

Complete Solutions to the Practice Exercises for Section 18 begin on page 367.

1. (B) 33	**3.** (D) 45	**5.** (D) 27	**7.** 288	**9.** 28
11. 68	**13.** 10	**15.** 7	**17.** 8	**19.** 7
21. 18	**23.** 90	**25.** 20	**27.** 300	**29.** (D) 60
31. (C) 40	**33.** (E) 144	**35.** 100	**37.** 480	**39.** 10
41. (A) 50%	**43.** (D) 80%	**45.** 90%	**47.** 10%	**49.** (C) 41
51. (B) 40%	**53.** (D) 600	**55.** 20	**57.** $8360	**59.** 6600

Section 19, page 164

Complete Solutions to the Practice Exercises for Section 19 begin on page 368.

1. (B) $224	**3.** (C) $149.80	**5.** (D) $1.30	**7.** $47,840
9. 6%	**11.** $5.04	**13.** $28,196	**15.** 50%
17. (E) $19,600	**19.** (D) $17.00	**21.** (C) $22.20	**23.** 174,600
25. 25%	**27.** $14,080	**29.** 20%	**31.** $36,000

Review Exercises for Unit IV, page 166

1. $\frac{4}{5}$ **2.** $\frac{17}{25}$ **3.** $\frac{7}{4}$ **4.** .7

5. .025   **6.** .1225   **7.** *i.* $\frac{9}{20}$ *ii.* .45   **8.** 40%

9. 4%   **10.** .4%   **11.** 102.5%   **12.** 90%

13. 15%   **14.** 160%   **15.** $66\frac{2}{3}\%$ **16.** 225%

17. $341\frac{2}{3}\%$ **18.** (A) 14   **19.** (E) 90   **20.** 9

21. 500   **22.** 7   **23.** 4.5   **24.** 50

25. 40   **26.** (E) 80   **27.** (C) 52   **28.** 400

29. 200   **30.** (D) 75%   **31.** (D) 70%   **32.** 32%

33. 1%   **34.** (C) 90%   **35.** (C) $250   **36.** (D) $58.80

37. (D) $53,040   **38.** 225   **39.** 280,000   **40.** (C) 22.50

41. (A) $108,000   **42.** 15,360,000   **43.** 585,000   **44.** 25%

45. 20%   **46.** $48.60   **47.** $31,200

Review Exercises on Units I–III, page 169

1. forty-seven thousand two   **2.** 139,320   **3.** $2^6 \times 3^2 \times 5^2$

4. 162   **5.** 12   **6.** $\dfrac{3}{20}$ **7.** $\dfrac{4}{5}$

8. $4\dfrac{3}{8}$ or $\dfrac{35}{8}$ **9.** $\dfrac{147}{8}$ or $18\dfrac{3}{8}$ **10.** .027   **11.** .06

12. $\dfrac{1}{40}$ **13.** .000 112   **14.** 400   **15.** 21 inches

16. 40

Practice Exam on Unit IV, page 169
Complete Solutions begin on page 369.

Practice Midterm Exam A on Units I–IV, page 171
Complete Solutions begin on page 370.

Practice Midterm Exam B on Units I–IV, page 173
Complete Solutions begin on page 370.

Section 20, page 181
Complete Solutions to the Practice Exercises for Section 20 begin on page 371.

1. 8 hours 52 minutes   **3.** 8 hours 10 minutes   **5.** 4 hours 20 minutes

7. 4 hours 52 minutes   **9.** 3 hours 39 minutes   **11.** 4 minutes 51 seconds

13. 2 hours 25 minutes   **15.** 1 hour 45 minutes   **17.** 1 hour 45 minutes

19. 1:50 P.M.   **21.** 20   **23.** 12 pounds

25. 17 kilograms 500 grams   **27.** 13 grams 27 centigrams   **29.** 3 pounds 2 ounces

31. 2 pounds 4 ounces   **33.** 5 kilograms 225 grams   **35.** 3 grams 55 centigrams

37. 15 pounds 1 ounce   **39.** 4 pounds 11 ounces   **41.** 6 kilograms 500 grams

43. 9 pounds 6 ounces   **45.** 9 feet   **47.** 10 kilometers

49. 3 feet 11 inches   **51.** 4 feet 3 inches   **53.** 89 centimeters

55. 12   **57.** 10 inches   **59.** 55 pounds

61. 22.9 centimeters   **63.** 88.0 kilometers   **65.** 36.0 miles

Section 21, page 191
Complete Solutions to the Practice Exercises for Section 21 begin on page 372.

1. (C) 18 ft.^{2} **3.** (D) 700 in.^{2} **5.** 54 in.^{2} **7.** 220 in.^{2}

9. 300 in.^{2} **11.** 135 ft.^{2} **13.** (A) 10 yd.   **15.** 3 ft.

17. (B) 49 cm.^{2}	**19.** 6.25 cm.^{2}	**21.** 9 ft.^{2}	**23.** (D) $450
25. (D) $180	**27.** $84	**29.** $210	**31.** $9
33. (B) 26 cm.	**35.** (C) 22 ft.	**37.** 308 yd.	**39.** (E) 4 in.
41. 6 in.	**43.** 24 in.^{2}	**45.** 200 cm.^{2}	**47.** 36 cm.
49. 25 in.^{2}	**51.** (C) $300	**53.** (D) $1400	**55.** $25
57. $420			

Section 22, page 207

Complete Solutions to the Practice Exercises for Section 22 begin on page 373.

1. 12 in.^{2} **3.** 3 ft.^{2} **5.** $\frac{45}{2}$ in.^{2} or $22\frac{1}{2}$ in.^{2}

7. (C) 8 in.	**9.** 18 cm.	**11.** 13 cm.	**13.** 31 ft.
15. 6 in.	**17.** 12 in.	**19.** 2 ft.	**21.** 3 in.
23. 12.6 ft.	**25.** 31.4 in.	**27.** (C) 14 in.	**29.** 153.9 cm.^{2}
31. 50.2 ft.^{2}	**33.** 50.2 in.^{2}	**35.** 201.0 in.^{2}	**37.** 50.2 m.^{2}
39. 63.6 cm.^{2}	**41.** (C) 36π cm.^{2}	**43.** $(6 + \pi)$ cm.^{2}	**45.** 27π in.^{2}

47. *i.* 8 ft.^{3} *ii.* 28 ft.^{2} **49.** *i.* 108 in.^{3} *ii.* 144 in.^{2}

51. *i.* 24 m.^{3} *ii.* 52 m.^{2} **53.** *i.* 125 in.^{3} *ii.* 150 in.^{2}

55. 216 in.^{2} **57.** $1200.50

Section 23, page 214

Complete Solutions to the Practice Exercises for Section 23 begin on page 375.

1. (B) 2500	**3.** (D) 4500	**5.** (D) 1984
7. (D) 140,000	**9.** (E) 20,000	**11.** (D) $5,000,000
13. (C) $9,000,000	**15.** (D) $1,000,000	**17.** (B) $325
19. (D) $1250	**21.** (C) $900,000	**23.** (D) $1,584,000
25. (C) $288,000	**27.** (E) $100,000	**29.** (B) $200,000
31. (B) 2,750,000	**33.** (C) 1,250,000	**35.** (B) 500,000

Review Exercises for Unit V, page 218

1. 7 hours 20 minutes	**2.** 2 minutes 49 seconds	**3.** (D) 4 hr. 45 min.
4. (D) 18 minutes	**5.** 18 pounds 2 ounces	**6.** 750 grams
7. 10 pounds 4 ounces	**8.** 12.5 kilograms	**9.** 6 meters 10 centimeters
10. 2 feet 9 inches	**11.** 7 in.	**12.** 36 in.^{2}
13. (E) 25 ft.	**14.** (A) 12 ft.^{2}	**15.** 121 cm.^{2}

16. $150   17. 280 ft.²   18. $980

19. 26 ft.   20. 3 m.   21. $840

22. 75 cm.²   23. 30 inches   24. (B) 9 in.²

25. (C) 16 cm.   26. 24 in.   27. 10 cm.

28. 31 cm.   29. 25π cm.²   30. 7 in.

31. 44.0 in.   32. 49π in.²   33. 6 in.

34. 36π in.²   35. $(8 + \frac{\pi}{2})$, or $\frac{16 + \pi}{2}$, cm.²   36. (A) 1981

37. (B) 200   38. (C) 1750   39. (C) $7500

40. (B) taxes   41. (B) $1250   42. (C) $2,000,000

43. (D) $7,000,000   44. (D) $43,000,000

Review Exercises on Units I–IV, page 221

1. one hundred five thousand twenty   2. *i.* 14 *ii.* 7   3. $2^3 \times 3 \times 13$ 4. 600

5. $\frac{1}{6}$ 6. $\frac{14}{9}$ 7. $7\frac{3}{4}$ or $\frac{31}{4}$ 8. $\frac{5}{2}$ or $2\frac{1}{2}$ 9. $\frac{3}{1000}$

10. .222 222...   11. .01   12. .02   13. .025   14. 109%

15. $\frac{6}{25}$ 16. $32

Practice Exam on Unit V, page 222

Complete Solutions begin on page 375.

Section 24, page 237

Complete Solutions to the Practice Exercises for Section 24 begin on page 376.

1.–8.

9. 5   11. 15   13. .73   15. $\frac{1}{3}$ 17. -5

19. 2   21. 4   23. -20 25. -4 27. -3

29. 1   31. 12   33. -1 35. 7   37. -7

39. -7 41. 23   43. 0   45. -12 47. 12

49. 35   51. 36   53. 0   55. -24 57. -2

59. 2   61. -5 63. -4 65. undefined   67. undefined

69. undefined   71. $-16°$ Fahrenheit   73. 3 miles west   75. -14

Section 25, page 244

Complete Solutions to the Practice Exercises for Section 25 begin on page 377.

1. 4	**3.** 36	**5.** 100	**7.** −121	**9.** 0
11. −27	**13.** −81	**15.** −32	**17.** −128	**19.** −10,000,000
21. (B) 20	**23.** (D) −20	**25.** 49	**27.** −19	**29.** (C) 42
31. (A) 50	**33.** (B) −199	**35.** 22	**37.** 14	**39.** 75
41. 26	**43.** −36	**45.** −70	**47.** −14	**49.** 10
51. −2	**53.** −36	**55.** $100	**57.** 6 hours	**59.** $49
61. 4 yards				

Section 26, page 249

Complete Solutions to the Practice Exercises for Section 26 begin on page 377.

1. 17	**3.** 61	**5.** 40	**7.** 72
9. 92	**11.** 54.5	**13.** 78.5	**15.** 47
17. 86	**19.** 96	**21.** 12	**23.** 22 centimeters
25. (A) 22	**27.** (C) 43	**29.** (C) 44	**31.** (D) 72
33. (B) 79	**35.** (C) 48.75	**37.** (B) $16,000	

Review Exercises for Unit VI, page 251.

1.–4.

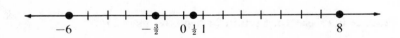

5. 10	**6.** $\frac{1}{4}$	**7.** 0	**8.** −10	**9.** 6
10. −20	**11.** −8	**12.** 1	**13.** 2	**14.** −2
15. 0	**16.** 27	**17.** undefined	**18.** undefined	**19.** −13
20. 18	**21.** −8°	**22.** −1	**23.** 36	**24.** −125
25. −1	**26.** −64	**27.** (D) −12	**28.** (A) 77	**29.** (C) −6
30. (B) −27	**31.** 1	**32.** 75	**33.** 127	**34.** −5
35. 2	**36.** 36	**37.** −6	**38.** 5	**39.** $2
40. 300	**41.** 41	**42.** 23.5	**43.** 71.5	**44.** 36
45. 70				

Review Exercises on Units I–V, page 253

1. $2^3 \times 3 \times 11$ 2. 2000   3. $\dfrac{3}{35}$ 4. $\dfrac{13}{100}$ 5. $3\dfrac{7}{8}$

6. .0297   7. .135   8. $\dfrac{28}{25}$ 9. 14%   10. .105

11. 540,800   12. 1 hour 36 minutes   13. 6 inches   14. 6 pounds 4 ounces

15. $18   16. 18 inches   17. 78.5 in.²   18. 96 cm.²

Practice Exam on Unit VI, page 253

Complete Solutions begin on page 378.

Section 27, page 259

Complete Solutions to the Practice Exercises for Section 27 begin on page 378.

1. $x + 3$ 3. $6x$ 5. $x + 2$ 7. $x + 12$

9. x^2 11. $1 + 3x$ or $3x + 1$ 13. $\dfrac{1}{2}x - 6$ or $\dfrac{x}{2} - 6$ 15. (D) $5n + 10d$

17. (E) $25q + 50h$ 19. $5x + 10y$ 21. $5x + 20y + 50z$ 23. $25x$

25. (A) $50h$ 27. mh 29. $80x + 60y$ 31. $20x + 17y$

Section 28, page 266

Complete Solutions to the Practice Exercises for Section 28 begin on page 379.

1. (B) 12   3. (C) 0   5. (D) -15 7. 16

9. -9 11. 10,000   13. 12   15. -8

17. (D) 16   19. 35   21. 980   23. (D) -2

25. (C) -4 27. -1 29. 33   31. 77

33. -1 35. (B) 24   37. (E) 0   39. -23

41. 25   43. 33   45. 24   47. -6

49. *i.* 49 in.²   *ii.* 144 in.²   51. *i.* 1200 miles   *ii.* 3600 miles

53. *i.* 80 ft.   *ii.* 128 ft.   *iii.* 0 ft. (ground level)   55. 42 cm.²

57. 972π in.³   59. 40π in.³   61. 60π in.²   63. 66π in.²

Section 29, page 278

Complete Solutions to the Practice Exercises for Section 29 begin on page 379.

1. 20   **3.** 1   **5.** 16   **7.** x^2 and $4x$

9. $3b^2$ and -5 **11.** unlike terms   **13.** like terms   **15.** like terms

17. unlike terms   **19.** like terms   **21.** $12a$ **23.** x

25. 0   **27.** $12xy$ **29.** (A) $6x + 10y$ **31.** (D) $7x$

33. $a - 1$ **35.** $8x + 2y$ **37.** $5x - y$ **39.** $6x^2 + 2x + 3$

41. (E) $5x + 3y$ **43.** (D) $4p + 9q$ **45.** $3x^2 - 2x + 7$ **47.** $-4a + 2b + 3c$

49. $x^2 - 7x - 1$ **51.** (D) $10x - 2$ **53.** (E) $3x - 1$

Section 30, page 286

Complete Solutions to the Practice Exercises for Section 30 begin on page 380.

1. yes   **3.** no   **5.** no   **7.** no   **9.** no

11. no   **13.** 10   **15.** 5   **17.** 12   **19.** -18

21. -4 **23.** -8 **25.** 7   **27.** -6 **29.** -11

31. -6 **33.** -1 **35.** 5   **37.** (D) 3   **39.** (A) 6

41. 5   **43.** 2   **45.** (C) 1   **47.** (D) 2   **49.** 2

51. -25 **53.** $\dfrac{5}{2}$ **55.** 8   **57.** $-\dfrac{3}{5}$

Section 31, page 294

Complete Solutions to the Practice Exercises for Section 31 begin on page 381.

1. $\dfrac{5}{1}$ **3.** $\dfrac{7}{2}$ **5.** $\dfrac{3}{5}$ **7.** (C) 10   **9.** (E) 49

11. (C) 6   **13.** 10   **15.** $\dfrac{3}{5}$ **17.** $\dfrac{3}{2}$ **19.** 9

21. 6   **23.** 10   **25.** (D) 6   **27.** 30   **29.** (D) 5

31. 5   **33.** 2   **35.** 4   **37.** -1 **39.** $\dfrac{1}{2}$

41. (A) 6   **43.** (E) 20   **45.** 10   **47.** -24 **49.** 36

51. 15   **53.** -6

Section 32, page 300

Complete Solutions to the Practice Exercises for Section 32 begin on page 381.

1. (D) 900   **3.** (B) 375   **5.** 8 meters   **7.** 4000

9. 95   **11.** (C) 15 inches   **13.** (C) 14   **15.** 50

17. 198   **19.** (B) 200   **21.** 240

Section 33, page 311

Complete Solutions to the Practice Exercises for Section 33 begin on page 382.

1. 4   **3.** 7   **5.** 11   **7.** 0

9. (B) 20   **11.** (C) 90   **13.** 800   **15.** (D) 6

17. 5   **19.** 8   **21.** (B) $\frac{1}{4}$ **23.** $\frac{2}{9}$

25. $\frac{5}{8}$ **27.** $\frac{10}{9}$ **29.** (D) 3   **31.** (B) 3

33. $\sqrt{13}$ **35.** (C) 5   **37.** (D) 6   **39.** $\sqrt{41}$

41. (A) 10   **43.** (D) 12   **45.** (D) 7   **47.** (E) $\sqrt{125}$ (or $5\sqrt{5}$)

49. (A) 2

Review Exercises for Unit VII, page 314

1. $x - 10$ **2.** $3x + 7$ **3.** 85 cents   **4.** (C) $2x + 3y$ **5.** 13

6. 8   **7.** (D) 7   **8.** 24 ft.   **9.** 12   **10.** 10

11. $\boxed{3x^2}$, $\boxed{4x}$, $\boxed{-1}$ **12.** no   **13.** yes   **14.** $9a$ **15.** $7b^2 + 2b$

16. $8x + 8y$ **17.** $a - 2b + 1$ **18.** $13x + 3y$ **19.** $6xy + 2$ **20.** $5a - b - 4$

21. no   **22.** yes   **23.** 4   **24.** 15   **25.** 25

26. 100   **27.** 1   **28.** 6   **29.** 7   **30.** 10

31. (B) 10   **32.** (A) 0   **33.** (D) 8   **34.** 3200   **35.** $7\frac{1}{2}$ pounds

36. 99   **37.** 11   **38.** 5   **39.** 10   **40.** 14

41. (E) $\sqrt{13}$

Review Exercises on Units I–VI, page 316

1. $2^4 \times 3^3 \times 5^3$ **2.** 240   **3.** 10^4 **4.** $\dfrac{1}{6}$ **5.** $\dfrac{-3}{20}$

6. .012 96   **7.** .000 024   **8.** $\dfrac{3}{250}$ **9.** $2\dfrac{1}{2}$ **10.** 38

11. 10 cm.   **12.** $\dfrac{25}{4}\pi$ in.^{2} **13.** 2520 ft.^{3} **14.** 10:45 P.M.   **15.** 4

16. 9   **17.** 50   **18.** 31.4

Practice Exam on Unit VII, page 317
Complete Solutions begin on page 383.

Practice Final Exam A, page 319
Complete Solutions begin on page 384.

Practice Final Exam B, page 324
Complete Solutions begin on page 387.

Practice Final Exam C, page 329
Complete Solutions begin on page 390.

COMPLETE SOLUTIONS TO THE PRACTICE EXERCISES AND PRACTICE EXAMS

Practice Exercises for Section 1, pages 11–13

1. 53,504

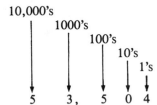

2. 6,090,015

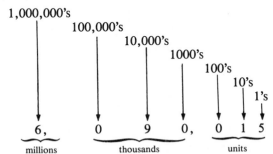

3. 943,000   **4.** thirty thousand four hundred thirty

5. twelve million five hundred twenty-nine

$$\underbrace{1\ 2,}_{\text{millions}}\ \underbrace{0\ 0\ 0,}_{\text{thousands}}\ \underbrace{5\ 2\ 9}_{\text{units}}$$

6. twelve million five hundred twenty-nine thousand

$$\underbrace{1\ 2,}_{\text{millions}}\ \underbrace{5\ 2\ 9,}_{\text{thousands}}\ \underbrace{0\ 0\ 0}_{\text{units}}$$

7. (B) $20,600

Practice Exercises for Section 2, pages 16–24

1. 781

$$\begin{array}{r} 1 \\ 453 \\ +328 \\ \hline 781 \end{array}$$

3 + 8 = 11, or 10 + 1.
Carry the 10 to the
10's column.

2. 15,113

Step 1:

1	
602	9
389	4
519	0
	3

9 + 4 + 0 = 13.
Carry the 1 to the
10's column.

Step 2:

2	1	
60	2	9
38	9	4
51	9	0
	1	3

1 + 2 + 9 + 9 = 21.
Carry the 2 to the
100's column.

3. 406

$$\begin{array}{r} {\scriptstyle 1\ 15} \\ 7\,\cancel{2}\,\cancel{5} \\ -3\ 1\ 9 \\ \hline 4\ 0\ 6 \end{array}$$

Step 3:

1	2	1
6	0	29
3	8	94
5	1	90
	1	13

2 + 0 + 8 + 1 = 11.
Carry the 1.

4. 389   Check:

$$\begin{array}{r} 389 \\ +218 \\ \hline 607 \end{array}$$

Step 4:

	1	21
6	029	
3	894	
5	190	
15	113	

1 + 6 + 3 + 5 = 15.
The sum is 15,113.

353

5. 10   **6.** 175,085

$$
\begin{array}{r}
4\ 8\ 5 \\
\times 3\ 6\ 1 \\
\hline
4\ 8\ 5 \quad \text{------} \quad 485 \times 1 \quad = \quad 4\ 8\ 5 \\
291\ 0\,\boxed{0} \quad \text{------} \quad 485 \times 6\,\boxed{0} \quad = \quad 291\ 0\,\boxed{0} \\
+1455\,\boxed{0\ 0} \quad \text{------} \quad 485 \times 3\,\boxed{00} \quad = 1455\,\boxed{0\ 0} \\
\hline
1750\ 8\ 5
\end{array}
$$

In the final step, add the individual products.

7. 18,567,360

$$
\begin{array}{r}
6048 \\
\times 3070 \\
\hline
423360 \\
18144000 \\
\hline
18567360
\end{array}
$$

8. 70

$$\underbrace{5 \times 2}_{10} \times 7 = 70$$

9. 35

$$
\begin{array}{r}
35 \\
23\overline{)805} \\
\underline{69} \\
115 \\
\underline{115} \\
0
\end{array}
$$

--- $80 \div 23 > 3$ $(3 \times 23 = 69)$
--- Subtract 69 from 80 and
--- bring down the next digit, 5.
  $115 \div 23 = 5.$
  The remainder is zero; therefore the quotient is exactly 35.

10. 46

$$
\begin{array}{r}
46 \\
57\overline{)2622} \\
\underline{228} \\
342 \\
\underline{342}
\end{array}
$$

Check:
$$
\begin{array}{r}
57 \quad \text{divisor} \\
\times 46 \quad \text{quotient} \\
\hline
342 \\
228 \\
\hline
2622 \quad \text{dividend}
\end{array}
$$

11. *i.* The quotient is 24.
ii. The remainder is 10.

$$
\begin{array}{r}
24 \quad \text{quotient} \\
17\overline{)418} \\
\underline{34} \\
78 \\
\underline{68} \\
10 \quad \text{remainder}
\end{array}
$$

12. *i.* 90   *ii.* 29

Check:
$$
\begin{array}{r}
58 \quad \text{divisor} \\
\times 90 \quad \text{quotient} \\
\hline
5220 \\
+29 \quad \text{remainder} \\
\hline
5249 \quad \text{dividend}
\end{array}
$$

13. 45 $\boxed{0}$

14. 63 $\boxed{00}$ **15.** 508, $\boxed{000}$ **16.** $172

Multiply 86 programs times 2 dollars per program to obtain $172.

Practice Exercises for Section 3, pages 28–31

1. 9

$$3^2 = 3 \times 3 = 9$$

2. 400

$$20^2 = 20 \times 20 = 400$$

3. 64

$$4^3 = \underbrace{4 \times 4}_{16} \times 4 = 64$$

4. 1,000,000

$$100^3 = \underbrace{100 \times 100}_{10,000} \times 100 = 1,000,000$$

5. 81

$$3^4 = \underbrace{3 \times 3}_{9} \times \underbrace{3 \times 3}_{9} = 81$$

6. 10,000,000

7. 625   **8.** *i.* 3
ii. 9
iii. the ninth power

9. 392

$$2^3 \times 7^2 = \underbrace{2 \times 2 \times 2}_{8} \times \underbrace{7 \times 7}_{49}$$
$$= 8 \times 49$$
$$= 392$$

10. 75,000

$$3 \times 5^2 \times 10^3$$
$$= 3 \times 25 \times 1000$$
$$= 75,000$$

Practice Exercises for Section 4, pages 34–36

1. 1, 2, and 4

    $4 = 1 \times 4$ (Both 1 and 4 are factors of 4.)

    $4 = 2 \times 2$ (2 is a factor of 4.)

2. 1 and 11

    $11 = 1 \times 11$ (11 cannot be factored in any other way.)

3. 1, 2, 3, 4, 6, and 12

    $12 = 1 \times 12$

    or $12 = 2 \times 6$

    or $12 = 3 \times 4$

4. *i.* (B), (D), and (E) are multiples of 8.

    ii. (A) and (B) are divisors of 8.

    (A) $8 = 2 \times 4$ (4 is a divisor of 8.)

    (B) $8 = 1 \times 8$ (8 is both a multiple of 8 and a divisor of 8.)

    (D) $16 = 2 \times 8$ (16 is a multiple of 8.)

    (E) $48 = 6 \times 8$ (48 is a multiple of 8.)

5. *i.* (D) 50 and (E) 230 are multiples of 10.

    ii. (A) 1 and (B) 5 are divisors of 10.

6. 21 is a composite

    because $21 = 3 \times 7$.

7. 23 is a prime

    because its only factors are 1 and 23.

8. 31 is a prime.

9. 39 is a composite.

10. $32 = 2 \times 2 \times 2 \times 2 \times 2 = 2^5$

11. $44 = 2^2 \times 11$

12. $84 = 2^2 \times 3 \times 7$

Practice Exercises for Section 5, pages 39–44

1. 558 is even

    because its 1's digit, 8, is even.

2. 6943 is odd

    because its 1's digit, 3, is odd.

3. 7530 is divisible by 2.

4. 68,421 is not divisible by 2

    because its 1's digit is odd.

5. 105 is divisible by 5

    because its 1's digit is 5.

6. 559 is not divisible by 5.

7. 8040 is divisible by 5

    because its 1's digit is 0.

8. (A) 60, (C) 4400, and (E) 1010 are divisible by 10.

9. (B) 5400 and (C) 54,000 are divisible by 100

    because they both end in two zeros.

10. (C) 201,000 and (D) 1000 are divisible by 1000.

11. 10^5 or 100,000

    Because 5,800,000 ends in five zeros, it is divisible by 10^5 or 100,000.

12. (A) 72, (B) 96, and (D) 8076 are divisible by 3

    because in each case the sum of the digits is a multiple of 3.

    (A) $7 + 2 = 9$; (B) $9 + 6 = 15$;

    (D) $8 + 0 + 7 + 6 = 21$

13. (A) 171, (B) 567, and (D) 6777 are divisible by 9.

    (A) $1 + 7 + 1 = 9$; (B) $5 + 6 + 7 = 18$;

    (D) $6 + 7 + 7 + 7 = 27$

14. $625 = 5^4$

15. $4050 = 2 \times 3^4 \times 5^2$

    $4050 = 10 \times 405$

    $= (2 \times 5) \times (9 \times 45)$

    $= (2 \times 5) \times (3 \times 3 \times 3 \times 3 \times 5)$

    $= 2 \times 3^4 \times 5^2$

16. $147 = 3 \times 7^2$

    $147 = 3 \times 49 = 3 \times 7 \times 7 = 3 \times 7^2$

Practice Exercises for Section 6, pages 46–51

1. 4

The factors of 4 are 1, 2, and 4; the factors of 8 are 1, 2, 4, and 8. The common factors of 4 and 8 are 1, 2, and 4.
$gcd (4, 8) = 4$

2. 5

3. 7

4. 6

$24 = 2^3 \times 3; \quad 42 = 2 \times 3 \times 7$
The common prime factors are 2 and 3. There is only one factor of each in 42. Therefore, the greatest common divisor is found by multiplying 2×3.
$gcd (24, 42) = 6$

5. 12

$72 = 2^3 \times 3^2; \quad 132 = 2^2 \times 3 \times 11$
Multiply the smallest powers of the common prime factors:
$gcd (72, 132) = 2^2 \times 3 = 12$

6. 36

$108 = 2^2 \times 3^3; \quad 144 = 2^4 \times 3^2$
$gcd (108, 144) = 2^2 \times 3^2 = 36$

7. 5

$10 = 2 \times 5; \quad 25 = 5^2; \quad 50 = 2 \times 5^2$
The only common prime factor is 5. The smallest power to which it appears is 5^1.
Thus $gcd (10, 25, 50) = 5$

8. 8

$16 = 2^4; \quad 40 = 2^3 \times 5; \quad 96 = 2^5 \times 3$
$gcd (16, 40, 96) = 2^3 = 8$

9. 1

$28 = 2^2 \times 7; \quad 25 = 5^2; \quad 50 = 2 \times 5^2$
There are no common prime factors.
Thus $gcd (28, 25, 50) = 1$

10. 12

The multiples of 3 are 3, 6, 9, ⑫, 15, etc. The multiples of 4 are 4, 8, ⑫, 16, etc. Therefore, $lcm (3, 4) = 12$.

11. 60

The multiples of 10 are 10, 20, 30, 40, 50, ⑥⓪, etc. The multiples of 12 are 12, 24, 36, 48, ⑥⓪, etc. $lcm (10, 12) = 60$.

12. 30

13. 160

$20 = 2^2 \times 5; \quad 32 = 2^5$
The highest power of 2 that occurs is 2^5; the highest power of 5 is 5 itself. Therefore, $lcm (20, 32) = 2^5 \times 5 = 32 \times 5 = 160$.

14. 90

$18 = 2 \times 3^2; \quad 45 = 3^2 \times 5$
$lcm (18, 45) = 2 \times 3^2 \times 5 = 90$.

15. 432

$48 = 2^4 \times 3; \quad 54 = 2 \times 3^3$
The highest power of 2 that occurs is 2^4; the highest power of 3 is 3^3. Therefore, $lcm (48, 54) = 2^4 \times 3^3 = 16 \times 27 = 432$.

16. 144

$9 = 3^2; \quad 12 = 2^2 \times 3; \quad 16 = 2^4$
The highest power of 2 that occurs in these factorizations is 2^4. The highest power of 3 is 3^2.
Therefore, $lcm (9, 12, 16) = 2^4 \times 3^2 = 144$.

17. 300

$20 = 2^2 \times 5; \quad 25 = 5^2;$
$30 = 2 \times 3 \times 5$
$lcm (20, 25, 30) = 2^2 \times 3 \times 5^2 = 300$

18. 336

$24 = 2^3 \times 3; \quad 42 = 2 \times 3 \times 7;$
$48 = 2^4 \times 3$
$lcm (24, 42, 48) = 2^4 \times 3 \times 7$
$= 16 \times 21 = 336$

Practice Exam on Unit I, page 54

1. 42,106

2. two hundred twenty thousand fifteen

3. 2,732,865

$$
\begin{array}{r}
40\ 8\ 5 \\
\times 6\ 6\ 9 \\
\hline
367\ 6\ 5 \\
2451\ 0\ \boxed{0} \\
24510\ \boxed{0\ 0} \\
\hline
27328\ 6\ 5
\end{array}
$$

------- $9 \times 4085 = \quad 367\ 6\ 5$

----- $6\,\boxed{0} \times 4085 = \ 2451\ 0\,\boxed{0}$

--- $6\,\boxed{0\ 0} \times 4085 = 24510\,\boxed{0\ 0}$

4. *i.* 16   *ii.* 13

$$
\begin{array}{r}
16 \ \text{quotient} \\
36\overline{)589} \\
\underline{36\ \ } \\
229 \\
\underline{216} \\
13 \ \text{remainder}
\end{array}
$$

5. 18

$72 \div 4 = 18$

6. 121

$11^2 = 11 \times 11 = 121$

7. 256

$4^4 = \underbrace{4 \times 4}_{16} \times \underbrace{4 \times 4}_{16} = 16 \times 16 = 256$

8. 175

$$
\begin{aligned}
5^2 \times 7 &= 5 \times 5 \times 7 \\
&= 25 \times 7 \\
&= 175
\end{aligned}
$$

9. $84 = 2^2 \times 3 \times 7$

10. yes

The 1's digit of

$840\,\textcircled{5}$ is 5.

11. 10^3 or 1000

12. yes

5589 is divisible by 9 because the sum of its digits is divisible by 9. $(5 + 5 + 8 + 9 = 27)$

13. $17,820 = 2^2 \times 3^4 \times 5 \times 11$

14. 16

$$
\begin{aligned}
48 &= 2^4 \times 3 \\
64 &= 2^6 \\
gcd\ (48,\ 64) &= 2^4 = 16
\end{aligned}
$$

15. 300

$$
\begin{aligned}
12 &= 2^2 \times 3 \\
20 &= 2^2 \times 5 \\
25 &= 5^2 \\
lcm\ (12,\ 20,\ 25) &= 2^2 \times 3 \times 5^2 = 300
\end{aligned}
$$

Practice Exercises for Section 7, pages 57–63

1. $\dfrac{1}{4}$

$\dfrac{5}{20} = \dfrac{\cancel{5} \times 1}{\cancel{5} \times 4} = \dfrac{1}{4}$

2. $\dfrac{4}{5}$

$gcd\ (12,\ 15) = 3$

$\dfrac{12}{15} = \dfrac{\cancel{3} \times 4}{\cancel{3} \times 5} = \dfrac{4}{5}$

3. $\dfrac{6}{11}$

$\dfrac{24}{44} = \dfrac{\cancel{4} \times 6}{\cancel{4} \times 11} = \dfrac{6}{11}$

4. $\dfrac{8}{12}$

$\dfrac{2}{3} = \dfrac{4 \times 2}{4 \times 3} = \dfrac{8}{12}$

5. *i.* $\dfrac{14}{20}$ *ii.* $\dfrac{35}{50}$

6. $\dfrac{5}{7}$

$\dfrac{3}{7} + \dfrac{2}{7} = \dfrac{3 + 2}{7} = \dfrac{5}{7}$

7. $\dfrac{1}{3}$

$\dfrac{5}{9} - \dfrac{2}{9} = \dfrac{5 - 2}{9} = \dfrac{3}{9} = \dfrac{3 \times 1}{3 \times 3} = \dfrac{1}{3}$

8. 1

$\dfrac{5}{12} + \dfrac{7}{12} = \dfrac{12}{12} = 1$

9. *i.* 12   *ii.* $\dfrac{3}{4} = \dfrac{9}{12}$; $\dfrac{1}{6} = \dfrac{2}{12}$

i. $4 = 2^2$ and $6 = 2 \times 3$. There-fore, *lcm* $(4, 6) = 2^2 \times 3 = 12$.

The *lcd* of $\dfrac{3}{4}$ and $\dfrac{1}{6}$ is 12.

ii. $\dfrac{3}{4} = \dfrac{3 \times 3}{3 \times 4} = \dfrac{9}{12}$;

$\dfrac{1}{6} = \dfrac{2 \times 1}{2 \times 6} = \dfrac{2}{12}$

12. $\dfrac{3}{20}$

The *lcd* of $\dfrac{9}{10}$ and $\dfrac{3}{4}$ is 20.

$\dfrac{9}{10} = \dfrac{2 \times 9}{2 \times 10} = \dfrac{18}{20}$;

$\dfrac{3}{4} = \dfrac{5 \times 3}{5 \times 4} = \dfrac{15}{20}$

$\dfrac{9}{10} - \dfrac{3}{4} = \dfrac{18}{20} - \dfrac{15}{20} = \dfrac{3}{20}$

14. $\dfrac{38}{45}$

lcm $(5, 9, 3) = 45$

$\dfrac{2}{5} + \dfrac{1}{9} + \dfrac{1}{3} = \dfrac{18}{45} + \dfrac{5}{45} + \dfrac{15}{45}$

$= \dfrac{18 + 5 + 15}{45}$

$= \dfrac{38}{45}$

10. *i.* 40   *ii.* $\dfrac{5}{8} = \dfrac{25}{40}$; $\dfrac{3}{20} = \dfrac{6}{40}$

11. $\dfrac{11}{12}$

lcm $(3, 4) = 12$

$\dfrac{2}{3} = \dfrac{4 \times 2}{4 \times 3} = \dfrac{8}{12}$;

$\dfrac{1}{4} = \dfrac{3 \times 1}{3 \times 4} = \dfrac{3}{12}$

$\dfrac{2}{3} + \dfrac{1}{4} = \dfrac{8}{12} + \dfrac{3}{12} = \dfrac{8 + 3}{12} = \dfrac{11}{12}$

13. $\dfrac{19}{50}$

$\dfrac{3}{10} + \dfrac{2}{25} = \dfrac{5 \times 3}{5 \times 10} + \dfrac{2 \times 2}{2 \times 25}$

$= \dfrac{15}{50} + \dfrac{4}{50} = \dfrac{19}{50}$

15. (D) $\dfrac{11}{12}$

$\dfrac{7}{12} - \dfrac{1}{2} + \dfrac{5}{6} = \dfrac{7}{12} - \dfrac{6}{12} + \dfrac{10}{12}$

$= \dfrac{7 - 6 + 10}{12}$

$= \dfrac{11}{12}$

Practice Exercises for Section 8, pages 68–73

1. $\dfrac{6}{35}$

$\dfrac{3}{7} \times \dfrac{2}{5} = \dfrac{3 \times 2}{7 \times 5} = \dfrac{6}{35}$

2. $\dfrac{1}{6}$

$\dfrac{\overset{1}{\cancel{5}}}{\underset{2}{\cancel{8}}} \times \dfrac{\overset{1}{\cancel{4}}}{\underset{3}{\cancel{15}}} = \dfrac{1 \times 1}{2 \times 3} = \dfrac{1}{6}$

3. $\dfrac{3}{4}$

The numerator 9 and the denominator 6 have the factor 3 in common. Divide them both by 3.

$\dfrac{\overset{3}{\cancel{9}}}{10} \times \dfrac{5}{\underset{2}{\cancel{6}}}$. Similarly, the numerator 5 and the denominator 10 have the factor 5 in common.

Divide them both by 5 and multiply. Thus, $\dfrac{\overset{3}{\cancel{9}}}{\underset{2}{\cancel{10}}} \times \dfrac{\overset{1}{\cancel{5}}}{\underset{2}{\cancel{6}}} = \dfrac{3 \times 1}{2 \times 2} = \dfrac{3}{4}$.

4. $\dfrac{1}{30}$

Divide the numerator 2 and the denominator 4 by the common factor 2. Divide the numerator 3 and the denominator 9 by the common factor 3. Thus, $\dfrac{\overset{1}{\cancel{2}}}{\underset{3}{\cancel{9}}} \times \dfrac{1}{\underset{2}{\cancel{4}}} \times \dfrac{\overset{1}{\cancel{3}}}{5} = \dfrac{1 \times 1 \times 1}{3 \times 2 \times 5} = \dfrac{1}{30}.$

5. $\dfrac{3}{4}$

Express the whole number 9 as a fraction, $\dfrac{9}{1}$.

$\dfrac{\overset{1}{\cancel{5}}}{\underset{2}{\cancel{6}}} \times \dfrac{1}{\underset{2}{\cancel{10}}} \times \dfrac{\overset{3}{\cancel{9}}}{1} = \dfrac{1 \times 1 \times 3}{2 \times 2 \times 1} = \dfrac{3}{4}$

6. $\dfrac{1}{20}$

7. $\dfrac{9}{16}$

$\left(\dfrac{3}{4}\right)^2 = \dfrac{3^2}{4^2} = \dfrac{9}{16}$

8. $\dfrac{1}{1000}$

$\left(\dfrac{1}{10}\right)^3 = \dfrac{1^3}{10^3} = \dfrac{1 \times 1 \times 1}{10 \times 10 \times 10} = \dfrac{1}{1000}$

9. $\dfrac{1}{16}$ **10.** $\dfrac{5}{6}$

$\dfrac{2}{3} \div \dfrac{4}{5} = \dfrac{2}{3} \times \dfrac{5}{\underset{2}{\cancel{4}}} = \dfrac{1 \times 5}{3 \times 2} = \dfrac{5}{6}$

11. $\dfrac{9}{7}$

$\dfrac{3}{4} \div \dfrac{7}{12} = \dfrac{3}{\underset{1}{\cancel{4}}} \times \dfrac{\overset{3}{\cancel{12}}}{7} = \dfrac{3 \times 3}{1 \times 7} = \dfrac{9}{7}$

12. $\dfrac{16}{5}$ Check: $\dfrac{\overset{4}{\cancel{16}}}{5} \times \dfrac{3}{\underset{5}{\cancel{20}}} = \dfrac{4 \times 3}{5 \times 5} = \dfrac{12}{25}$

(quotient) × (divisor) =   (dividend)

13. (B) $\dfrac{1}{3}$

$\dfrac{\overset{1}{\cancel{5}}}{\underset{3}{\cancel{9}}} \times \dfrac{\overset{1}{\cancel{3}}}{\cancel{5}} = \dfrac{1 \times 1}{3 \times 1} = \dfrac{1}{3}$

14. (D) $\dfrac{7}{9}$

$\dfrac{7}{12} \div \dfrac{3}{4} = \dfrac{7}{\underset{3}{\cancel{12}}} \times \dfrac{\overset{1}{\cancel{4}}}{3} = \dfrac{7 \times 1}{3 \times 3} = \dfrac{7}{9}$

15. (C) $\dfrac{1}{54}$

$\dfrac{1}{9}$ "of" $\dfrac{1}{6}$ means $\dfrac{1}{9} \times \dfrac{1}{6} = \dfrac{1}{54}.$

16. (E) 6

To find what number times $\dfrac{1}{8}$ equals $\dfrac{3}{4}$, use division.

Thus, $\dfrac{3}{4} \div \dfrac{1}{8} = \dfrac{3}{4} \times \dfrac{8}{1} = 6.$

Practice Exercises for Section 9, pages 78–83

1. $\dfrac{7}{2}$

$3\dfrac{1}{2} = 3 + \dfrac{1}{2}$

$= \dfrac{3}{1} + \dfrac{1}{2}$

$= \dfrac{6}{2} + \dfrac{1}{2} = \dfrac{7}{2}$

2. $\dfrac{11}{4}$

$2\dfrac{3}{4} = 2 + \dfrac{3}{4} = \dfrac{8}{4} + \dfrac{3}{4} = \dfrac{11}{4}$

3. $\dfrac{26}{3}$

$8\dfrac{2}{3} = 8 + \dfrac{2}{3} = \dfrac{24}{3} + \dfrac{2}{3} = \dfrac{26}{3}$

4. $4\frac{1}{2}$ **5.** $3\frac{2}{3}$ **6.** $2\frac{6}{7}$ **7.** $7\frac{5}{7}$

$$\begin{array}{r} 4 \quad \text{quotient} \\ 2\overline{)9} \\ \underline{8} \\ 1 \quad \text{remainder} \end{array}$$

Thus, $\dfrac{9}{2} = 4\dfrac{1}{2}$.

8. $1\frac{7}{8}$

The *lcd* of $\dfrac{1}{4}$ and $\dfrac{3}{8}$ is 8. $3\dfrac{1}{4} = 3\dfrac{2}{8}$. $\dfrac{2}{8}$ is smaller than $\dfrac{3}{8}$. Therefore, borrow 1 from

the whole part and convert it to eighths. $3\dfrac{2}{8} = 2 + 1 + \dfrac{2}{8} = 2 + \dfrac{8}{8} + \dfrac{2}{8}$.

$$\begin{array}{r} 2\dfrac{10}{8} \\ -1\dfrac{3}{8} \\ \hline 1\dfrac{7}{8} \end{array}$$

9. $11\frac{5}{12}$

The *lcd* of $\dfrac{2}{3}$ and $\dfrac{3}{4}$ is 12. Convert both to twelfths.

$$\begin{array}{r} 6\dfrac{2}{3} = 6\dfrac{8}{12} \\ +4\dfrac{3}{4} = 4\dfrac{9}{12} \\ \hline 10\dfrac{17}{12} \end{array}$$

Change the improper fraction $\dfrac{17}{12}$ to $1\dfrac{5}{12}$. Thus, $10\dfrac{17}{12} = 10 + 1\dfrac{5}{12} = 11\dfrac{5}{12}$.

10. $7\frac{1}{2}$ **11.** $1\frac{17}{19}$

$$2\frac{1}{4} \times 3\frac{1}{3} = \frac{\overset{3}{\cancel{9}}}{\underset{2}{\cancel{4}}} \times \frac{\overset{5}{\cancel{10}}}{\underset{1}{\cancel{3}}}$$

$$4\frac{1}{2} \div 2\frac{3}{8} = \frac{9}{2} \div \frac{19}{8}$$

$$= \frac{3 \times 5}{2 \times 1} = \frac{15}{2} = 7\frac{1}{2}$$

$$= \frac{9}{\underset{1}{\cancel{2}}} \times \frac{\overset{4}{\cancel{8}}}{19} = \frac{9 \times 4}{1 \times 19} = \frac{36}{19} = 1\frac{17}{19}$$

12. 99   **13.** 10

$$8 \times 4\frac{1}{2} \times 2\frac{3}{4} = \frac{\overset{2}{\cancel{8}}}{1} \times \frac{9}{2} \times \frac{11}{\underset{1}{\cancel{4}}}$$

$$2\frac{1}{2} \times 4 = \frac{5}{\underset{1}{\cancel{2}}} \times \frac{\overset{2}{\cancel{4}}}{1} = 10$$

$$= \frac{\cancel{2} \times 9 \times 11}{\underset{1}{\cancel{2}}} = 99$$

Practice Exercises for Section 10, pages 88–91

1. (A) $\dfrac{2}{7}$ **2.** $\dfrac{2}{3}$

$$\frac{2}{3} = \frac{8}{12}; \quad \frac{3}{4} = \frac{9}{12}$$

$\dfrac{8}{12} < \dfrac{9}{12}$ and therefore $\dfrac{2}{3} < \dfrac{3}{4}$.

Note also that $\dfrac{2}{3} < \dfrac{3}{4}$

because $2 \times 4 < 3 \times 3$.

3. $\dfrac{5}{8}$

$$\frac{5}{8} < \frac{7}{10}$$

because $5 \times 10 < 8 \times 7$.

4. $\dfrac{5}{12}$

5. (A) $\dfrac{1}{6}$

(A), (B): $\dfrac{1}{6} < \dfrac{5}{6}$ because $1 < 5$

(A), (C): $\dfrac{1}{6} < \dfrac{3}{10}$ because $1 \times 10 < 3 \times 6$

(A), (D): $\dfrac{1}{6} < \dfrac{7}{10}$ because $1 \times 10 < 7 \times 6$

(A), (E): $\dfrac{1}{6} < \dfrac{9}{10}$ because $1 \times 10 < 9 \times 6$

6. (B) $\dfrac{2}{5}$ **7.** (D) $\dfrac{3}{5}$

8. (E) $\dfrac{11}{12}$

Fractions (A), (B), and (C) all have the denominator 10. (C), $\dfrac{9}{10}$, is the largest.

For (D) and (E), $\dfrac{11}{12} > \dfrac{7}{12}$. $\dfrac{11}{12} > \dfrac{9}{10}$

because $11 \times 10 > 9 \times 12$, or $110 > 108$.

9. (B) $\dfrac{7}{8}$ **10.** $\dfrac{3}{4}$ of a ton weighs less.

$\dfrac{3}{4} < \dfrac{4}{5}$ because $3 \times 5 < 4 \times 4$

11. $\dfrac{5}{8}$ mile

$\dfrac{5}{8} < \dfrac{3}{4}$ because $5 \times 4 < 3 \times 8$;

$\dfrac{5}{8} < \dfrac{2}{3}$ because $5 \times 3 < 2 \times 8$

Practice Exam on Unit II, page 98

1. $\dfrac{5}{8}$

$gcd\,(20, 32) = 4$. Divide both numerator and denominator by 4. $\dfrac{20}{32} = \dfrac{4 \times 5}{4 \times 8} = \dfrac{5}{8}$.

3. $\dfrac{1}{16}$

The *lcd* is 16.

$$\frac{3}{8} - \frac{5}{16} = \frac{6}{16} - \frac{5}{16} = \frac{6 - 5}{16} = \frac{1}{16}$$

4. 1

The *lcd* is 10.

$$\frac{7}{10} - \frac{1}{5} + \frac{1}{2} = \frac{7}{10} - \frac{2 \times 1}{2 \times 5} + \frac{5 \times 1}{5 \times 2}$$

$$= \frac{7}{10} - \frac{2}{10} + \frac{5}{10}$$

$$= \frac{10}{10} = 1$$

2. $\dfrac{23}{36}$

The *lcd* of $\dfrac{5}{12}$ and $\dfrac{2}{9}$ is the *lcm* of 12 and 9.

$12 = 2^2 \times 3$; $\quad 9 = 3^2$.

$lcm\,(12, 9) = 2^2 \times 3^2 = 36$

$$\frac{5}{12} + \frac{2}{9} = \frac{3 \times 5}{3 \times 12} + \frac{4 \times 2}{4 \times 9}$$

$$= \frac{15}{36} + \frac{8}{36} = \frac{15 + 8}{36} = \frac{23}{36}$$

5. $\dfrac{2}{9}$

Divide the common factor 3 from the numerator 3 and the denominator 27; divide the common factor 10 from the numerator 20 and the denominator 10. Thus,

$$\overset{1}{\cancel{3}} \times \frac{\overset{2}{\cancel{20}}}{\underset{1}{\cancel{10}}} \cdot \overset{}{\underset{9}{\cancel{27}}} = \frac{1 \times 2}{1 \times 9} = \frac{2}{9}$$

6. $\dfrac{7}{8}$

To divide fractions, invert the divisor, $\dfrac{10}{21}$; then multiply.

$$\dfrac{5}{12} \div \dfrac{10}{21} = \dfrac{\cancel{5}^{1}}{\cancel{12}_{4}} \times \dfrac{\cancel{21}^{7}}{\cancel{10}_{2}} = \dfrac{1 \times 7}{4 \times 2} = \dfrac{7}{8}$$

7. $\dfrac{23}{4}$

$$5\dfrac{3}{4} = \dfrac{5}{1} + \dfrac{3}{4} = \dfrac{4 \times 5}{4 \times 1} + \dfrac{3}{4} = \dfrac{20}{4} + \dfrac{3}{4} = \dfrac{23}{4}$$

8. $3\dfrac{1}{3}$

$$\dfrac{10}{3} = \dfrac{9}{3} + \dfrac{1}{3} = 3 + \dfrac{1}{3} = 3\dfrac{1}{3}$$

9. $1\dfrac{3}{4}$

The *lcd* of 2 and 4 is 4. $6\dfrac{1}{2} = 6\dfrac{2}{4}$. But $\dfrac{2}{4}$ is smaller than $\dfrac{3}{4}$; therefore, borrow 1 from the whole part, 6, before subtracting.

Thus, $6\dfrac{1}{2} = 6\dfrac{2}{4} = 5 + 1 + \dfrac{2}{4} = 5 + \dfrac{4}{4} + \dfrac{2}{4} =$

$$\begin{array}{r} 5\dfrac{6}{4} \\ -\,4\dfrac{3}{4} \\ \hline 1\dfrac{3}{4} \end{array}$$

10. $3\dfrac{1}{15}$ or $\dfrac{46}{15}$

Convert the mixed numbers to improper fractions. To divide, invert the divisor; then multiply.

Thus, $4\dfrac{3}{5} \div 1\dfrac{1}{2} = \dfrac{23}{5} \div \dfrac{3}{2} = \dfrac{23}{5} \times \dfrac{2}{3} = \dfrac{46}{15}$.

Change the improper fraction $\dfrac{46}{15}$ to the mixed number $3\dfrac{1}{15}$.

11. (C) $\dfrac{5}{16}$

(A), (B): $\dfrac{3}{4} > \dfrac{3}{8}$ because when the numerators are the same, the smaller fraction corresponds to the larger denominator.

(B), (C): $\dfrac{3}{8} > \dfrac{5}{16}$ because $3 \times 16 > 5 \times 8$

(C), (D): $\dfrac{5}{16} < \dfrac{7}{20}$ because $5 \times 20 < 7 \times 16$

(C), (E): $\dfrac{5}{16} < \dfrac{13}{22}$ because $5 \times 22 < 13 \times 16$

12. (C) $\dfrac{2}{3}$

13. (D) 8 miles

How many times will $\dfrac{1}{3}$ of an inch go into $2\dfrac{2}{3}$ inches?

The operation called for is division. Thus,

$$2\dfrac{2}{3} \div \dfrac{1}{3} = \dfrac{8}{3} \div \dfrac{1}{3} = \dfrac{8}{3} \times \dfrac{3}{1} = 8.$$

14. (C) $57\dfrac{5}{8}$

$$\begin{array}{r} 56\dfrac{1}{8} = 56\dfrac{1}{8} \\ +\,1\dfrac{1}{2} = 1\dfrac{4}{8} \\ \hline 57\dfrac{5}{8} \end{array}$$

Practice Exercises for Section 11, pages 101–106

1. $\dfrac{2}{10} + \dfrac{1}{100}$

$.21 = \dfrac{21}{100} = \dfrac{20 + 1}{100}$

$\qquad = \dfrac{20}{100} + \dfrac{1}{100}$

$\qquad = \dfrac{2}{10} + \dfrac{1}{100}$

2. $\dfrac{3}{10} + \dfrac{5}{100} + \dfrac{1}{1000}$

$.351 = \dfrac{351}{1000} = \dfrac{300 + 50 + 1}{1000}$

$\qquad = \dfrac{300}{1000} + \dfrac{50}{1000} + \dfrac{1}{1000}$

$\qquad = \dfrac{3}{10} + \dfrac{5}{100} + \dfrac{1}{1000}$

3. $\dfrac{8}{10} + \dfrac{7}{100} + \dfrac{7}{1000}$ **4.** *i.* 5 *ii.* 1 *iii.* 0 *iv.* 5 **5.** 0 **6.** .629 **7.** .5

8. .005 **9.** four tenths **10.** seventeen hundredths **11.** 5.1 **12.** 63.09 **13.** 25.17

14. .75

To change $\dfrac{3}{4}$ to a decimal, divide the numerator 3 by the denominator 4. Thus,

$$\begin{array}{r} .75 \\ 4\overline{)3.00} \\ 2\,8 \\ \hline 20 \\ 20 \\ \hline \end{array}$$

Thus, $\dfrac{3}{4} = .75$

15. .35

$$\begin{array}{r} .35 \\ 20\overline{)7.00} \\ 6\,0 \\ \hline 1\,00 \\ 1\,00 \\ \hline \end{array}$$

Thus, $\dfrac{7}{20} = .35$

16. $\dfrac{6}{25}$

$.24 = \dfrac{24}{100}$ (Divide numerator and denominator by the *gcd* 4.)

$\qquad = \dfrac{4 \times 6}{4 \times 25}$

$\qquad = \dfrac{6}{25}$

17. $\dfrac{1}{250}$

$.004 = \dfrac{4}{1000} = \dfrac{1}{250}$

18. .444 444...

$$\begin{array}{r} .444... \\ 9\overline{)4.000} \end{array}$$

19. .833 333...

$$\begin{array}{r} .833... \\ 6\overline{)5.000} \end{array}$$

Practice Exercises for Section 12, pages 109–113

1. (C) .1 **2.** (E) .32

Because all choices have the same tenths digit, compare their hundredths digits. Choice (E) has the smallest hundredths digit.

3. (B) .46 **4.** (E) .131 **5.** (B) .407

Note that .407 has the smallest hundredth's digit.

6. (D) 2.7 **7.** (A) 1.01 **8.** (D) 10.366 **9.** (C) .491 **10.** (C) 1.10

Practice Exercises for Section 13, pages 116–118

1. 1.8 **2.** 17.3 **3.** 32.27 **4.** .3 **5.** 6.8

$$\begin{array}{r} .6 \\ .5 \\ + \ .7 \\ \hline 1.8 \end{array}$$

6. 7.97   **7.** 35.14   **8.** 4.12   **9.** $162.10   **10.** $74.85

$$
\begin{array}{r}
45.50 \\
-37.53 \\
\hline
7.97
\end{array}
$$

Practice Exercises for Section 14, pages 120–125

1. 2.36   **2.** 2753

To multiply by 100, or 10^2, move the decimal point 2 digits to the right. Thus, 27 53.

3. 2.8

.0028 × 1000 = 002.8   (Move the decimal point three digits to the right.)
= 2.8   (When zeros are to the left of the decimal point, they may be dropped.)

4. 38.4   **5.** .076   **6.** .0321   **7.** .25   **8.** .027

$$
\begin{array}{r}
.38 \leftarrow \ \ 2 \text{ decimal digits} \\
\times\ .2 \leftarrow +1 \text{ decimal digit} \\
\hline
.076 \leftarrow \ \ 3 \text{ decimal digits}
\end{array}
$$

9. .0001   **10.** $11.60   **11.** .585   **12.** .027 36   **13.** .072 14

2.736 ÷ 100 = .02 736

(Move the decimal point
2 digits to the left.)

14. 80   **15.** 139

$$\frac{48}{.6} = \frac{48 \times 10}{.6 \times 10} = \frac{480}{6} = 80$$

$$\frac{6.95}{.05} = \frac{6.95 \times 100}{.05 \times 100} = \frac{695}{5} = 139$$

16. 5300   **17.** 7.4 pounds

To divide by thousandths, multiply both divisor and dividend by 1000. This is equivalent to moving both decimal points 3 digits to the right.

$$\frac{18.50}{2.50} = \frac{18.50 \times 10}{2.50 \times 10} = \frac{185}{25}$$

$$
\begin{array}{r}
5300. \\
.032\,\overline{)169.600}
\end{array}
\text{ is the same as }
\begin{array}{r}
32.\,\overline{)169600.}
\end{array}
$$

$$
\begin{array}{r}
160 \\
\hline
96 \\
96 \\
\hline
00
\end{array}
$$

$$
\begin{array}{r}
7.4 \\
25\,\overline{)185.0} \\
175 \\
\hline
10\ 0
\end{array}
$$

18. (C) 3 inches   **19.** (D) 12 hours

Multiply .05 in. × 60.

"What number times $7.50 yields $90?"
The answer is found by dividing 90 ÷ 7.50.

$$
\begin{array}{r}
60 \leftarrow \ \ 0 \text{ decimal digits} \\
\times\,.05 \leftarrow +2 \text{ decimal digits} \\
\hline
3.00 \leftarrow \ \ 2 \text{ decimal digits}
\end{array}
$$

$$\frac{90}{7.5} = \frac{90 \times 10}{7.5 \times 10} = \frac{900}{75} = 12$$

Practice Exercises for Section 15, pages 128–130

1. (A) \$2.30

    apples: $4 \times \$.35 = \$1.40$
    oranges: $3 \times \$.30 = \underline{\$\ .90}$
                         \$2.30 (Total Cost)

2. (E) \$255

The first day's rental costs \$55. There are $(6 - 1)$, or 5, additional days at \$40 per day. The total cost is

$$\$55 + 5 \times \$40 = \$55 + \$200$$
$$= \$255$$

3. \$4.00

$24 \times \$.30 = $ \$7.20 (Total Sales)
              $\underline{-\ \$3.20}$ (Cost)
              \$4.00 (Profit)

4. (D) \$20.20

The total amount of sales is

$$(40 \times \$.85) + (20 \times \$.45) = \$34 + \$9$$
$$= \$43$$

The total number of loaves baked is $40 + 20$, or 60 loaves, at a cost of \$.38 per loaf. The total cost then is $60 \times \$.38 = \$22.80$

    \$43.00 (Total Sales)
$\underline{-\ \$22.80}$ (Cost)
    \$20.20 (Profit)

Practice Exercises for Section 16, pages 134–138

1. 8000

The hundred's digit of 8174 is 1, which is less than 5. Retain the 1000's digit, 8, and replace the last 3 digits with 0's.

2. 13,000   **3.** 76,000,000   **4.** 49,000,000

5. 7,400,000

To round 7,391,504 to the nearest 100,000, note that the 100,000's digit is 3. The following digit, 9, is more than 5. Therefore, increase the 3 to 4 and replace all the following digits with 0's.

6. .9   **7.** .6   **8.** .66   **9.** .71

10. .814

To round .8135 to the nearest thousandth, note that the thousandth's digit is 3. Because the following digit is 5, increase the 3 to 4.

11. .491   **12.** 3   **13.** 100   **14.** 58,000,000   **15.** 600

16. 2.33

$$\begin{array}{r} 2.333\ldots \\ 3.\overline{)7.000\ldots} \end{array}$$

To the nearest hundredth, 2.333 rounds to 2.33.

17. .11

$$\begin{array}{r} .113\ldots \\ 8.\overline{)0.910\ldots} \end{array}$$

To the nearest hundredth, .113 rounds to .11.

Practice Exam on Unit III, page 144

1. .097   **2.** .175   **3.** $\dfrac{26}{25}$ **4.** (A) .049

$$\begin{array}{r} .175 \\ 4.0\overline{)7.00} \end{array}$$

$$1.04 = \frac{104}{100} = \frac{26 \times 4}{25 \times 4} = \frac{26}{25}$$

5. 102

$$
\begin{array}{r}
27.2 \\
36.9 \\
15.91 \\
+ \ 21.99 \\
\hline
102.00
\end{array}
$$

6. 3.1

$$
\begin{array}{r}
13.08 \\
- \ 9.98 \\
\hline
3.10
\end{array}
$$

7. 4.5

To multiply by 100, move the decimal point 2 digits to the right.

8. .00417

To divide by 1000, or 10^3, move the decimal point 3 digits to the left.

9. .0882

$$
\begin{array}{r}
.042 \ \leftarrow \quad 3 \text{ decimal digits} \\
\times \ 2.1 \ \leftarrow \ +1 \text{ decimal digit} \\
\hline
42 \\
84 \\
\hline
.0882 \ \leftarrow \quad 4 \text{ decimal digits}
\end{array}
$$

10. 101

Multiply both divisor and dividend by 1000. Thus, $.505 \div .005 = 505 \div 5 = 101$

11. $9300

$$
\begin{array}{ll}
800 \times \$7.50 = \$6000 & \text{(orchestra seats)} \\
550 \times \$6.00 = \underline{\$3300} & \text{(balcony seats)} \\
\$9300 & \text{(total)}
\end{array}
$$

12. (D) 200 hours

Divide: $\$900 \div \$4.50$ per hour

$$
\frac{\$900}{\$4.50} = \frac{900 \times 10}{4.5 \times 10} = \frac{9000}{45} = 200 \text{ hours}
$$

13. 474,000

14. *i* .1   *ii* .09   *iii* .091

15. .03

$$
\begin{array}{r}
.028 \\
42 \overline{\smash{)}01.200} \\
84 \\
\hline
360 \\
336 \\
\hline
24
\end{array}
$$

To round to the nearest hundredth, carry the division to the thousandths digit; then round the quotient to the hundredths digit.

Thus, .028 rounds to .03.

Practice Exercises for Section 17, pages 146–150

1. $\dfrac{3}{5}$

$60\% = \dfrac{60}{100}$ Divide numerator and denominator by 20.

$= \dfrac{3}{5}$

2. $\dfrac{21}{25}$

$84\% = \dfrac{84}{100}$ Divide by 4.

$= \dfrac{21}{25}$

3. $\dfrac{5}{4}$, or $1\dfrac{1}{4}$

$125\% = \dfrac{125}{100} = \dfrac{5}{4}$

4. .67

$67\% = 67.\% = .67$

5. 1.8

$180\% = 180.\% = 1.80$

6. .1025

Because $\dfrac{1}{4} = .25$, it follows that

$10\dfrac{1}{4}\% = 10.25\% = .10 \ 25.$

7. 82%

$.82 = 82.\%, \text{ or } 82\%$

8. 60%

Insert a zero after .6 so that there are two decimal digits. Thus, $.6 = .60 = 60.\%,$ or 60%.

9. 290%

$2.9 = 2.90 = 2 \ 90.\%$

10. 100.3%

11. 30%

$$\frac{3}{10} = .3 = .30 = \underset{\curvearrowright}{30}\ \%$$

12. 160%

$$5\overline{)8.00}^{\,1.60} \qquad \text{Thus, } \frac{8}{5} = 1.60 = 1\underset{\curvearrowright}{\,60}\ \%.$$

13. $83\frac{1}{3}\%$

$$6\overline{)5.00}^{\,.83} \qquad \frac{5}{6} = .83\frac{2}{6} = .83\frac{1}{3} = 83\frac{1}{3}\%$$
$$\underline{4\ 8}$$
$$\ \ 20$$
$$\ \ \underline{18}$$
$$\ \ \ 2 \quad \text{remainder}$$

14. 420%

$$4\frac{1}{5} = 4 + \frac{1}{5}$$
$$= 4 + .20$$
$$= 4.20$$
$$= 420\%$$

15. 175%

$$1\frac{3}{4} = 1.75 = 175\%$$

Practice Exercises for Section 18, pages 153–157

1. (B) 15

$$25\% \text{ of } 60 = .25 \times 60 = 15.00$$

2. 67.5

$$45\% \text{ of } 150 = .45 \times 150 = 67.50 = 67.5$$

3. 1.05

$$7\% \text{ of } 15 = .07 \times 15 = 1.05$$

4. 9

$$5 \times n = 45 \qquad \text{Divide both sides of the equation by 5.}$$
$$\frac{5 \times n}{5} = \frac{45}{5}$$
$$n = 9$$

5. $\frac{11}{2}$ or $5\frac{1}{2}$

$$\frac{8 \times n}{8} = \frac{44}{8}$$
$$n = \frac{44}{8}$$
$$= \frac{11}{2}, \text{ or } 5\frac{1}{2}$$

6. 25

Change 20% to .2.
$$20\% \times n = .2 \times n$$
$$.2 \times n = 5 \qquad \text{To eliminate decimals, multiply both sides by 10.}$$
$$10 \times .2 \times n = 10 \times 5$$
$$2 \times n = 50 \qquad \text{Divide by 2 on both sides.}$$
$$\frac{2 \times n}{2} = \frac{50}{2}$$
$$n = 25$$

7. 20

$$90\% \times n = 18$$
$$.9 \times n = 18$$
$$9 \times n = 180$$
$$n = \frac{180}{9}$$
$$= 20$$

8. (B) 80

Let n stand for the unknown number.
$$40 = 50\% \text{ of } n$$
$$40 = .5 \times n \qquad \text{Multiply both sides by 10.}$$
$$400 = 5 \times n \qquad \text{Divide by 5.}$$
$$\frac{400}{5} = n$$
$$n = 80$$

9. 150

$$90 = 60\% \text{ of } n$$
$$90 = .6 \times n$$
$$900 = 6 \times n$$
$$n = \frac{900}{6} = 150$$

10. 25

$$200\% \text{ of } n = 50$$
$$2 \times n = 50$$
$$n = \frac{50}{2} = 25$$

11. 40%

$$\frac{20}{50} \text{ represents the fractional part of 50 that 20 is.}$$

Change $\frac{20}{50}$ to a decimal, then to a percent.

$$\frac{20}{50} = \frac{2}{5} = .4 = 40\%$$

12. 250%

What percent of 20 is 50?

$\underbrace{\qquad\qquad}_{n} \times 20 = 50$ Divide both sides by 20.

$$n = \frac{50}{20} = 2.5$$

Change 2.5 to percent. Thus,

$2.5 = 2\underset{\smile}{50}\%$

13. 37.5% or $37\frac{1}{2}\%$

$\frac{12}{32} = \frac{3}{8}$ $8\overline{)3.000}$ .375

Thus, $\frac{3}{8} = .375 = 37.5\%$

Because $.5 = \frac{1}{2}$, $37.5\% = 37\frac{1}{2}\%$

14. (C) 6 gallons

Let n be the number of gallons of grape juice. Translate the words into an equation.

How many gallons are 40% of 15 (gallons)?

$\underbrace{\qquad\qquad}_{n} = .40 \times 15$

$n = 6.00 = 6$ (gallons)

15. 550

Let n be the total number of students.

60% of n = 330

$.6 \times n = 330$ Multiply both sides by 10.

$6 \times n = 3300$ Divide both sides by 6.

$$n = \frac{3300}{6} = 550$$

Practice Exercises for Section 19, pages 162–164

1. (C) $12.96

The total of $12.00 plus 8% of $12.00 is 108% of $12.00, or $1.08 \times \$12$.

$\begin{array}{r} 1.08 \\ \times\ \ \ 12 \\ \hline 12.96 \end{array}$

2. 52,000

The new population is 100% + 4%, or 104% of the old population.

104% of 50,000 = $1.04 \times 50{,}000$

$= 52{,}000$

3. (D) 96 cents

$150\% \times 64 = 1.5 \times 64$

$= 96$

4. $33\frac{1}{3}\%$

The actual increase in enrollment is 12,000 − 9000 = 3000 students. Divide the increase by the original enrollment. Thus,

$$\frac{3000}{9000} = \frac{1}{3} = 33\frac{1}{3}\%$$

5. 7%

$\begin{array}{r} \$17.12 \\ -\ 16.00 \\ \hline 1.12 \end{array}$

$16.00\overline{)1.12}$ .07 = 7%

6. 12%

$\begin{array}{r} \$56,000 \\ -\ 50,000 \\ \hline \$\ 6\ 000 \end{array}$

$\frac{6000}{50,000} = \frac{3}{25} = .12 = 12\%$

7. (D) $6.75

100% − 10% = 90%. Therefore, the reduced price is 90% of the original price.

90% of $7.50 = $.9 \times \$7.50$

$= \$6.75$

8. 3,500,000

$100\% - 12\frac{1}{2}\% = 87\frac{1}{2}\% = 87.5\%$

The new population is 87.5% of the old population.

$87.5\% \times 4{,}000{,}000 = .875 \times 4{,}000{,}000$

$= 3{,}500{,}000$

9. (C) 15%

The actual amount paid in taxes is $40,000 − $34,000 = $6,000. To find what percent of his salary this is, divide 6000 by 40,000. Thus,

$$\frac{6000}{40,000} = \frac{6}{40} = \frac{3}{20} = .15$$

Expressed as a percent, .15 becomes 15%.

10. 37,500 copies

$$100\% - 25\% = 75\%$$
$$75\% \text{ of } 50,000 = .75 \times 50,000$$
$$= 37,500.00$$
$$= 37,500$$

Practice Exam on Unit IV, pages 169–170

1. $\frac{13}{25}$

$$52\% = \frac{52}{100}$$ Divide numerator and denominator by 4.

$$= \frac{13}{25}$$

2. .05

$$5\% = \frac{5}{100} = .05$$

3. 87.5%, or $87\frac{1}{2}\%$

Divide the numerator, 7, by the denominator, 8.

$$8\overline{)7.000}^{.875}$$

Thus, $\frac{7}{8} = .875 = \underset{\smile}{87.5}\%$

4. .4%

To change from a decimal to percent, move the decimal point 2 digits to the right and insert the percent sign.

$$.004 = \underset{\smile}{00.4}\%$$

5. 220%

$$2\frac{1}{5} = 2.2 = 2\underset{\smile}{20.}\%$$

6. 17

$$25\% \text{ of } 68$$
$$= .25 \times 68$$
$$= 17.00$$

7. 50

$80\% \times n$ is the same as $.8 \times n$.

$$.8 \times n = 40$$ Multiply both sides by 10.
$$10 \times .8 \times n = 10 \times 40$$
$$8 \times n = 400$$ Divide both sides by 8.
$$n = \frac{400}{8} = 50$$

8. (D) 500

$$40\% \text{ of } n = 200$$
$$.4 \times n = 200$$ Multiply by 10.
$$4 \times n = 2000$$ Divide by 4.
$$n = \frac{2000}{4} = 500$$

9. (B) 90%

$$\frac{54}{60} = \frac{9}{10} = .9 = 90\%$$

10. (A) $28.80

An increase of 20% means that the new price is 120% of the old price. Thus,

$$120\% \text{ of } \$24.00 = 1.2 \times \$24.00$$
$$= \$28.80$$

11. 7,600,000

$$100\% - 5\% = 95\%$$
$$95\% \text{ of } 8,000,000 = .95 \times 8,000,000$$
$$= 7,600,000$$

12. $25,500

$100\% - 15\% = 85\%$. Take-home pay is 85% of the total income. Thus,

$$85\% \text{ of } \$30,000 = .85 \times \$30,000$$
$$= \$25,500$$

13. $17.28

$$100\% + 8\% = 108\%$$
$$108\% \text{ of } \$16 = 1.08 \times \$16$$
$$= \$17.28$$

Practice Midterm Exam A, pages 171–172

1. (C) 220,500   **2.** (B) 318   **3.** (C) 312   **4.** 81   **5.** 80   **6.** $1440 = 2^5 \times 3^2 \times 5$

$$
\begin{array}{r}
312 \\
18\overline{)5616} \\
54 \\
\overline{21} \\
18 \\
\overline{36} \\
36 \\
\overline{}
\end{array}
$$

7. (D) $\dfrac{4}{7}$

8. $\dfrac{23}{45}$

$$\frac{2}{5} + \frac{1}{9} = \frac{18}{45} + \frac{5}{45} = \frac{23}{45}$$

9. $1\dfrac{7}{10}$

$$4\frac{1}{5} - 2\frac{1}{2} = 4\frac{2}{10} - 2\frac{5}{10} = 3\frac{12}{10} - 2\frac{5}{10} = 1\frac{7}{10}$$

10. $\dfrac{13}{7}$ or $1\dfrac{6}{7}$

$$3\frac{1}{4} \div 1\frac{3}{4} = \frac{13}{4} \div \frac{7}{4} = \frac{13}{4} \times \frac{4}{7} = \frac{13}{7} \text{ or } 1\frac{6}{7}$$

11. $\dfrac{4}{25}$

$$\left(\frac{2}{5}\right)^2 = \frac{2}{5} \times \frac{2}{5} = \frac{4}{25}$$

12. (D) .0079   **13.** 14.65   **14.** (B) .77

$$
\begin{array}{r}
27.10 \\
-\ 12.45 \\
\overline{14.65}
\end{array}
$$

$$
\begin{array}{r}
.769 \approx .77 \\
13\overline{)10.000} \\
9\ 1 \\
\overline{90} \\
78 \\
\overline{120} \\
117 \\
\overline{}
\end{array}
$$

15. $35.40

$$\$2.95 \times 12 = \$35.40$$

16. 45

$$.6\cancel{0} \times 75 = 45.\cancel{0} = 45$$

17. 80

$$
\begin{aligned}
60 &= .75 \times a \\
6000 &= 75a \\
\frac{6000}{75} &= a \\
80 &= a
\end{aligned}
$$

18. .58 333...

$$
\begin{array}{r}
.58333\ldots \\
12\overline{)7.00000}
\end{array}
$$

19. .000 008

20. $145.60

$$
\begin{array}{r}
\$140 \\
\times 1.04 \\
\overline{5\ 60} \\
140\ 0 \\
\overline{\$145.60}
\end{array}
$$

Practice Midterm Exam B, pages 173–174

1. (C) 5000.2   **2.** 53   **3.** 64   **4.** $13,200 = 2^4 \times 3 \times 5^2 \times 11$

$$
\begin{array}{r}
53 \\
35\overline{)1855} \\
175 \\
\overline{105} \\
105 \\
\overline{}
\end{array}
$$

5. 56

$$
\begin{aligned}
8 &= 2^3 \\
28 &= 2^2 \times 7 \\
lcm\ (8, 56) &= 2^3 \times 7 = 8 \times 7 = 56
\end{aligned}
$$

6. $\dfrac{7}{10}$

$$\frac{1}{10} + \frac{3}{5} = \frac{1}{10} + \frac{6}{10} = \frac{7}{10}$$

7. $3\dfrac{5}{8}$

$$4\frac{7}{8} - 1\frac{1}{4} = 4\frac{7}{8} - 1\frac{2}{8} = 3\frac{5}{8}$$

8. (E) $\dfrac{3}{13}$ **9.** .428 571 428 571...

$$\begin{array}{r} .428571428571 \\ 7\overline{\smash{\big)}3.000000000000} \end{array}$$

10. (E) .004 49

11. $\dfrac{40}{13}$ or $3\dfrac{1}{13}$

$$10 \div 3\frac{1}{4} = \frac{10}{1} \div \frac{13}{4} = \frac{10}{1} \times \frac{4}{13} = \frac{40}{13} \text{ or } 3\frac{1}{13}$$

12. (C) 2.16

$$\begin{array}{r} 4.5 \leftarrow \\ \times .48 \leftarrow \\ \hline 360 \\ 1\ 80 \\ \hline 2.16\cancel{0} \leftarrow \end{array} \quad \begin{array}{l} 1 \text{ decimal digit} \\ +2 \text{ decimal digits} \\ \\ \\ 3 \text{ decimal digits} \end{array}$$

13. .0144

$$\begin{array}{r} .12 \leftarrow \\ \times .12 \leftarrow \\ \hline .0144 \leftarrow \end{array} \quad \begin{array}{l} 2 \text{ decimal digits} \\ +2 \text{ decimal digits} \\ 4 \text{ decimal digits} \end{array}$$

14. (C) 150

$$\frac{\$18.00}{\$.12} = \frac{1800}{12} = 150$$

15. 22.95

$$\begin{array}{r} 27.90 \\ -\ 4.95 \\ \hline 22.95 \end{array}$$

16. (C) $\dfrac{22}{25}$

$$88\% = \frac{88}{100} = \frac{22}{25}$$

17. 3000

$$\begin{aligned} 900 &= .3\cancel{0}x \\ 9000 &= 3x \\ 3000 &= x \end{aligned}$$

18. 10

$$\begin{aligned} .7\cancel{0}x &= 7 \\ 7x &= 70 \\ x &= 10 \end{aligned}$$

19. (C) $12

$$\begin{aligned} 100\% - 25\% &= 75\% \\ \$16 \times .75 &= \$12.00 \end{aligned}$$

20. $300

$$\begin{array}{ll} \$100 & \leftarrow \text{ first month} \\ \underline{5 \times \$40} & \leftarrow \text{ next five months} \\ \$200 & \end{array}$$

$$\$100 + \$200 = \$300$$

Practice Exercises for Section 20, pages 177–181

1. 4 hours 20 minutes

$$\begin{array}{l} 2 \text{ hours } 35 \text{ minutes} \\ \underline{+1 \text{ hour } \ 45 \text{ minutes}} \\ 3 \text{ hours } \cancel{80 \text{ minutes}} \quad [80 \text{ minutes} = \\ \underline{1 \text{ hour } \ 20 \text{ minutes}} \quad \quad 1 \text{ hour } 20 \text{ minutes}] \\ 4 \text{ hours } 20 \text{ minutes} \end{array}$$

2. 2 minutes 35 seconds

$$\begin{array}{l} 10 \text{ minutes} \\ 20 \text{ seconds} = \quad \begin{array}{r} 9 \text{ minutes } 80 \text{ seconds} \\ \underline{-7 \text{ minutes } 45 \text{ seconds}} \\ 2 \text{ minutes } 35 \text{ seconds} \end{array} \end{array}$$

3. 2 hours 15 minutes

To find the amount of time from 8:45 to 11:00 P.M.,

subtract: $\begin{array}{rcl} 11 \text{ hours} & = & 10 \text{ hours } 60 \text{ minutes} \\ -\ 8 \text{ hours } 45 \text{ minutes} & = & -\ 8 \text{ hours } 45 \text{ minutes} \\ \hline & & 2 \text{ hours } 15 \text{ minutes} \end{array}$

4. 15

$$1\frac{1}{2} \text{ hours} = 1\frac{1}{2} \times 60 \text{ minutes}$$
$$= 90 \text{ minutes}$$
$$90 \text{ minutes} \div 6 \text{ minutes} = 15$$

5. 10 pounds 4 ounces

$$\begin{array}{l} 6 \text{ pounds } 11 \text{ ounces} \\ \underline{+3 \text{ pounds } \ 9 \text{ ounces}} \\ 9 \text{ pounds } 20 \text{ ounces} = 10 \text{ pounds } 4 \text{ ounces} \end{array}$$

[Note that 20 ounces = 1 pound 4 ounces.]

6. 9 pounds 11 ounces

1 pound = 16 ounces. To do the subtraction, borrow 1 pound from 20 pounds.

$$\begin{array}{l} 20 \text{ pounds} \\ 5 \text{ ounces} = \quad \begin{array}{r} 19 \text{ pounds } 21 \text{ ounces} \\ \underline{-10 \text{ pounds } 10 \text{ ounces}} \\ 9 \text{ pounds } 11 \text{ ounces} \end{array} \end{array}$$

7. 6 kilograms

$$\begin{array}{ll} & 2\text{ kilograms} \quad 750\text{ grams} \\ +\,&3\text{ kilograms} \quad 250\text{ grams} \\ \hline & 5\text{ kilograms } 1000\text{ grams} = 5\text{ kilograms} + 1\text{ kilogram} \\ & \qquad\qquad\qquad\qquad\qquad = 6\text{ kilograms} \end{array}$$

[Note that 1000 grams = 1 kilogram.]

8. 800 grams

4 kilograms
$= 4 \times 1000$ grams
$= 4000$ grams

4000 grams $\div$ 5 = 800 grams

9. 9 feet 6 inches

1 foot = 12 inches

$$\begin{array}{ll} & 5\text{ feet } 10\text{ inches} \\ +\,&3\text{ feet } \;\;8\text{ inches} \\ \hline & 8\text{ feet } 18\text{ inches} \\ & = 8\text{ feet } + 12\text{ inches } + 6\text{ inches} \\ & = 9\text{ feet } 6\text{ inches} \end{array}$$

10. 3 meters 90 centimeters

6 meters
$$\begin{array}{lll} 20\text{ centimeters} = & 5\text{ meters} & 120\text{ centimeters} \\ = & -\,2\text{ meters} & \;\;30\text{ centimeters} \\ \hline & 3\text{ meters} & \;\;90\text{ centimeters} \end{array}$$

11. 10 inches

3 feet 4 inches = 3(12 inches) + 4 inches
$\qquad\qquad\qquad = 36$ inches + 4 inches
$\qquad\qquad\qquad = 40$ inches

40 inches $\div$ 4 = 10 inches

12. 2724 grams

1 pound $\approx$ 454 grams. Therefore,
6 pounds $\approx$ 6(454 grams) = 2724 grams

13. 145,280 centigrams

1 pound $\approx$ 454 grams. Therefore,
3.2 pounds $\approx$ 3.2 $\times$ 454 grams

$$\begin{array}{r} 454\text{ grams} \\ \times\;\; 3.2 \\ \hline 90\;8 \\ 1362 \\ \hline 1452.8\text{ grams} \end{array} = 1452.8 \times 100\text{ centigrams}$$
$$\qquad\qquad\qquad\qquad = 145,2\,\underline{80}\text{ centigrams}$$

14. 33 pounds

1 kilogram $\approx$ 2.2 pounds. Therefore,
15 kilograms $\approx$ 15 $\times$ 2.2 pounds

$$\begin{array}{r} 15 \\ \times 2.2\text{ pounds} \\ \hline 3\;0 \\ 30 \\ \hline 33.\cancel{0}\text{ pounds} \end{array}$$

15. 61.0 centimeters

1 inch $\approx$ 2.54 centimeters. Therefore,
2 feet = 2 $\times$ 12 inches
$\qquad = 24$ inches
$\qquad \approx 24 \times 2.54$ centimeters = 60.96 centimeters

Rounded to the nearest tenth, 60.96 centimeters
becomes 61.0 centimeters.

16. 19.2 kilometers

1 mile $\approx$ 1.6 kilometers
12 miles $\approx$ 12 $\times$ 1.6 kilometers
$\qquad\quad = 19.2$ kilometers

Practice Exercises for Section 21, pages 187–191

1. 70 square inches

Area = length $\times$ width
$\qquad = 10$ in. $\times$ 7 in. = 70 in.^{2}

2. 8 square feet

The base of the trunk is the bottom. To find the area of the base, it is not necessary to know the height.

Area = length $\times$ width
$\qquad = 4$ feet $\times$ 2 feet = 8 ft.^{2}

3. 8 feet

120 ft.^{2} = 15 ft. $\times$ height   Divide both sides
$\dfrac{120\text{ ft.}^2}{15\text{ ft.}}$ = height   by 15 ft.

height = 8 feet

4. 121 square inches

Area = (11 in.)^2 = 121 in.^{2}

5. $648

The area of the bedspread is
12 feet × 9 feet = 108 ft.²

At $\frac{\$6}{\text{ft.}^2}$, the cost is $108 \, \cancel{\text{ft.}^2} \times \frac{\$6}{\cancel{\text{ft.}^2}} = \$648$.

6. (D) $11.25

Area of the wall = 25 yards × 3 yards = 75 yd.²

15 cents per square yard = $\frac{\$.15}{\text{yd.}^2}$

Total cost = $75 \, \cancel{\text{yd.}^2} \times \frac{\$.15}{\cancel{\text{yd.}^2}} = \$11.25$

7. 320 feet

Perimeter = 2 × length + 2 × width
= 2 × 100 feet + 2 × 60 feet
= 200 feet + 120 feet = 320 feet

8. *i.* 150 meters   *ii.* 37,500 square meters

i. Perimeter = 2 × length + 2 × width

800 meters = 2 × 250 meters + 2 × width

$$
\begin{array}{rl}
800 \text{ meters} = & 500 \text{ meters} + 2 \times \text{width} \\
-500 \text{ meters} & -500 \text{ meters} \\
\hline
300 \text{ meters} = & 2 \times \text{width} \\
150 \text{ meters} = & \text{width}
\end{array}
$$

Subtract 500 meters from both sides.
Divide both sides by 2.

ii. Area = length × width
= 250 meters × 150 meters = 37,500 meters²

9. 120 centimeters

Perimeter of a square = 4 × side length
= 4 × 30 centimeters
= 120 centimeters

10. $440

Perimeter = 2 × 300 feet + 2 × 250 feet
= 600 feet + 500 feet = 1100 feet

Cost = $1100 \, \cancel{\text{ft.}} \times \frac{\$.40}{\cancel{\text{ft.}}} = \$440.00$

Practice Exercises for Section 22, pages 198–207

1. (A) 20 in.²

Area of a triangle = $\frac{1}{2}$ × base × height

= $\frac{1}{2}$ × 10 in. × 4 in.

= 20 in.²

2. 10 centimeters

25 cm.² = $\frac{1}{2}$ × 5 cm. × height   Multiply both sides by 2.

50 cm.² = 5 cm. × height   Divide both sides by 5 cm.

$\frac{50 \text{ cm.}^2}{5 \text{ cm.}}$ = height = 10 cm.

3. 16 centimeters

To find the perimeter, add the lengths of all sides.

3 cm. + 6 cm. + 7 cm. = 16 cm.

4. 14 centimeters

diameter = 2 × radius
= 2 × 7 centimeters
= 14 centimeters

5. 3.5 centimeters

7 centimeters = 2 × radius   To find the radius, divide by 2.

$\frac{7}{2}$ centimeters = radius = 3.5 centimeters

6. 9π inches

Circumference = π × diameter
= π × 9 inches
= 9π inches

7. 50.2 inches

Circumference = $2\pi \times$ radius
$= 2\pi \times 8$ inches
≈ 16 in. $\times 3.14$
$= 50.24$ in.

To the nearest tenth, 50.24 in. rounds to 50.2 in.

8. 3 feet

6π feet $= 2\pi \times$ radius   Divide both sides by 2π.

$\dfrac{6\pi \text{ feet}}{2\pi} =$ radius $= 3$ feet

9. 12.6 in.²

Area $= \pi \times$ radius² $= \pi(2 \text{ in.})^2 = 4\pi$ in.²
$4\pi \approx 4 \times 3.14 = 12.56$, or 12.6 to the nearest tenth.

10. 78.5 cm.²

If the diameter is 10 cm., then the radius is 5 cm.

Area $= \pi(5 \text{ cm.})^2 = 25\pi$ cm.²
$\approx 25(3.14)$ cm.²
$= 78.5$ cm.²

11. 50.2 ft.²

To find the radius, given the circumference, use

Circumference $= 2\pi \times$ radius
8π feet $= 2\pi \times$ radius   Divide both sides by 2π.
$\dfrac{8\pi \text{ feet}}{2\pi} =$ radius $= 4$ feet

Area $= \pi \times$ radius² $= \pi \times (4 \text{ feet})^2$
$= 16\pi$ ft.²
$\approx 16(3.14)$ ft.² $= 50.24$ ft.²
≈ 50.2 ft.², to the nearest tenth

12. (D) 6.25π in.²

Radius $= \dfrac{\text{Circumference}}{2\pi}$

$= \dfrac{5\pi \text{ in.}}{2\pi}$

$\cdot\ = 2.5$ in.

Area $= \pi \times$ radius²
$= \pi(2.5 \text{ in.})^2$
$= 6.25\pi$ in.²

13. $(24 + 8\pi)$ in.²

Area of triangle $= \dfrac{1}{2}$ base $\times$ height

$= \dfrac{1}{2}(6 \text{ in.})(8 \text{ in.})$

$= 24$ in.²

Radius of circle $= \dfrac{1}{2} \times 8$ in.

$= 4$ in.

Area of semicircle $= \dfrac{1}{2}$ area of circle

$= \dfrac{1}{2}\pi(4 \text{ in.})^2$

$= 8\pi$ in.²

Total area $= 24$ in.² $+ 8\pi$ in.²
$= (24 + 8\pi)$ in.²

14. *i.* 12 cubic feet   *ii.* 38 square feet

i. Volume $=$ length $\times$ width $\times$ height
$= 3$ ft. $\times 1$ ft. $\times 4$ ft.
$= 12$ ft.³

ii. Surface area $= 2 \times 3$ ft. $\times 1$ ft.
$+ 2 \times 3$ ft. $\times 4$ ft.
$+ 2 \times 1$ ft. $\times 4$ ft.
$= 6$ ft.² $+ 24$ ft.² $+ 8$ ft.²
$= 38$ ft.²

15. *i.* 64 in.³   *ii.* 96 in.²

i. Volume $= (4 \text{ in.})^3 = 4$ in. $\times 4$ in. $\times 4$ in. $= 64$ in.³

ii. Surface area $= 6 \times (4 \text{ in.})^2 = 6 \times 16$ in.² $= 96$ in.²

Practice Exercises for Section 23, pages 211–213

1. (C) 7500

According to the graph,

June sales	= 2000
July sales	= 3000
August sales	= 2500 (midway between 2000 and 3000)
Total	= 7500

2. (D) June and September

3. (A) $1,920,000

Housing represents 12% of the total, or

$.12 \times \$16,000,000 = \$1,920,000$

4. (C) schools

$\dfrac{\$4,000,000}{\$16,000,000} = \dfrac{1}{4} = 25\%$. Therefore, 4 million dollars is 25% of 16 million dollars, and the item that receives 25% is schools.

5. (A) 1980

6. (E) 1984

Practice Exam on Unit V, page 222

1. 1 hour 45 minutes

10 hours 5 minutes =

$$\begin{array}{r} 9 \text{ hours } 65 \text{ minutes} \\ -8 \text{ hours } 20 \text{ minutes} \\ \hline 1 \text{ hour } 45 \text{ minutes} \end{array}$$

2. 3 pounds 9 ounces

1 pound = 16 ounces

6 pounds 3 ounces =

$$\begin{array}{r} 5 \text{ pounds } 19 \text{ ounces} \\ -2 \text{ pounds } 10 \text{ ounces} \\ \hline 3 \text{ pounds } 9 \text{ ounces} \end{array}$$

3. 10 feet 5 inches

1 foot = 12 inches

$$\begin{array}{r} 4 \text{ feet } 8 \text{ inches} \\ +5 \text{ feet } 9 \text{ inches} \\ \hline 9 \text{ feet } 17 \text{ inches} \\ 1 \text{ foot } 5 \text{ inches} \\ \hline 10 \text{ feet } 5 \text{ inches} \end{array}$$

(But 17 inches = 1 foot 5 inches)

4. (D) 44 cm.

Area of a rectangle = length × width. Therefore,

$$\text{width} = \frac{\text{area}}{\text{length}} = \frac{40 \text{ cm.}^2}{20 \text{ cm.}} = 2 \text{ cm.}$$

Perimeter = 2 × length + 2 × width
$$= 2 \times 20 \text{ cm.} + 2 \times 2 \text{ cm.}$$
$$= 44 \text{ cm.}$$

5. (A) $720

The height of the room is irrelevant.
Area of floor = 15 feet × 12 feet = 180 ft.²

$$\text{Cost} = 180 \text{ ft.} \times \frac{\$4}{\text{ft.}^2} = \$720$$

6. 20 cm.²

$$\text{Area of triangle} = \frac{1}{2} \text{base} \times \text{height}$$
$$= \frac{1}{2} \times 5 \text{ cm.} \times 8 \text{ cm.}$$
$$= 20 \text{ cm.}^2$$

7. $\dfrac{81\pi}{4}$ in.²

Circumference = 2π × radius. Therefore,

$$\text{radius} = \frac{\text{Circumference}}{2\pi}$$
$$= \frac{9\pi \text{ in.}}{2\pi} = \frac{9}{2} \text{ in.}$$

Area of circle = π × radius² = $\pi \left(\dfrac{9}{2} \text{ in.}\right)^2$

$$= \pi \times \frac{81}{4} \text{ in.}^2 = \frac{81\pi}{4} \text{ in.}^2$$

8. 432 cubic feet

Volume = length × width × height
$$= 8 \text{ ft.} \times 6 \text{ ft.} \times 9 \text{ ft.}$$
$$= 432 \text{ ft.}^3$$

9. (D) $5,000,000

Salaries are 25% of expenditures, or
$.25 \times \$20,000,000 = \$5,000,000$

10. (D) maintenance

$$\frac{\$3,000,000}{\$20,000,000} = \frac{3}{20} = 15\%$$

Therefore, 3 million is 15% of 20 million. 15% is spent on maintenance [choice (D)].

11. (C) $7,000,000

Salaries and scholarships combined represent 25% + 10%, or 35% of expenditures.

.35 × $20,000,000 = $7,000,000

Practice Exercises for Section 24, pages 225–237

1–5.

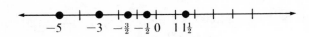

6. 10

7. 15

Because −15 is negative, $|-15| = -(-15) = 15$.

8. $\frac{1}{2}$

9. −3

$|-5| > |2|$. Therefore, the sum of −5 and 2 is negative.
Thus, $-5 + 2 = -(5 - 2) = -3$

10. −10

11. −19

For convenience, rearrange the terms in order to add the positive and negative numbers separately.

$$
\begin{array}{rrr}
12 & -9 & \text{Finally,} \quad -44 \\
13 & -15 & +25 \\
\overline{25} & -20 & \overline{-19} \\
& \overline{-44} &
\end{array}
$$

Add the positive sum and the negative sum, by taking their difference with the sign of the larger absolute value.

12. −8

$4 - 12 = 4 + (-12)$
$\quad = -(12 - 4) = -8$

13. 8

$(-4) - (-12) = -4 + 12$
$\quad = 12 - 4 = 8$

14. −15

$-9 - 6 = -9 + (-6) = -(9 + 6) = -15$

15. 10

$0 - (-10) = 0 + 10 = 10$

16. −14

17. 12

Two negative factors yield a positive product.

$(-4)(-3) = 12$

18. 0

19. −8

Three negative factors yield a negative product.

$(-2)(-4)(-1) = 8\,(-1) = -8$
$\underline{\quad} +8 \underline{\quad}$

20. −2

When numerator and denominator have different signs, the quotient is negative.

Therefore, $\frac{10}{-5} = -2$.

21. 3

$$\frac{-18}{-6} = \frac{18}{6} = 3$$

22. 0

$$\frac{0}{-3} = 0$$

because $0 \cdot (-3) = 0$

23. undefined

24. $-10°F$

$$-6 - 4 = -6 + (-4) = -10$$

Practice Exercises for Section 25, pages 240–244

1. 25

$$(-5)^2 = (-5)(-5) = 25$$

2. -25

$$-5^2 = -(5 \times 5) = -25$$

3. 121   **4.** -64

$$(-4)^3 = (-4)(-4)(-4) = -64$$
(Odd powers of negative numbers are negative.)

5. 256

$$(-4)^4 = (-4)(-4)(-4)(-4) = 256$$
(Even powers of negative numbers are positive.)

6. -64

$$-2^6 = -(2)(2)(2)(2)(2)(2) = -64$$

7. 128

$$-(-2)^7 = -(-128) = 128$$

8. (D) -8

$$4(1 - 3) = 4(-2) = -8$$

9. (A) 9

$$
\begin{aligned}
10(6 - 3) - 7(2 + 1) &\quad \text{Perform operations in parentheses,} \\
= 10(3) - 7(3) &\quad \text{then multiply,} \\
= 30 - 21 &\quad \text{then add.} \\
= 30 + (-21) = 9
\end{aligned}
$$

10. 17

$$
\begin{aligned}
5^2 - 2^3 &= 5 \times 5 - 2 \times 2 \times 2 \\
&= 25 - 8 = 17
\end{aligned}
$$

11. -8

$$8 - 4^2 = 8 - 16 = -8$$

12. 4

$$(9 - 7)^2 = 2^2 = 4$$

13. -2

$$\frac{9 - 1}{4} - 2^2 = \frac{8}{4} - 4 = 2 - 4 = -2$$

14. 5 pounds per week

$$\frac{250 \text{ pounds} - 210 \text{ pounds}}{8 \text{ weeks}} = \frac{40 \text{ pounds}}{8 \text{ weeks}}$$

$$= \frac{5 \text{ pounds}}{1 \text{ week}}$$

Practice Exercises for Section 26, pages 248–249

1. 13

$$\frac{10 + 12 + 17}{3} = \frac{39}{3} = 13$$

2. 30

$$\frac{24 + 26 + 30 + 40}{4} = \frac{120}{4} = 30$$

3. 56.5

$$\frac{46 + 50 + 62 + 68}{4}$$

$$= \frac{226}{4} = 56.5$$

4. 15.5

$$12 + 16 + 14 + 20 = 62$$

$$\frac{62 \text{ games}}{4 \text{ years}} = 15.5 \text{ games per year}$$

5. (C) 20

$$\frac{15 + 19 + 26}{3} = \frac{60}{3} = 20$$

6. (B) 7.5

$$\frac{6 + 7 + 8 + 9}{4} = \frac{30}{4} = 7.5$$

Practice Exam on Unit VI, pages 253–254

1. −11

$(-3) + (-8) = -(3 + 8) = -11$

The sum of negative numbers is negative.

2. 20

$(-5) \times (-4) = 20$

The product of two negative numbers is positive.

3. −6   **4.** 0   **5.** −13

$$\begin{array}{r} 14 \\ +\,15 \\ \hline 29 \end{array} \qquad \begin{array}{r} -9 \\ -12 \\ -21 \\ \hline -42 \end{array}$$

$$\begin{array}{r} -42 \\ +\,29 \\ \hline -13 \end{array}$$

6. −3°F

$2 - 5 = 2 + (-5) = -3$

7. 31

$$2^2 - (-3)^3 = 4 - (-27)$$
$$= 4 + 27 = 31$$

8. 1

$$\frac{8 - 2}{(-2)(-3)} = \frac{6}{6} = 1$$

9. 2

$$\left(\frac{14 - 10}{2}\right)^2 + \frac{8 - 4}{-2} = \left(\frac{4}{2}\right)^2 + \frac{4}{-2}$$
$$= 2^2 + (-2)$$
$$= 4 + (-2)$$
$$= 2$$

10. −12

$$-2(5 - 3 + 1) - 3\big(4 + (-2)\big)$$
$$= -2(3) - 3(2)$$
$$= -6 - 6$$
$$= -12$$

11. 35.6

$$\frac{19 + 29 + 39 + 42 + 49}{5} = \frac{178}{5} = 35.6$$

The average of the five numbers is 35.6.

12. (C) 20

$$\frac{18 + 15 + 18 + 29}{4} = \frac{80}{4} = 20$$

13. $19

$$2(\$20) - [2(\$3) + 3(\$5)]$$
$$= \$40 - (\$6 + \$15)$$
$$= \$40 - \$21$$
$$= \$19$$

Practice Exercises for Section 27, pages 258–259

1. $x + 9$

9 more than the number

$$9 \quad + \quad x \quad , \quad \text{or } x + 9$$

2. $x - 4$ **3.** $6 \div x$, or $\dfrac{6}{x}$ **4.** $2x + 1$

5. $3(x + 2)$ **6.** $5n + 10d + 25q$

7. $2a + 3b$

Number of items × Cost per item = Total cost

$$\begin{array}{ccccc} a & \times & 2 & = & 2a \\ b & \times & 3 & = & +3b \end{array}$$

The total cost of the two-dollar items and the three-dollar items is $(2a + 3b)$ dollars.

8. (A) $5x$ miles

Rate × Time = Distance

$$\frac{x \text{ miles}}{\text{hr.}} \times 5 \text{ hr.} = 5x \text{ miles}$$

Practice Exercises for Section 28, pages 263–266

1. (D) 47

When $y = 5$,
$$10y - 3 = 10(5) - 3$$
$$= 50 - 3 = 47$$

2. 5

When $x = 4$,
$$x^2 - 3x + 1 = 4^2 - 3(4) + 1$$
$$= 16 - 12 + 1 = 5$$

3. −2

When $t = -1$,
$$t^3 + 2t^2 + t - 2$$
$$= (-1)^3 + 2(-1)^2 + (-1) - 2$$
$$= -1 + 2(1) - 1 - 2 = -2$$

4. 13

When $a = 3$,
$$2a + 5 = 2^3 + 5$$
$$= 8 + 5 = 13$$

5. (B) 14

Substitute 5 for a and 2 for b:
$$2ab - 3b = 2(5)(2) - 3(2) = 20 - 6 = 14$$

6. 8

When $x = 1$ and $y = -1$,
$$x^2 - 3xy + 4 = 1^2 - 3(1)(-1) + 4$$
$$= 1 + 3 + 4 = 8$$

7. 48

$y = 3ab^2 = \underbrace{3(4)}_{12} \times \underbrace{(2)^2}_{4}$ (Evaluate 2^2 before multiplying.)
$$= 12 \times 4$$
$$= 48$$

8. −13

$$x^2y - 1 = (-2)^2(-3) - 1$$
$$= 4(-3) - 1$$
$$= -12 - 1$$
$$= -13$$

9. 30 square inches

$$A = \frac{b_1 + b_2}{2} \times h$$
$$= \frac{4 \text{ in.} + 8 \text{ in.}}{2} \times 5 \text{ in.}$$
$$= \frac{12 \text{ in.}}{2} \times 5 \text{ in.}$$
$$= 6 \text{ in.} \times 5 \text{ in.} = 30 \text{ in.}^2$$

10. 4 hours

$$t = \frac{d}{r} = \frac{200 \text{ miles}}{50 \text{ mph}} = 4 \text{ hours}$$

Practice Exercises for Section 29, pages 272–278

1. 12   **2.** 5   **3.** −5   **4.** x^2 and $5x$ **5.** $3a^4$, $-5a$, and 1

6. like terms

$4xy$ and $-4yx$ contain the same variables with the same exponents, although in a different order.

7. unlike terms

x^2y and xy^2 differ in their exponents on both x and y.

8. unlike terms

$10ab^3$ and $10ab^5$ have different exponents on b.

9. $8a$

$$5a + 3a = (5 + 3)a = 8a$$

10. $6b^2$

$$12b^2 - 6b^2 = (12 - 6)b^2 = 6b^2$$

11. $10a - 7b$

Unlike terms cannot be combined.

12. $5x - 1$

$$\begin{array}{r} 2x - 3 \\ 3x + 2 \\ \hline 5x - 1 \end{array}$$

13. $9a + 2b - 1$

14. (B) $5x + 2y$

$$-(5x + y) = -5x - y$$

Add:

$$10x + 3y$$
$$\underline{-5x - y}$$
$$5x + 2y$$

15. $10b$

$$\begin{array}{rcrcr} 2a & + & 7b & = & 2a + 7b \\ -(2a & - & 3b) & = & -2a + 3b \\ \hline & & & & 0 + 10b \end{array}$$

16. $3x + 5$

4 less than $3x + 9$ means
$3x + 9 - 4$, or $3x + 5$.

Practice Exercises for Section 30, pages 282–286

1. yes

Substituting 5 for x, a true statement results:

$4(5) = 20$ (true)

2. yes

$$2(3) - 2 \overset{?}{=} 3 + 1$$
$$6 - 2 = 4 \text{ (true)}$$

3. no

Substituting -2 for t, a false statement results:

$$10(-2) + 4 \overset{?}{=} 8(-2) + 2$$
$$-20 + 4 = -16 + 2$$
$$-16 = -14 \text{ (false)}$$

-2 is not a root.

4. $x = 6$

$5x = 30$ Divide both sides by 5.

$$\frac{5x}{5} = \frac{30}{5}$$
$$x = 6$$

5. $y = -7$

$-4y = 28$ Divide both sides by -4.

$$\frac{-4y}{-4} = \frac{28}{-4}$$
$$y = -7$$

6. $z = -18$

$\dfrac{z}{3} = -6$ Multiply both sides by 3.

$$\frac{z}{3} \cdot 3 = -6(3)$$
$$z = -18$$

7. $x = 4$

$x + 8 = 12$ Subtract 8 from both sides.
$x + 8 - 8 = 12 - 8$
$x = 4$

8. $y = 7$

$y - 3 = 4$ Add 3 to both sides.
$y - 3 + 3 = 4 + 3$
$y = 7$

9. $t = -5$

$2 + t = -3$ Subtract 2 from both sides.
$2 + t - 2 = -3 - 2$
$t = -5$

10. (A) 2

$3t + 1 = 7$ Subtract 1 from both sides.

$3t + 1 - 1 = 7 - 1$

$3t = 6$ Divide both sides by 3.

$$t = \frac{6}{3} = 2$$

11. $y = 1$

$5y - 3 = 2$ Add 3 to both sides.
$5y = 5$ Divide both sides by 5.
$y = 1$

12. (E) 12

$3x + 12 = 4x$ Subtract $3x$ from both sides.

$3x - 3x + 12 = 4x - 3x$
$12 = x$

13. $t = 3$

$6t - 2 = 2t + 10$ Subtract $2t$.
$4t - 2 = 10$ Add 2.
$4t = 12$ Divide by 4.

$$t = \frac{12}{4} = 3$$

Practice Exercises for Section 31, pages 290–294

1. $\dfrac{2}{1}$ **2.** $\dfrac{5}{4}$ **3.** $\dfrac{3}{4}$ **4.** (D) 10

$\dfrac{x}{15} = \dfrac{2}{3}$ Multiply both sides by 15.

$$\overset{}{\cancel{15}} \cdot \dfrac{x}{\cancel{15}} = \overset{5}{\cancel{15}} \cdot \dfrac{2}{\underset{1}{\cancel{3}}}$$

$$x = 10$$

5. −5

$\dfrac{y}{20} = \dfrac{-1}{4}$ Multiply both sides by 20.

$$y = 20 \left(\dfrac{-1}{4}\right) = -5$$

6. 9   **7.** (C) 3

$\dfrac{3}{u} = \dfrac{5}{15}$ Cross-multiply.

$3(15) = 5u$ Divide by 5.

$9 = u$

$\dfrac{2x - 1}{10} = \dfrac{1}{2}$ Multiply both sides by 10.

$2x - 1 = \dfrac{1}{2}(10)$ Add 1 to both sides.

$2x = 5 + 1$ Divide by 2.

$x = \dfrac{6}{2} = 3$

8. 12   **9.** 0   **10.** (C) 6

$\dfrac{y}{3} = \dfrac{y + 8}{5}$

$5y = 3(y + 8)$

$5y = 3y + 24$

$2y = 24$

$y = 12$

$\dfrac{2 - t}{7} = \dfrac{t + 4}{14}$

$14(2 - t) = 7(t + 4)$

$28 - 14t = 7t + 28$

$0 = 21t$

$0 = t$

$\dfrac{x}{3} + 1 = \dfrac{x}{6} + 2$

$6\left(\dfrac{x}{3} + 1\right) = 6\left(\dfrac{x}{6} + 2\right)$

$2x + 6 = x + 12$

$2x - x + 6 = x - x + 12$

$x + 6 = 12$

$x = 6$

11. 20   **12.** 16

$\dfrac{y}{5} - 1 = \dfrac{y}{2} - 7$

$2y - 10 = 5y - 70$

$2y - 10 + 70 = 5y - 70 + 70$

$2y + 60 = 5y$

$2y - 2y + 60 = 5y - 2y$

$60 = 3y$

$20 = y$

$\dfrac{t + 4}{2} = \dfrac{t}{8} + 8$

Multiply both sides by the *lcd*, 8.

$4t + 16 = t + 64$

$4t - t + 16 = t - t + 64$

$3t + 16 = 64$

$3t + 16 - 16 = 64 - 16$

$3t = 48$

$t = 16$

Practice Exercises for Section 32, pages 298–300

1. 160

Let x be the number of tickets that must be sold. At 5 dollars per ticket, the total income will be $5x$.

Income − Expenses = Profit

$5x \quad - \quad 550 \quad = 250$

$5x = 250 + 550$

$5x = 800$

$x = \dfrac{800}{5}$

$= 160$

2. 39 inches

Let x be the length of the shorter piece (in inches). Then $3x$ is the length of the longer piece. The total length of the two pieces is 52 inches.

$$x + 3x = 52$$
$$4x = 52$$
$$x = 13 \text{ inches (shorter piece)}$$
$$3x = 3(13) = 39 \text{ inches (longer piece)}$$

3. 180

Let x be the total number of women surveyed. 126 represents 7 out of 10, or $\frac{7}{10}$ of x.

This suggests the proportion $\quad \frac{126}{x} = \frac{7}{10}$.

Cross-multiplying, $\quad 1260 = 7x$

$$\frac{1260}{7} = x = 180$$

4. 200

25% of $400 = .25 \times 400 = 100$ minority workers. Let $x =$ the number of additional minority workers to be hired. Then $100 + x$ will be the new total of minority workers and $400 + x$ will be the new total of all workers in the factory. The following equation states that the new ratio of minority workers to all workers is 50%, or $\frac{1}{2}$.

$$\frac{100 + x}{400 + x} = \frac{1}{2} \quad \text{Cross-multiply.}$$
$$2(100 + x) = 400 + x$$
$$200 + 2x = 400 + x \quad \text{Subtract 200 and subtract } x \text{ from both sides.}$$
$$200 - 200 + 2x - x = 400 - 200 + x - x$$
$$x = 200$$

5. 96

Let g represent the score on the fourth exam. If the average of the four grades is 90, then

$$\frac{84 + 92 + 88 + g}{4} = 90 \quad \text{Add the first 3 grades and multiply both sides by 4.}$$

$$264 + g = 360 \quad \text{Subtract 264 from both sides.}$$
$$g = 360 - 264 = 96$$

Practice Exercises for Section 33, pages 305–311

1. 8

$\sqrt{64} = 8$ because 8 is positive and $8^2 = 64$.

2. 10

$\sqrt{100} = 10$ because 10 is positive and $10^2 = 100$.

3. (B) 40

$$\sqrt{1600} = \sqrt{16 \times 100}$$
$$= \sqrt{16} \times \sqrt{100}$$
$$= 4 \times 10$$
$$= 40$$

4. 9

$$\sqrt{27} \times \sqrt{3} = \sqrt{27 \times 3} = \sqrt{81} = 9$$

or

$$\sqrt{27} \times \sqrt{3} = \sqrt{9 \times 3} \times \sqrt{3} = \sqrt{9} \times \underbrace{\sqrt{3} \times \sqrt{3}}_{3} = 3 \times 3 = 9$$

5. $\frac{3}{2}$

$$\sqrt{\frac{9}{4}} = \frac{\sqrt{9}}{\sqrt{4}} = \frac{3}{2}$$

6. (C) 4

$$\sqrt{25 - 9} = \sqrt{16} = 4$$

7. 10

$$\sqrt{6^2 + 8^2}$$
$$= \sqrt{36 + 64}$$
$$= \sqrt{100}$$
$$= 10$$

8. 5

$$\sqrt{5^2 - 0^2}$$
$$= \sqrt{5^2}$$
$$= 5$$

9. 15

$$c = \sqrt{9^2 + 12^2} = \sqrt{81 + 144} = \sqrt{225} = 15$$

(Note that $\sqrt{225} = \sqrt{9 \times 25} = \sqrt{9}\sqrt{25} = 3 \times 5 = 15$.)

10. $\sqrt{21}$

The hypotenuse is 5. Thus,

$$2^2 + b^2 = 5^2$$
$$b^2 = 5^2 - 2^2$$
$$b^2 = 25 - 4 = 21$$
$$b = \sqrt{21}$$

Practice Exam on Unit VII, page 317

1. (A) $2x - 5$

Twice the number, x, is $2x$. Five less than $2x$ is obtained by subtracting 5 from $2x$. Thus, $2x - 5$.

2. (E) $2x + 5y$

$$\underbrace{\text{Number of items}} \times \underbrace{\text{Cost per item}} = \text{Cost}$$
$$\begin{array}{ccccc} x & \times & 2 \text{ dollars} & = & 2x \\ + \, y & \times & 5 \text{ dollars} & = & + \, 5y \end{array}$$

The total cost is $2x + 5y$.

3. -25

Substitute -2 for each occurrence of x.

$$\begin{aligned} 4x^3 - 2x + 3 &= 4(-2)^3 - 2(-2) + 3 \\ &= 4(-8) + 4 + 3 \\ &= -32 + 7 \\ &= -25 \end{aligned}$$

4. 12

$x = -1$ and $y = -3$. Thus,

$$\begin{aligned} 5xy - 3x^2 &= 5(-1)(-3) - 3(-1)^2 \\ &= 15 - 3(1) \\ &= 12 \end{aligned}$$

5. 12

$(4x^2)(3y^2) = 12x^2y^2$. The coefficient is 12.

6. $9a$

Combine like terms.

$$\begin{aligned} 5a - 2b + 4a + 2b &= (5a + 4a) - 2b + 2b \\ &= 9a + 0 = 9a \end{aligned}$$

7. $6x - y + 2$

$$\begin{array}{r} \text{Add:} \quad 4x + 3y - 1 \\ 2x - 4y + 3 \\ \hline 6x - \quad y + 2 \end{array}$$

8. $x + y + 5$

$$x + y - 1 + 6 = x + y + 5$$

9. $x = -28$

$$\frac{x}{4} = -7$$
$$4\left(\frac{x}{4}\right) = 4(-7)$$
$$x = -28$$

10. $t = 10$

$$\begin{array}{rcr} 4t - 8 &=& 2t + 12 \\ -2t + 8 && -2t + 8 \\ \hline 2t &=& 20 \end{array}$$
$$t = \frac{20}{2} = 10$$

11. $t = 11$

$$\frac{t - 1}{5} = \frac{t + 1}{6} \quad \text{Cross-multiply.}$$
$$6(t - 1) = 5(t + 1)$$
$$\begin{array}{rcr} 6t - 6 &=& 5t + 5 \\ -5t + 6 &=& -5t + 6 \\ \hline t &=& 11 \end{array}$$

12. 6

Let x = the number of pounds of cheese for 45 guests. Set up a proportion equating the ratios of pounds of cheese to numbers of guests. Thus,

$\dfrac{x}{45} = \dfrac{4}{30}$ Multiply both sides by 45.

$x = \dfrac{4}{30}(45) = 6$ (pounds of cheese)

13. 8

$\sqrt{10^2 - 6^2}$
$= \sqrt{100 - 36}$
$= \sqrt{64}$
$= 8$

14. (D) $\sqrt{45}$ or $3\sqrt{5}$

$a^2 + 2^2 = 7^2$
$a^2 = 7^2 - 2^2$
$a^2 = 49 - 4$
$a^2 = 45$
$a = \sqrt{45}$

(Note that $\sqrt{45} = \sqrt{9 \times 5}$
$= \sqrt{9} \times \sqrt{5}$
$= 3\sqrt{5}$.)

Practice Final Exam A, pages 319–324

1. (D) 420,500

2. (A) 249

3. (B) 407

$$
\begin{array}{r}
407 \\
16\overline{)6512} \\
\underline{64} \\
112 \\
\underline{112} \\
\end{array}
$$

4. (D) 64

$4^3 = 4 \times 4 \times 4$
$= 4 \times 16 = 64$

5. (C) 2×3^4

$162 = 2 \times 81 = 2 \times 9 \times 9$
$ = 2 \times (3 \times 3) \times (3 \times 3) = 2 \times 3^4$

6. (B) $\dfrac{3}{5}$

7. (C) $\dfrac{35}{36}$

The *lcd* is 36. $\dfrac{2}{9} = \dfrac{8}{36}$, $\dfrac{3}{4} = \dfrac{27}{36}$; $\dfrac{2}{9} + \dfrac{3}{4} = \dfrac{8}{36} + \dfrac{27}{36} = \dfrac{35}{36}$

8. (B) $2\dfrac{1}{6}$

$4\dfrac{1}{2} = 4\dfrac{3}{6}$, $2\dfrac{1}{3} = 2\dfrac{2}{6}$.

$$
\begin{array}{r}
4\dfrac{3}{6} \\
-2\dfrac{2}{6} \\
\hline
2\dfrac{1}{6} \\
\end{array}
$$

9. (A) 3

$8\dfrac{1}{4} = \dfrac{33}{4}$, $2\dfrac{3}{4} = \dfrac{11}{4}$

$8\dfrac{1}{4} \div 2\dfrac{3}{4} = \dfrac{33}{4} \div \dfrac{11}{4}$

$= \dfrac{\overset{3}{\cancel{33}}}{\cancel{4}} \times \dfrac{\overset{1}{\cancel{4}}}{\underset{1}{\cancel{11}}} = 3$

10. (B) $\dfrac{4}{9}$

$\left(\dfrac{2}{3}\right)^2 = \dfrac{2}{3} \times \dfrac{2}{3} = \dfrac{2 \times 2}{3 \times 3} = \dfrac{4}{9}$

11. (B) 26.072

$$
\begin{array}{r}
15.093 \\
+2.979 \\
+8. \\
\hline
26.072 \\
\end{array}
$$

12. (C) .0309

13. (B) 79.84

$$
\begin{array}{r}
87.80 \\
-7.96 \\
\hline
79.84 \\
\end{array}
$$

14. (A) .92

$$
\begin{array}{r}
.923 \\
13\overline{\smash{\big)}12.000} \\
\underline{11\ 7}\ \ \\
30\ \ \\
\underline{26}\ \ \\
40 \\
\underline{39} \\
\end{array}
$$

To the nearest hundredth,

$$\frac{12}{13} \approx .92$$

15. (C) .222 222...

Divide numerator by denominator.

$$
\begin{array}{r}
.222... \\
9\overline{\smash{\big)}2.000\ 00} \\
\underline{1\ 8}\ \ \ \\
20\ \ \\
\underline{18}\ \ \\
20 \\
\end{array}
$$

16. (D) .0016

$$(.04)^2 = \underbrace{(.04)}_{\substack{2\ \text{decimal}\\ \text{digits}}} \times \underbrace{(.04)}_{\substack{2\ \text{decimal}\\ \text{digits}}} = \underbrace{.0016}_{\substack{4\ \text{decimal}\\ \text{digits}}}$$

17. (C) $127.20

$$
\begin{array}{r}
\$\ \ 7.95 \leftarrow \ 2\ \text{decimal digits}\\
\times\ \ 16 \leftarrow +0\ \text{decimal digits}\\
\hline
47\ 70\\
79\ 5\ \ \\
\hline
\$127.20 \leftarrow \ 2\ \text{decimal digits}\\
\end{array}
$$

18. (D) 45

60% *of* 75
= .60 × 75
= 45.00
= 45

19. (D) 75

Let x be the number in question.

60 is 80% of a certain number.

$$
\begin{aligned}
60 &= .80 \times x \quad \text{Multiply both sides by 10.}\\
600 &= 8x\\
\frac{600}{8} &= x\\
75 &= x
\end{aligned}
$$

20. (D) $40.50

A 10% reduction on a $45 dress means .10 × $45, which equals $4.50.

$$
\begin{array}{r}
\$45.00 \leftarrow \text{original price}\\
-\ 4.50 \leftarrow \text{reduction}\\
\hline
\$40.50 \leftarrow \text{sale price}\\
\end{array}
$$

21. (B) $2707

$$
\begin{aligned}
\text{PROFIT} &= \text{SALES} - \text{COSTS}\\
&= 952 \times \$3 - (\$97 + \$52)\\
&= \$2856 - \$149\\
&= \$2707
\end{aligned}
$$

22. (D) 8 pounds 14 ounces

$$
\begin{array}{r}
{\scriptstyle 26\ \text{pounds}}\quad {\scriptstyle 28\ \text{ounces}}\\
\cancel{27\ \text{pounds}}\ \cancel{12\ \text{ounces}}\\
-\ 18\ \text{pounds}\ 14\ \text{ounces}\\
\hline
8\ \text{pounds}\ 14\ \text{ounces}\\
\end{array}
$$

23. (B) 880 in.²

$$
\begin{aligned}
\text{Area} &= \text{length} \times \text{width}\\
&= 40\ \text{in.} \times 22\ \text{in.}\\
&= 880\ \text{in.}^2
\end{aligned}
$$

24. (C) 12 in.²

$$
\begin{aligned}
\text{Area of triangle} &= \frac{1}{2}\text{base} \times \text{height}\\
&= \frac{1}{2}(6\ \text{in.})(4\ \text{in.})\\
&= \frac{1}{2}(24\ \text{in.}^2)\\
&= 12\ \text{in.}^2
\end{aligned}
$$

25. (C) 12,500

$$
\begin{array}{r}
3\ 000 \leftarrow \text{September}\\
2\ 000 \leftarrow \text{October}\\
3\ 500 \leftarrow \text{November}\\
4\ 000 \leftarrow \text{December}\\
\hline
12,500\\
\end{array}
$$

26. (B) 22

$$
\begin{aligned}
2(-5)^2 &+ 4(-7)\\
&= 2(25) + (-28)\\
&= 50 - 28\\
&= 22
\end{aligned}
$$

27. (A) −6

$$
\begin{aligned}
5(2 - 4) &+ (3 - 1)^2 \quad \text{Perform operations in parentheses first.}\\
&= 5(-2) + (2)^2 \quad \text{Apply exponent, then multiply.}\\
&= -10 + 4 \quad\quad\ \ \text{Lastly, add.}\\
&= -6
\end{aligned}
$$

28. (C) 27

$$\frac{(10-1)^2}{3} = \frac{9^2}{3} = \frac{81}{3} = 27$$

29. (C) 83

$$\frac{78 + 81 + 82 + 91}{4} = \frac{332}{4} = 83$$

30. (A) $10x$

$$\begin{array}{r} 4x + 9y \\ 6x - 9y \\ \hline 10x \end{array}$$

31. (A) $4x^2 - 2x - 1$

$$\begin{array}{r} 3x^2 - 2x + 4 \\ +\ x^2 \qquad\ -\ 5 \\ \hline 4x^2 - 2x - 1 \end{array}$$

32. (C) $5f + 10t$

Each ten-dollar bill is worth 10 dollars;
    t ten-dollar bills are worth $t \times 10$ dollars or $10t$ dollars.
Each five-dollar bill is worth 5 dollars;
    f five-dollar bills are worth $f \times 5$ dollars or $5f$ dollars.
The total value, in dollars, is $10t + 5f$ or $5f + 10t$.

33. (D) 4

$$\begin{aligned} 4x + 3 &= 5x - 1 \\ 4x - 4x + 3 &= 5x - 4x - 1 \\ 3 &= x - 1 \\ 4 &= x \end{aligned}$$

34. (A) 76

Substitute 4 for x and -3 for y in the expression $x^2 - 5xy$.

$$\begin{aligned} 4^2 &- 5(4)(-3) \\ &= 16 - 20(-3) \\ &= 16 + 60 = 76 \end{aligned}$$

35. (C) $-\dfrac{15}{4}$

$$\frac{x}{5} = \frac{-3}{4} \quad \text{Cross-multiply.}$$
$$4x = (-3) \times 5$$
$$4x = -15$$
$$x = \frac{-15}{4} \text{ or } -\frac{15}{4}$$

36. (C) -32

Substitute -1 for a and 2 for b. Then

$$\begin{aligned} c &= 8ab^2 \\ &= 8(-1)2^2 \\ &= (-8) \times 4 \\ &= -32 \end{aligned}$$

37. (D) 525

$$\frac{\text{hits}}{\text{at-bats}} = \frac{\text{hits}}{\text{at-bats}}$$
$$\frac{2 \text{ hits}}{7 \text{ at-bats}} = \frac{150 \text{ hits}}{x \text{ at-bats}}$$
$$\frac{2}{7} = \frac{150}{x} \quad \text{Cross-multiply.}$$
$$2x = 7 \times 150$$
$$2x = 1050$$
$$x = 525$$

38. (D) $\sqrt{17}$

$$\begin{aligned} a^2 + b^2 &= x^2 \\ 4^2 + 1^2 &= x^2 \\ 16 + 1 &= x^2 \\ 17 &= x^2 \\ \sqrt{17} &= x \end{aligned}$$

39. (D) 3

$$\sqrt{4^2} - \sqrt{1^2} = 4 - 1 = 3$$

40. (B) 3000

6000 games at \$5 per game costs \$30,000. The general overhead cost of \$15,000 brings the total costs to \$45,000. He receives \$15 per game. To break even, he must sell

$$\frac{\$45,000}{\$15} = 3000 \text{ (games)}$$

Practice Final Exam B, pages 324–329

1. (C) 60,405

2. (C) 319

$$\begin{array}{r} 319 \\ 19\overline{\smash{\big)}6061} \\ \underline{57} \\ 36 \\ \underline{19} \\ 171 \\ \underline{171} \end{array}$$

3. (C) 2700

$$3^3 \times 10^2 = 27 \times 100$$
$$= 2700$$

4. (A) $2^7 \times 5$

$$640 = 64 \times 10$$
$$= 2^6 \times 10$$
$$= 2^6 \times 2 \times 5$$
$$= 2^7 \times 5$$

5. (A) 29

Choices (B), (D), and (E) are all divisible by 3. Choice (C) is divisible by 7. $49 = 7 \times 7$. Choice (A), 29, has only itself and 1 as factors and is therefore prime.

6. (D) 100

$$20 = 2^2 \times 5; \quad 25 = 5^2$$
$$lcm\,(20, 25) = 2^2 \times 5^2$$
$$= 4 \times 25$$
$$= 100$$

7. (E) $\dfrac{29}{40}$

The *lcd* is 40.

$$\frac{1}{8} = \frac{5}{40}, \quad \frac{3}{5} = \frac{24}{40}$$
$$\frac{1}{8} + \frac{3}{5} = \frac{5}{40} + \frac{24}{40} = \frac{29}{40}$$

8. (D) 4

$$2\frac{3}{4}$$
$$1\frac{1}{4}$$
$$3\frac{4}{4} = 3 + \frac{4}{4}$$
$$= 3 + 1$$
$$= 4$$

9. (A) $\dfrac{1}{64}$

$$\left(\frac{1}{4}\right)^3 = \frac{1^3}{4^3} = \frac{1}{64}$$

10. (C) 1.11

$$\begin{array}{r} 3.09 \\ -1.98 \\ \hline 1.11 \end{array}$$

11. (D) .008

$$(.2)^3 = \left(\frac{2}{10}\right)^3 = \frac{2^3}{10^3} = \frac{8}{1000} = .008$$

12. (E) $53

$$\text{PROFIT} = \text{SALES} - \text{COSTS}$$
$$= (70 \times \$1.50) - (80 \times \$.65)$$
$$= \$105.00 - \$52.00$$
$$= \$53$$

13. (C) .72

$$9\% \text{ of } 8$$
$$= .09 \times 8$$
$$= .72$$

14. (B) $24

A 25% reduction on a $32 sweater means .25 × $32, which equals $8.

$$\begin{array}{r} \$32 \leftarrow \text{normal price} \\ -\ 8 \leftarrow \text{reduction} \\ \hline \$24 \leftarrow \text{new price} \end{array}$$

15. (A) $4.20

$$\frac{\text{cost}}{\text{ounces}} = \frac{\text{cost}}{\text{ounces}}$$
$$\frac{\$2.80}{8 \text{ ounces}} = \frac{\$x}{12 \text{ ounces}}$$
$$\frac{2.80}{8} = \frac{x}{12} \quad \text{Cross-multiply.}$$
$$12 \times 2.80 = 8x$$
$$33.60 = 8x$$
$$4.20 = x$$

The cost of a 12-ounce slice is $4.20.

16. (C) 40

$$\frac{\$6.00}{\$.15} = \frac{600}{15} = 40$$

17. (B) 2

$$4\frac{1}{2} \div 2\frac{1}{4} = \frac{9}{2} \div \frac{9}{4}$$
$$= \frac{\overset{1}{\cancel{9}}}{\underset{1}{\cancel{2}}} \times \frac{\overset{2}{\cancel{4}}}{\underset{1}{\cancel{9}}}$$
$$= 2$$

18. (B) $\dfrac{1}{10}$ **19. (C)** 2.009 **20. (D)** 20.118

$$\begin{array}{r} 5.08 \\ 3.038 \\ \underline{12.} \\ 20.118 \end{array}$$

21. (B) $90.10

$$\begin{aligned} 6\% \ of \ \$85 &= .06 \times \$85 \\ &= \$5.10 \end{aligned}$$

$$\begin{array}{rl} \$85 & \leftarrow \text{ price} \\ \underline{+5.10} & \leftarrow \text{ sales tax} \\ \$90.10 & \leftarrow \text{ total price} \end{array}$$

22. (B) $154

$$\frac{\text{earnings}}{\text{hours}} = \frac{\text{earnings}}{\text{hours}}$$

$$\frac{\$44}{8 \text{ hours}} = \frac{\$x}{28 \text{ hours}}$$

$$\frac{44}{8} = \frac{x}{28}$$

$$44 \times 28 = 8x$$
$$1232 = 8x$$
$$154 = x$$

He earns $154 for 28 hours of work.

23. (C) $45

$$\text{Cost} = \text{Area} \times \text{Cost per yd.}^2$$

$$= (6 \text{ yd.} \times 5 \text{ yd.}) \times \frac{\$1.50}{\text{yd.}^2}$$

$$= 30 \overset{1}{\cancel{\text{yd.}^2}} \times \frac{\$1.50}{\underset{1}{\cancel{\text{yd.}^2}}}$$

$$= \$45.00$$

24. (D) 5 hours 15 minutes

$$\begin{array}{cc} 9 \text{ hours} & 66 \text{ minutes} \\ \cancel{10 \text{ hours}} & \cancel{6 \text{ minutes}} \\ -\ 4 \text{ hours} & 51 \text{ minutes} \\ \hline 5 \text{ hours} & 15 \text{ minutes} \end{array}$$

25. (C) 2 oz.

$$1 \text{ pound} = 16 \text{ ounces}$$

$$16 \text{ ounces} \div 8$$

$$= \frac{16}{8} \text{ ounces}$$

$$= 2 \text{ ounces}$$

26. (B) 4 inches

$$\text{Area} = \frac{1}{2} \text{ base} \times \text{height}$$

$$12 \text{ in.}^2 = \frac{1}{2} \times (6 \text{ in.}) \times \text{height}$$

$$12 \text{ in.}^2 = (3 \text{ in.}) \times \text{height} \quad \text{(Divide by 3 in.)}$$

$$\text{height} = \frac{12 \text{ in.}^2}{3 \text{ in.}} = 4 \text{ in.}$$

27. (D) 48

$$\begin{array}{rl} 10 & \leftarrow \text{ January} \\ 13 & \leftarrow \text{ February} \\ 10 & \leftarrow \text{ March} \\ \underline{15} & \leftarrow \text{ April} \\ 48 & \end{array}$$

28. (A) −80

$$\begin{aligned} 5(2^2) &- (2 \times 5)^2 \\ &= 5(4) - 10^2 \\ &= 20 - 100 = -80 \end{aligned}$$

29. (D) 81

$$\frac{72 + 85 + 80 + 87}{4} = \frac{324}{4} = 81$$

30. (A) 4

The quotient of two negative numbers is positive.

$$\frac{-12}{-3} = +\frac{12}{3} = 4$$

31. (B) 1

$$\begin{aligned} (-2)(-5) &- (-3)^2 \\ &= 10 - 9 = 1 \end{aligned}$$

32. (D) $150

$$\underbrace{\text{price}}_{x} \times \underbrace{\text{tax rate}}_{.04} = \underbrace{\text{tax}}_{\$6}$$

$$4x = \$600$$
$$x = \$150$$

33. (D) $a + 9$

$$P - Q = P + (-Q)$$

$$\begin{array}{rl} P\text{:} & 5a + 4 \\ Q\text{:} & 4a - 5 \\ -Q\text{:} & -4a + 5 \end{array}$$

$$\begin{array}{r} 5a + 4 \\ \underline{-4a + 5} \\ a + 9 \end{array}$$

34. (D) $9x + 7y$

One pair of gloves costs 9 dollars.
x pairs of gloves cost $x \cdot 9$ dollars
   or $9x$ dollars.
One scarf costs 7 dollars.
y scarves cost $y \cdot 7$ dollars
   or $7y$ dollars.
The total cost, in dollars, is
$9x + 7y$.

35. (B) -21

Substitute 4 for a and -1 for b in the expression $5ab - b^2$.

$5 \times 4 \times (-1) - (-1)^2 = -20 - 1 = -21$

36. (C) $\dfrac{24}{5}$

$$5x - 2 = 22$$
$$5x - 2 + 2 = 22 + 2$$
$$5x = 24$$
$$x = \dfrac{24}{5}$$

37. (C) 5

$$\dfrac{x + 3}{2} = \dfrac{3x - 3}{3} \quad \text{Cross-multiply.}$$
$$3(x + 3) = 2(3x - 3)$$
$$3x + 9 = 6x - 6$$
$$9 = 3x - 6$$
$$15 = 3x$$
$$5 = x$$

39. (D) 8

$$\sqrt{10^2 - 6^2} = \sqrt{100 - 36} = \sqrt{64} = 8$$

40. (D) 302

Let x be the number of tickets sold.

Profit = Sales − Expenses
$$\$1156 = x \cdot \$8 - \quad \$1260$$
$$\$2416 = \$8x$$
$$2416 = 8x$$
$$302 = x$$

38. (C) $10\dfrac{1}{2}$ inches

$$\dfrac{\text{width}}{\text{length}} = \dfrac{\text{width}}{\text{length}}$$

$$\dfrac{5 \text{ inches}}{7 \text{ inches}} = \dfrac{7\dfrac{1}{2} \text{ inches}}{x \text{ inches}}$$

$$\dfrac{5}{7} = \dfrac{7.5}{x} \quad \text{Cross-multiply.}$$
$$5x = 7 \times 7.5$$
$$5x = 52.5$$
$$\dfrac{5x}{5} = \dfrac{52.5}{5}$$
$$x = 10.5 \text{ or } 10\dfrac{1}{2}$$

The length of the enlargement is $10\dfrac{1}{2}$ inches.

Practice Final Exam C, pages 329–334

1. (C) 80,000.4   **2.** (B) 504   **3.** (C) 125   **4.** (E) 72

$$5^3 = 5 \times 5 \times 5 = 5 \times 25 = 125 \qquad 2^3 \times 3^2 = 8 \times 9$$
$$= 72$$

5. (A) $2^4 \times 3^2$ **6.** (C) 96

$$144 = 12 \times 12$$ $$12 = 2^2 \times 3, \quad 32 = 2^5$$
$$= (2^2 \times 3) \times (2^2 \times 3)$$ $$lcm\,(12, 32) = 2^5 \times 3 = 96$$
$$= 2^4 \times 3^2$$

7. (D) $-\dfrac{1}{40}$ **8.** (B) $\dfrac{2}{3}$ **9.** (A) $\dfrac{2}{3}$

The *lcd* is 40. $\dfrac{3}{8} = \dfrac{15}{40}, \quad \dfrac{2}{5} = \dfrac{16}{40}$

$$\dfrac{3}{8} - \dfrac{2}{5} = \dfrac{15}{40} - \dfrac{16}{40} = -\dfrac{1}{40}$$

$$\overset{1}{\cancel{\dfrac{3}{5}}} \times \overset{2}{\underset{3}{\cancel{\dfrac{10}{9}}}} = \dfrac{2}{3} \qquad \dfrac{3}{10} \div \dfrac{9}{20} = \overset{1}{\underset{1}{\cancel{\dfrac{3}{10}}}} \times \overset{2}{\underset{2}{\cancel{\dfrac{20}{9}}}}$$

$$= \dfrac{2}{3}$$

10. (B) 5   **11.** (D) $\dfrac{5}{12}$ **12.** (D) .0289   **13.** (B) 22.812

$$3\dfrac{3}{4}$$ 6.003
$$+\,1\dfrac{1}{4}$$ 5.809
                                                                             11.
$$4\dfrac{4}{4} = 4 + \dfrac{4}{4} = 4 + 1 = 5$$ ———
                                                                             22.812

14. (E) .0064   **15.** (D) $20

$$(.08)^2 = \underbrace{(.08)}_{\substack{2\ \text{decimal} \\ \text{digits}}} \times \underbrace{(.08)}_{\substack{2\ \text{decimal} \\ \text{digits}}} = \underbrace{.0064}_{\substack{4\ \text{decimal} \\ \text{digits}}}$$

1 gallon costs $1.25.
16 gallons cost 16 × $1.25, or $20.00.

16. (C) $1.67   **17.** (D) .31

1 pound of potatoes costs $.22.
5 pounds of potatoes cost 5 × $.22 or $1.10.
1 pound of onions costs $.19.
3 pounds of onions cost 3 × $.19 or $.57.

$$\begin{array}{r} \$1.10 \\ +\ .57 \\ \hline \$1.67 \end{array}$$

$$\begin{array}{r} .307 \\ 13\overline{)4.000} \\ 3\ 9 \\ \hline 100 \\ 91 \end{array}$$

The third decimal digit is 5 or
more. To the nearest hundredth,

$$\dfrac{4}{13} \approx .31$$

18. (E) 64

Let x be the number.

48 is 75% of a certain number.

$$48 = .75 \times x \qquad \text{Multiply both sides}$$
$$4800 = 75x \qquad \text{by 100.}$$
$$\frac{4800}{75} = x$$
$$64 = x$$

19. (B) 36 tons

$$30\% \times 120 \text{ tons}$$
$$= .30 \times 120 \text{ tons}$$
$$= 36.00 \text{ tons}$$

20. (C) \$45

A 25% discount
on \$60 means
$.25 \times \$60$ or \$15.

$$\begin{array}{rl} \$60 & \leftarrow \text{ original price} \\ -15 & \leftarrow \text{ discount} \\ \hline \$45 & \leftarrow \text{ new price} \end{array}$$

21. (D) 3 pounds 15 ounces

$$\begin{array}{ll} 9 \text{ pounds} & 27 \text{ ounces} \\ \cancel{10 \text{ pounds } 11 \text{ ounces}} \\ -6 \text{ pounds} & 12 \text{ ounces} \\ \hline 3 \text{ pounds} & 15 \text{ ounces} \end{array}$$

22. (E) \$768

$$\text{Cost} = \text{Area} \times \text{Cost per ft.}^2$$
$$= (16 \text{ ft.} \times 12 \text{ ft.}) \times \frac{\$4}{\text{ft.}^2}$$
$$= 192 \overset{1}{\cancel{\text{ft}^2}} \times \frac{\$4}{\underset{1}{\cancel{\text{ft}^2}}}$$
$$= \$768$$

23. (D) 4

$$\frac{4^2 - 2^3}{2} = \frac{16 - 8}{2}$$
$$= \frac{8}{2}$$
$$= 4$$

24. (A) 12

$$(6 - 2)^2 - \left(\frac{4}{2}\right)^2 = 4^2 - 2^2$$
$$= 16 - 4 = 12$$

25. (D) 21

$$(-2 - 1)^2 - 4(-3) = (-3)^2 - (-12)$$
$$= 9 + 12 = 21$$

26. (B) 11.5

$$\frac{10 + 11 + 12 + 13}{4} = \frac{46}{4} = 11.5$$

27. (A) 24 in.^{3}

$$\text{Volume} = \text{length} \times \text{width} \times \text{height}$$
$$= 4 \text{ in.} \times 3 \text{ in.} \times 2 \text{ in.} = 24 \text{ in.}^3$$

28. (C) 5 feet 10 inches

1 foot = 12 inches

$$\begin{array}{lll} 6 \text{ feet 2 inches} = & 5 \text{ feet 14 inches} \\ -\phantom{6 \text{ feet}}4 \text{ inches} = & -\phantom{5 \text{ feet 1}}4 \text{ inches} \\ \hline & 5 \text{ feet 10 inches} \end{array}$$

29. (A) 500

$$\begin{array}{rl} 4000 & \leftarrow \text{ 1984 enrollment} \\ -3500 & \leftarrow \text{ 1983 enrollment} \\ \hline 500 & \leftarrow \text{ increase} \end{array}$$

30. (C) $4x - 5$

5 less than 4 times the number

$$-5 + 4 \quad \cdot \quad x \qquad \text{or } 4x - 5$$

31. (D) $-x^2 + 3x - 7$

$$P - Q = P + (-Q)$$
$$\begin{array}{ll} P: & x^2 - 6 \\ Q: & 2x^2 - 3x + 1 \\ -Q: & -2x^2 + 3x - 1 \end{array}$$
$$\begin{array}{l} x^2 - 6 \\ -2x^2 + 3x - 1 \\ \hline -x^2 + 3x - 7 \end{array}$$

32. (B) 8

$$
\begin{array}{rcl}
4x - 7 &=& 2x + 9 \\
+ 7 && + 7 \\
\hline
4x &=& 2x + 16 \\
-2x && -2x \\
\hline
2x &=& 16 \\
x &=& 8
\end{array}
$$

33. (E) 5

$\dfrac{t + 3}{4} = \dfrac{t + 1}{3}$ Cross-multiply.

$$
\begin{array}{rcl}
3(t + 3) &=& 4(t + 1) \\
3t + 9 &=& 4t + 4 \\
9 &=& t + 4 \\
5 &=& t
\end{array}
$$

34. (B) 24

Substitute -2 for p
and 3 for q.

$$
\begin{array}{l}
3(-2)^2 - 2(-2)3 \\
\quad = 3 \times 4 - (-12) \\
\quad = 12 + 12 \\
\quad = 24
\end{array}
$$

35. (D) $\dfrac{3}{8}$

Substitute 3 for x.

$\dfrac{3}{2^3} = \dfrac{3}{8}$

36. (D) $5n + 25q + 3$

The value of 1 nickel is 5 cents.
The value of n nickels is $n \times 5$ cents or $5n$ cents.
The value of 1 quarter is 25 cents.
The value of q quarters is $q \times 25$ cents or $25q$ cents.
The value of 3 pennies is 3 cents.
The total value, in cents, is $5n + 25q + 3$.

37. (C) 13.5 pounds

$$\dfrac{\text{number of pounds}}{\text{number of rats}} = \dfrac{\text{number of pounds}}{\text{number of rats}}$$

$$\dfrac{9 \text{ pounds}}{50 \text{ rats}} = \dfrac{x \text{ pounds}}{75 \text{ rats}}$$

$\dfrac{9}{50} = \dfrac{x}{75}$ Cross-multiply.

$$
\begin{array}{rcl}
9 \times 75 &=& 50x \\
675 &=& 50x \\
13.5 &=& x
\end{array}
$$

38. (D) 12

$\dfrac{x}{3} - 2 = \dfrac{x - 2}{5}$ Multiply both sides by 15.

$$15\left(\dfrac{x}{3} - 2\right) = \overset{3}{\cancel{15}} \cdot \dfrac{x - 2}{\underset{1}{\cancel{5}}}$$

$$
\begin{array}{rcl}
\dfrac{15x}{3} - 30 &=& 3(x - 2) \\
5x - 30 &=& 3x - 6 \\
2x - 30 &=& -6 \\
2x &=& 24 \\
x &=& 12
\end{array}
$$

39. (E) $\sqrt{15}$

$$
\begin{array}{rcl}
a^2 + b^2 &=& c^2 \\
x^2 + 1^2 &=& 4^2 \\
x^2 + 1 &=& 16 \\
x^2 &=& 15 \\
x &=& \sqrt{15}
\end{array}
$$

40. (C) 30

$25\% \text{ of } 60 = .25 \times 60 = 15$

Thus 15 of the guidance counsellors are women.
If x additional women are hired, then

$$\dfrac{\text{number of women counsellors}}{\text{total number of counsellors}} \underbrace{\text{will be } 50\%}.$$

$$\dfrac{15 + x}{60 + x} = \dfrac{1}{2}$$ Cross-multiply.

$$
\begin{array}{rcl}
2(15 + x) &=& 60 + x \\
30 + 2x &=& 60 + x \\
30 + x &=& 60 \\
x &=& 30
\end{array}
$$

INDEX